Environmental
and
Economic
Sustainability

T0179283

Environmental and Ecological Risk Assessment

Series Editor
Michael C. Newman
College of William and Mary
Virginia Institute of Marine Science
Gloucester Point, Virginia

Published Titles
Coastal and Estuarine Risk Assessment
Edited by
Michael C. Newman, Morris H. Roberts, Jr., and Robert C. Hale

Risk Assessment with Time to Event Models
Edited by
Mark Crane, Michael C. Newman, Peter F. Chapman, and John Fenlon

Species Sensitivity Distributions in Ecotoxicology
Edited by
Leo Posthuma, Glenn W. Suter II, and Theo P. Traas

**Regional Scale Ecological Risk Assessment:
Using the Relative Risk Method**
Edited by
Wayne G. Landis

**Economics and Ecological Risk Assessment:
Applications to Watershed Management**
Edited by
Randall J.F. Bruins

**Environmental Assessment of Estuarine Ecosystems:
A Case Study**
Edited by
Claude Amiard-Triquet and Philip S. Rainbow

Environmental and Economic Sustainability
Edited by
Paul E. Hardisty

Environmental and Economic Sustainability

Paul E. Hardisty

CRC Press
Taylor & Francis Group
Boca Raton London New York

CRC Press is an imprint of the
Taylor & Francis Group, an **informa** business

CRC Press
Taylor & Francis Group
6000 Broken Sound Parkway NW, Suite 300
Boca Raton, FL 33487-2742

First issued in paperback 2019

ISBN-13: 978-1-4200-5948-9 (hbk)
ISBN-13: 978-0-367-38389-3 (pbk)

Library of Congress Cataloging-in-Publication Data

Hardisty, Paul E.
    Environmental and economic sustainability / Paul E. Hardisty.
        p. cm. -- (Environmental and ecological risk assessment)
    Includes bibliographical references and index.
    ISBN 978-1-4200-5948-9
    1. Sustainable development. 2. Environmental protection. I. Title.

HC79.E5H3537 2010
338.9'27--dc22

2009052896

Visit the Taylor & Francis Web site at
http://www.taylorandfrancis.com

and the CRC Press Web site at
http://www.crcpress.com

Dedication

To the mentors in my life: Dad, Fred, Tad, and Peter. Thanks for your wisdom, guidance, criticism, and support.

Contents

Foreword

Until relatively recently, serious discussion of environmental issues at board level was the preserve of an enlightened few companies. For most, protection of the environment was considered to be only a legal compliance issue. However, recognition of the magnitude and severity of human impact on the global climate, coupled with society's demand for greater corporate social responsibility, has changed all that. Whilst climate change has dominated the environmental agenda in recent years, there is a growing awareness that preservation of the wider environment, dwindling resources and social well-being demand an integrated approach if future generations are to prosper.

Whilst this is a great philosophical conclusion to reach, we live in a world where the common global language is money. Hardisty's book shows us how to use the language of money to make decisions that are right for the environment, society, and, critically, the commercial world that we rely upon to increase our quality of life. This does not mean that we are being encouraged to somehow "sell out" the environment, but rather that by measuring and internalizing the value of the environment and resources to society, we will make decisions that are more sustainable for all.

Dr. Steve Wallace
Head of Climate Change and Environment
National Grid

Preface

At the United Nations Copenhagen Climate Conference in December 2009, I had the opportunity to meet with a senior scientist from the U.S. National Oceanographic and Atmospheric Administration (NOAA) in the U.S. pavilion. He was playing with a remote control device that was directing the data feed to four high-definition projectors aimed at a massive translucent sphere hanging from the ceiling. The sphere, of course, was Earth. He brought up satellite and radar imaging data on Arctic sea ice for every day going back several years and then let it run. We watched the sea ice go through its yearly cycle of winter expansion and summer contraction. He stopped the run at mid-September 2009 and described what we could see: an ice pack that was at its third smallest areal extent ever (2007 was the lowest; it dropped 35% below the long-term average in one year, with a slight recovery in 2008). Then, he explained the significance of the vast gray areas clearly visible against the white ice. "These are areas of thinning ice," he said. He went on to explain that the overall volume of Arctic ice is now less than one-third of what it was in the 1970s, and that 2009 was the lowest ever on record (so far).

The data are coming in quickly now. The World Meteorological Organization reported that the decade ending in 2009 was the warmest ever on record, and that each successive decade has been warmer than the last. The year 2009 was the fifth warmest on record. Twelve of the warmest years on record have occurred in the last 12 years. The natural climate has always been variable, but now the human-induced overprinting is becoming more and more dominant. And yet, our emissions continue to accelerate.

Climate change is not the only issue facing us in the twenty-first century. Water scarcity, the urgent need to produce more food for the billions we will add to the world's population over the next 40 years, the increasing disparity between rich and poor, the unraveling of many of the world's ecosystems, species loss, and the plight of the oceans are all equally deserving of our attention. We need to find and implement solutions to all of these (and other) challenges, and do it quickly, or face a perilous future.

Many of the fixes, particularly to global issues like climate change, may at first appear to be global in scale, solved only by international treaties and national policy. But, the combined effect of the millions of smaller-scale project and policy decisions made every day by businesses, industry, and organizations of all kinds is what makes global trends. At this smaller scale, a move toward more environmentally, socially, and economically sustainable choices, options, and policies can have a powerful effect.

This book, the result of over 15 years of research and practice, introduces the environmental and economic sustainability assessment (EESA), a process that helps decision makers at all levels balance the needs of society, the environment, and business over the long term by quantifying sustainability in a way that is physically based and objective. Ultimately, this book is about communication: including stakeholders in a transparent process that provides a robust view of how various options compare over a wide range of possible future conditions using a language that everyone understands—money.

In Copenhagen, the real climate change debate was mostly about money: who is going to pay and how much, how developing countries can access financing. Although everyone understands that we must act, they also realize that nothing can be done without funding—simply because money is how we measure *value* (whether we like it or not). Ultimately, the solutions to the problems of the twenty-first century will come from understanding and acknowledging the tremendous value that the environment provides, and reflecting that value within decision making at every level so that *society as a whole* is better off from each choice we make. Perhaps it will be the sum of all of those beneficial decisions, taken every day, at every level, that will help to change the world.

Paul E. Hardisty

About the Author

Paul E. Hardisty is executive director, Sustainability and EcoNomics™ for WorleyParsons, one of the world's largest engineering companies. For over twenty years, he has been advising industry and governments around the world on environmental strategy and sustainability. He is a visiting professor in environmental engineering at Imperial College, London, and adjunct professor at the University of Western Australia School of Business, where he teaches sustainability and climate change to MBA students. Paul is the author of numerous technical papers, books, and newspaper articles on environmental issues and a soon-to-be-released novel, which he describes as an eco-thriller. He is a contributor to President Gorbachev's Climate Change  Task Force, a member of the Waste Management Authority of Western Australia, and a director of Green Cross Australia. Paul lives in Western Australia with his wife, Heidi, and two sons, Zachary and Declan, and for fun competes in Ironman triathlons.

1 Introduction

THE EXPONENTIAL ERA

In the twenty-first century, the world is a place of unrelenting and ever-accelerating change. Financial turmoil sends the global economy from the heights of boom to unprecedented depression in a few short months; the price of oil skyrockets to over five times its previous long-term average and then tumbles down again in a matter of weeks (Figure 1.1); after taking a hundred thousand years to reach just over 6 billion, the world's population will grow by almost 4 billion in the next 40 years[1] (Figure 1.2); the extent of arctic sea ice, in steady decline since the middle of the last century, falls off alarmingly in 2007 and 2008;[2] emissions of greenhouse gases (GHGs) to the atmosphere are rising faster than ever before.[3]

We live in the exponential era—a time unique in history, when a confluence of overlapping and mutually reinforcing factors is propelling the world into unknown economic, social, and environmental territory at an accelerating rate.[4] Not only are there ever more people on the planet,[5] but quickening development, particularly in India and China, means that each of these people is demanding more of the world's resources. Technology spurs development, and our exploding technological prowess allows us to wield greater power over our environment and surroundings than ever before. A single man with a D8 caterpillar can now clear as much land in a day as his grandfather could have in a decade of hard manual labor. Our ability to assimilate, use, and process data and information is exploding, just as predicted by Gordon Moore, the founder of Intel. In the 1960s, he predicted that the number of transistors on a silicon chip would double every 18 months—and it has, inexorably, since then.[6] But, a rapidly rising global population, combined with accelerating development and resource use, surging energy demand, and an ever-expanding need for water and food, is also creating huge stress on the natural environment. This combination of forces, which some are now calling simply *global change*, is leading to chronic overfishing, large-scale clearing of native forest, an alarming and accelerating loss of global biodiversity, and increasingly stronger evidence of the impacts of climate change.[7] Many are now calling this a time of unprecedented global environmental crisis.[8]

CRISIS—WHICH CRISIS?

But other issues, equally worthy of the dubious distinction "crisis," abound. Poverty remains a blight on humanity. Today, according to the most recent statistics from the United Nations, approximately 45% of the world's population lives on less than US$1 per day.[9] In the United States or Europe, that much would not buy one decent meal. An astonishing 65% of the world lives on less than US$2 per day. And, the

1

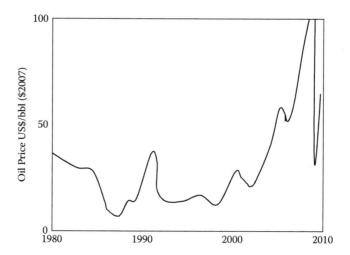

FIGURE 1.1   Actual oil price in U.S. dollars per barrel, 1975–2009, with 2% and 5% increase trend lines from 1988.

numbers of chronically poor are increasing despite the efforts of well-intentioned organizations and individuals around the world. But, the disparity in income is not the only measure of poverty. Never before in modern history has wealth been more concentrated in fewer hands: The richest 1% of the people on the planet control about half of the wealth. The poorest half of the population, over 3 billion people, owns less than 1% of the planet's wealth. This shocking inequality is also growing, accelerating in the wrong direction (20 years ago the top 1% controlled about a quarter of the wealth). Poverty can also be measured in other ways. Over 1 billion people on the planet lack access to safe, clean drinking water, and that number is rising. Lacking this most fundamental of goods, these people are *water poor*, and it affects every

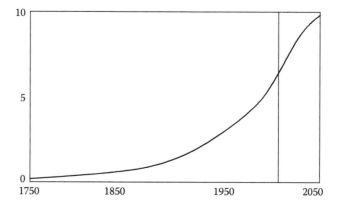

FIGURE 1.2   World population growth 1750–2050 based on data from U.N. Population Project and Cohen (1995).

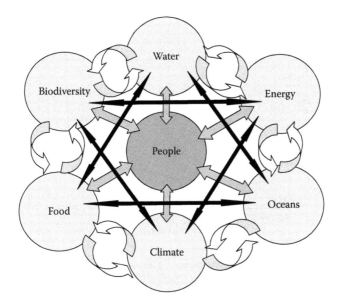

FIGURE 1.3   An interconnected world: humankind's all-affecting role on the planet.

part of their lives. These are all examples of increasingly unsustainable trends—they cannot continue indefinitely, as history has shown, without causing major ruptures in society.

ALL FEEDING OFF EACH OTHER

Many, if not most, of these crises are actually interlinked, interdependent, and mutually reinforcing. Figure 1.3 provides a basic schematic overview of the causative and consequential links between people and the world we inhabit. The interdependence is startling. An economic paradigm that focuses on gross domestic product (GDP) and does not explicitly account for the value of external issues (environment, society, depletion of natural capital) accelerates the use of natural resources of all kinds and concentrates wealth; concentration of economic wealth and income disparity create poverty; poverty causes environmental degradation as people are forced to destroy natural capital just to survive; environmental degradation further reinforces the poverty cycle as the land is degraded; and pollution leads to health impacts, further loss of income-generating potential, damage to the means of livelihood, and eventually social strife. Civil unrest among the disaffected and displaced leads to the rise of extremism and terrorism. And as the population grows, and each of these issues develops more rapidly, the need for solutions becomes even more urgent.[10]

CHEAP ENERGY, CLIMATE CHANGE, AND POVERTY

The widespread availability of "cheap" fossil energy has driven global economic growth, creating prosperity for many (but not most), but as a consequence has laden the atmosphere with billion of tonnes of GHGs, which are accelerating the natural

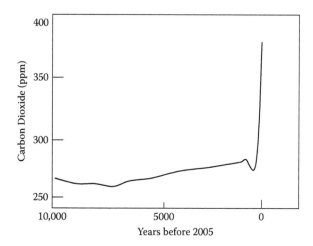

FIGURE 1.4 Atmospheric concentrations of CO_2 over the last 10,000 years (based on data from the Intergovernmental Panel on Climate Change, Fourth Assessment Report: The Physical Science Basics, 2007, Cambridge University Press, Cambridge, UK).

changes in the Earth's climate (Figure 1.4). Climate change is, among other things, essentially a story of the redistribution of water, increasingly through extreme weather events.[11] That means, in very general terms, more flooding in areas that are already wet and more drought in areas that are already arid.[12] Flood or drought—both lead to hardship, loss of economic activity, declining agricultural production, and damage to property. Climate change is predicted to have a disproportionate effect on the poorest people of the world and so will only reinforce poverty and the wealth and income disparities between haves and have-nots.[13] Even our efforts to protect ourselves against climate change, if executed using current business-as-usual decision making and technology, will act to reinforce climate change. In Australia, for instance, chronic drought due to changing rainfall patterns triggered by climate change[14] has led to the building of new desalination plants, with more planned. If powered by electricity from a predominantly coal-fired grid, these plants will add more GHGs to the atmosphere, exacerbating climate change. These anthropogenic feedback loops will simply reinforce the problem in a descending spiral. The harsher the impacts of climate change, the more energy we will need to protect ourselves and adapt, the worse climate change will get. One of the most pressing questions facing people and governments around the world today is: Which of these simultaneous crises do we deal with, and how?

A Crisis of Sustainability

These, and other issues such as the threat of terrorism, nuclear proliferation, AIDS, pandemics, and basic food security, are all essentially *crises of sustainability*—they cannot go on indefinitely. Societies, ecosystems, countries, sectors, industries, people—all are locked together on the same planet, subject to the same laws of physics and biology. One way or another, unsustainable behavior will eventually lead to correction or collapse in much the same way as the financial house of cards built on empty subprime assets imploded in late 2008, launching the world into an unprecedented economic downturn.

There are many examples of corrections to unsustainable behavior. Some are more brutal and permanent than others. Stock market "bubbles" burst, and sanity is reestablished. Unsustainable growth in the housing market in 2008 in the United States led directly to the subprime mortgage crisis and the ensuing economic corrections. The Easter Islanders, more than a century ago, destroyed their isolated island ecosystem through systemic unremitting deforestation, to the point where that once highly developed society collapsed into anarchy, cannibalism, and near extinction.[15] Overfishing in the Grand Banks off Newfoundland, Canada, through the 1970s and 1980s led to the complete collapse of the cod fishery—once the most prolific fish resource on the planet. The cod are gone, and they are not coming back (more on this in Chapter 2).[16] Climate change is another kind of correction, an irreversible one (from a human generational timescale perspective), wrought by the Earth system as it adapts to warming temperatures and higher concentrations of GHGs in the atmosphere, with the collateral damage in this case falling on humankind.[17] Further corrections and collapses are clearly on the horizon in this era of exponential change if we do not change our ways. The signals are all there. But so far, there is little evidence that we are changing.

Do We Want A Sustainable World?

Moving toward a truly sustainable world is clearly in everyone's best long-term interest. However, short-term thinking, and an economic and political system geared to a short-term perspective (annual bonuses, quarterly profit reports, 4-year electoral cycles, 5-year board appointments) prevent us from acting with a longer-term view in mind. The fundamentals of our economic system also entrench short-term thinking. The financial analysis that every enterprise, government, and individual uses to evaluate every transaction, project, and decision worldwide is based on the time value of money—the discount rate. The higher the discount rate, the greater the expected rate of return, the shorter the payback period and, not surprisingly, the less the future is worth. Short-term thinking is embedded deep within our fundamental systems and processes, on every level. Historically, true visionary long-term thinking has been the preserve of a rare and unusual breed of leaders, statesmen, and private citizens, who have occasionally been able to make a real difference: the Muirs, Mandela, Gorbachev, Ghandi. Of course, there are others, but modern society as a whole is not structurally geared to think for the long term. If it were, we would surely not be in the position we are in today.

But, we must be realistic. There are also people who do not want change, for whom movement to a more sustainable, equitable world is a threat. There is huge wealth invested in the status quo. Sometimes, sustainability is seen as a threat to return on investment and private enterprise, so it is sometimes the rich and the powerful, and the leaders of major industries, who are the most vocal detractors of efforts to move toward sustainability. In reality, of course, sustainability has as much, if not more, to offer shareholders and owners of industries and businesses as anyone else.

Sustainability is good for everyone. So far, for reasons we explore in more detail in this chapter, we have not been able to achieve it. We have not even come close and are in fact accelerating in the wrong direction. But, assuming that we do aspire to the benefits of a sustainable world, one that will continue to provide benefits and a good life for all of its inhabitants over the long term, who will make it happen?

INDUSTRY CAN AND MUST BE PART OF THE ANSWER

Industry cannot exist without its customers. It depends on a customer base able to purchase its products and use them to improve their own well-being without harming the places they inhabit, the societies within which they live, and the planet on which they depend. And ultimately, the more customers there are, and the wealthier they are in real terms, the more industry will prosper.

It is in the best interests of industry to be a prime mover in the shift toward a more sustainable world. Large private enterprise today wields immense financial, political, and technical muscle and know-how. It employs millions of highly educated people and has access to the leading technology, research, and tools. If ever there was a force that could lead the drive for real progress, it is here. The benefits to industry of more sustainable operations are massive. Sustainable operations are more efficient, use less of everything, waste less, produce more value, have lower environmental management costs and liabilities, enjoy better relationships with their neighbors and stakeholders, can obtain regulatory approvals more readily and smoothly, and can attract financing at better rates. True sustainability can deliver increased profits for business and industry, simultaneously delivering overall increases in human welfare: a classic win-win. But to achieve this, industry will have to change the way it makes decisions.

TRUE SUSTAINABILITY FOR THIS CENTURY

SUSTAINABILITY: DIFFERENT PERSPECTIVES, DIFFERENT MEANINGS

This book is about changing the fundamental way in which industry, business, and government assess and choose their projects and policies, and how they look at their relationships with the wider world. Since the 1990s, a lot as been written about the need to consider the environment and society when making policy, setting strategy, determining business objectives, and conceiving and designing projects of all types, from a new refinery to the refurbishment of a water treatment plant, from the analysis of options for ensuring a stable and secure energy future, to selecting which turbine to use in a new power plant. Much of this work and thought has crystallized around the concept of sustainability, a widely used but equally wide-ranging term, which at present seems to mean many different things to many different people. Business sustainability means keeping businesses alive and profitable so that they can continue to deliver the goods and services that society needs and wants. Sustainability in the environmental movement refers to the ongoing viability of the ecosystems that provide the basis for all life on Earth. In community terms, we consider sustainability of populations, cultures, and languages. All are equally valid.

WORDS, THOUGHTS, AND ACTION

However, as discussed in the following chapters, the concept of sustainability, in all of its guises, has clearly failed. Since the 1970s, despite a growing awareness of the importance of sustaining our businesses, environment, and communities over the long term, of making them secure and resilient over time, of managing these things in a way that guarantees long-term prosperity and ongoing value, the facts show that our actions have not lived up to our aspirations. We are not conducting ourselves in a sustainable way on any level.

Business as usual, with its focus on profit in the short term, on the maximization of private benefit without regard to external influences, continues to dominate the way we run our economies, our societies, our businesses, our environment, and even our households. This is not because of some moral defect or lack of ethics. It is simply because the fundamental decision-making systems that our society uses, on every level, every day, are rooted in concepts that belong in another century, another time, when the world was a very different place. Our ability to make wise and coherent decisions has not kept pace with our numbers, our technological capability, or our footprint on the planet (Figure 1.5). So, despite the fact that around the world most individuals *know* intrinsically that we must behave in a more sustainable way, and although they *want* to do the things that will benefit their children and future

FIGURE 1.5 Temperate rain forest on the West Coast of Canada. Is this worth anything to us, beyond simply the lumber? Surely it is. But, the current economic system ascribes little or no value to nature and the functioning ecosystems that support all life on the planet.

generations, what we *actually do* every day is increasingly opposed to this understanding, to this desire.

WHAT DO I GIVE UP? WHAT DO I GET?

Currently, of course, many organizations and governments around the world expend a large amount of time and effort trying to identify the various socioenvironmental issues associated with a project or decision, and then attempt to use that information in the decision-making process. This has been going on now in government and industry, particularly in the more developed Western nations, since the 1980s, to varying degrees. We undertake detailed environmental and social impact studies that examine the ramifications of project options, and use multicriteria assessments to try to rank and prioritize the many different socioenvironmental issues so that we can put them into a decision-making context. These processes have been developed because decision making is, at its most elemental level, a matter of *trade-offs*. Whatever we do, whatever policy we enact, we are always trading some of one benefit for some of another, a cost here for a gain there, having to weigh the lesser of evils, select the greater of goods. This is the central issue in every decision: What do I give up, and what do I get?

COMPARING APPLES, REFRIGERATORS, AND GIRAFFES

However, despite the growing realization that environment and society *matter*, decision making remains dominated by financial imperatives. The other socioenvironmental aspects, widely considered to be "external" to the decision, are referred to only in qualitative terms, as footnotes. The problem, of course, is one of language. Decision makers in our society speak the language of money. Projects, policies, and decisions are first and foremost framed in the context of cost, revenue, and profit. Environmental and community impacts and benefits are described in increasingly arcane technical terms in the accompanying reports, and quantified (if at all) in a bewildering variety of units: liters of water, hectares of wetland, thousands of people disaffected, probabilities of increased cancer rates. How can even the most enlightened decision maker, supported by the most knowledgeable technical people, hope to determine an optimal solution with so many different components to the analysis, each measured in a completely different unit of measure? It is akin to making trade-offs between apples, refrigerators, and giraffes. Of course, the answer is that they cannot. What is required to identify truly optimal approaches, strategies, policies, even technology selections, when there are many forces at work and many complex interlinked issues, is that *all* of the factors be expressed in the same unit of measure.

IN THE END, MONEY RULES

Current systems that attempt to make sense of environmental and social issues in decision making are largely based on qualitative, subjective, multicriteria approaches. There are many variations of this approach, and many publicly available and proprietary methods are available. Essentially, they all do the same thing. They rank and score a range of issues not normally included in conventional financially driven decision-making analysis. Scores are weighted and ranked, often by an expert panel (but just as often not), and a scaled index is created. For instance, some systems rank issues out of a score of one hundred, some a score out of ten. Others provide a range of qualitative descriptors expressed in pictorial fashion (smiling, neutral, or happy faces for instance). However, an essential problem remains with these methods—this score, whatever it is, must then be used by decision makers whose primary focus, whose training, whose understanding and calibration of value, and in almost all cases whose remuneration, is based on money. Such is the power of money in our world today that it is not surprising which of the two resulting measures receives the most weight and attention: of course, the financial.

Therein lies the essential dilemma, the reason that sustainability, despite being something that most espouse as desirable, has not made a real impact on our lives or on the current state of the planet. Although we have spent decades talking about achieving a sustainable balance in our world, in our communities, in our lives, over the same period we have seen the health of the world decline rapidly, bio-diversity disappear, emissions of climate-changing fossil fuels rise sharply, and the plight of the poor around the world worsen. This has not happened because we have *wanted* it to happen—quite the opposite. The notion that we can achieve prosperity for all, and increase our real wealth and happiness, is central to every-thing that organized human societies have sought for decades. The problem is that if money is the dominant measure of success and value in our society, while so many of things we really care about are *not* measured in monetary terms, then our decisions will invariably favor the monetary and the financial and ignore the nonfinancial, the unmonetized (Figure 1.5). Entrenched short-term perspectives, embodied in the concept of discounting, devalue the future and compound the problem.

THE DILEMMA FOR INDUSTRY

Humankind has achieved great material and scientific progress over the last two centuries. Since the dawn of the industrial revolution, increasingly rapid develop-ments in technology and engineering ability have brought unprecedented prosperity to millions of people around the world. Industries of all kinds have and continue to provide the products and tools that hold the key to our continuing future prosperity. Clearly, this newfound material wealth has not been shared equally among the peoples of the world, but globalization and the rise of the new economic powers in India, China, Brazil, and other countries are for the first time bringing the undeni-able benefits of development within the grasp of the majority of humankind. The stated goal of almost every national government the world over is now economic development, led by increased industrialization, access to resources, energy, and the benefits of modern technology.

But, there has been a price. As industrial development has grown, so has the state of the global environment suffered—and with environmental degradation has also come human consequences: displaced peoples, ruptured cultures, loss of traditional places and values, health impacts, death. Indeed, as the population of the planet has continued to grow, and the overall level of development and industrialization has accelerated, so have environmental and social costs multiplied. The old view, the view of the nineteenth and twentieth centuries, was that material and industrial advancement must come at a price. Environmental degradation was an unfortunate but necessary sacrifice for what was often termed simply: "progress."

But in the twenty-first century, this notion has become increasingly untenable and unpalatable. This is the dilemma faced by industry and business in this new century. Industries of all kinds, from the production of oil and gas and minerals, to the manufacture of goods, to the provision of water and power, are now wedged between the anvil of survival (the profit imperative) and the hammer of increasing environmental and social demands from communities, employees, customers, and investors. Increasingly, people everywhere are demanding that industry play a more active role in the quest for a sustainable future on Earth.

Everywhere, industry has started to respond to this challenge. Progress has been made. Yet, fundamentally, business and industry are not structurally set up to respond to altruism or what appear to be ethically-driven imperatives. Companies are designed to deliver shareholder return. They are, essentially, financial beings, with money coursing through their veins. The souls of their owners and employees can temper and guide their actions, but if they are cut off for too long from their life-giving flow of revenue and profit, they die. So, while spending some of their hard-earned profits on environmental and social protection is usually acceptable, and even in many cases desirable (good corporate citizens make good partners and suppliers), clearly if companies have to spend *too much*, they will wither. Again, it comes down to finding a balance. Industry needs to act to improve sustainability. But, how much is enough? And just as importantly, how much is too much?

SUSTAINABLE DECISIONS FOR THE TWENTY-FIRST CENTURY

Truly sustainable decisions must embrace the widest notion of what sustainability is and means. While there are many definitions of sustainability (Chapter 2 offers a brief history of sustainability), most have in common the idea that future generations should be able to live as well as we do now. Inherently, that means that the material well-being provided by the goods and services produced by industry should, in general terms, also be available to future generations. So must a thriving planetary ecosystem and its wondrous biodiversity. But, industry must also survive—it must be sustainable in its own right. In other words, industry must continue to produce the goods and services that people want, that improve their health, happiness and well-being, over the long term—over generations. Most shareholders would also want the companies that they invest in to be successful enough to stay in business, not just for years but for decades.

So, sustainable decisions must be ones that allow industry to operate successfully over the long term while maintaining (and at this point in history, necessarily also enhancing and repairing) the health of the environment on which we all depend, and simultaneously nurturing and sustaining communities and societies (Figure 1.6). Sustainability is unabashedly anthropocentric: It is about people. People want to live

FIGURE 1.6 Vancouver Harbor, British Columbia, Canada. Industry provides the commodities and services that, in many ways, define civilization and have greatly increased standards of living and prosperity around the world. But, we are now so many on the planet and our ability to impact our environment so large, that balancing our needs with those of the planet is increasingly difficult.

healthy, meaningful, and happy lives free from poverty and oppression. People also want and need a healthy, thriving environment, with flourishing biodiversity and a reasonably stable climate. Without the latter, they cannot have the first. And from these two, sustainable, lasting, durable societies can continue to grow.

MAKING SUSTAINABILITY RELEVANT TO BUSINESS AND INDUSTRY

At each stage of the product life cycle, from cradle to grave, industry, regulators, and the public face a wide range of choices on the level of environmental and social protection that is required. The costs of mitigation and protection measures can be considerable, and worldwide environmental expenditure by industry, for example, has risen steadily since the 1990s, partly in response to ever-tightening regulations and public pressure. Recognizing the need to protect their natural heritage for future generations, most governments have passed laws to protect the environment. Within this context, industrial development must in theory be carried out with due care for the environment and in adherence to national law and regulations. Of course, in practice, this is often not the case. In many less-developed nations, regulatory capacity and enforcement infrastructure are lacking, hindering enforcement of environmental protection laws. Corruption, as discussed in Chapter 8, can also weaken and usurp efforts to protect the environment.

Beyond this, many companies have for some time applied their own strict environmental and corporate responsibility (CR) policies uniformly wherever they operate, and some investment and lending institutions have developed voluntary environmental codes of practice, such as the equator principles.[18] However, it is usually the legally enforceable national standards that dictate to a large degree what industry spends on environmental and social protection and restoration. This total spending has recently been defined by the FASB (Federal Accounting Standards Board of the United States) as the environmental liability that firms must place on their balance sheets. For Western companies, and increasingly others, accurately quantified and verified accounting of environmental liabilities is also a requirement of the Securities and Exchange Commission (SEC), supported by the U.S. Sarbanes–Oxley legislation.[19]

Despite the importance placed by international financial regulatory bodies on these liabilities, the environmental regulations that actually define the cost of protection and cleanup have little or no economic basis. Environmental protection and cleanup of legacy issues can cost considerable amounts of money. Industry spends tens of billions of dollars each year on environmental protection worldwide. But, economists would argue that to justify any expenditure, there must be equal or greater benefits resulting from that expenditure. If the benefits accruing to society (including the proponent) from a project exceed the costs of implementation, then the project is worth doing. In environmental terms, therefore, what industry spends on environmental protection should have some relation to the value of the benefits that result. If costs vastly exceed benefits, then society loses—the funds could have been spent in a way that would benefit society more. Conversely, by explicitly valuing environmental assets and natural resources, appropriate restoration and protection expenditure levels can be determined. In this way, common goods such as air and water are much less likely to be treated as worthless sinks to be damaged or exploited without cost.

ENVIRONMENTAL AND SOCIAL ECONOMICS FOR INDUSTRY

Applying economic tools to the environment is a relatively new approach (when compared to our traditional economic and accounting systems) brought about by recent advancements in valuation of natural resources.[20] The underlying principle is that natural ecosystems provide services that benefit humankind, such as providing clean air to breathe, soil in which to grow our crops, animals and fish to eat, clean water to drink and irrigate our farms. Therefore, these natural resources are actually environmental assets and have measurable economic value.[21] So, when we prevent damage to a coral reef or remediate a contaminated aquifer, we can calculate a resultant benefit to compare with the cost of the action. Cost–benefit analysis can then be used to compare the net benefit (total benefit minus total cost) of various environmental and social protection options, either for a specific problem or when considering a range of strategic alternatives. This process allows all stakeholders to have their interests evaluated in a common unit of measure (money) on the same economic basis (Figure 1.7). By explicitly valuing the environment and including it in the overall economic analysis, decisions that are optimal for all of society are revealed.[22] And, if we also take the definition of sustainability at its most basic and literal—that what we do should continue to provide real benefits

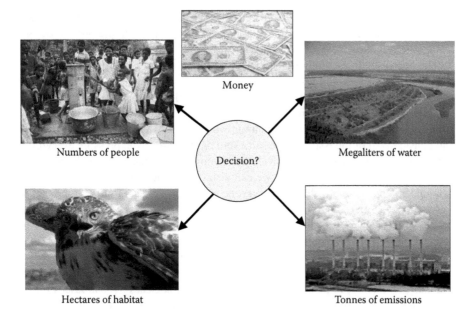

FIGURE 1.7   Sustainable decisions invariably involve trade-offs between issues with widely varying units.

over long periods of time—we end up with what is essentially an economic definition of sustainability:

> If over the long term a proposition delivers more benefit than cost over its complete life cycle, when all environmental, social, and economic factors are taken into account, using a socially acceptable discount rate, then the proposition is sustainable.

If costs are greater than benefits, then the proposition is unsustainable—society will not want to continue to fund and support the project over the long term because it simply costs more than it is worth. Even with government subsidy, eventually society will recognize that the expenditure of time, effort, energy, and materials is simply *not worth it*.

The advantages of this approach are many and are discussed and illustrated in full in the chapters that follow. Actions that drive organizations to overspend on issues that provide meager environmental benefits can be avoided, and situations for which significant expenditure on environmental protection and reclamation efforts is genuinely warranted can be identified. Equally, many of the things that we now do in the name of sustainability, we do because we *think* that they are good—we do them because they are the "right thing to do." But, as shown in the examples in the following chapters, many of the things that we do because "everyone knows that this is good," when examined in a rigorous, objective, full life-cycle environmental, social, and economic analysis, are not. The costs, energy, embedded carbon, hidden environmental damage, and social disruption involved erode and finally exceed the economic benefits produced. Equally, many opportunities for significant sustainability improvement are ignored or bypassed because they are not considered "green" or are singularly unattractive—things like heat recovery systems or simple process design changes that reduce energy consumption.

The approach discussed in this book reveals the true overall long-term attractiveness of options and compares them over a wide range of future conditions, allowing decision makers to identify robustly and consistently superior and more sustainable choices. This decision-making support approach removes the subjectivity of consensus-based ranking systems (commonly used in multicriteria-based approaches) and considers all of the issues in the same unit of measure. By expressing everything in monetary terms, we not only provide a common unit of measure that allows tradeoffs to be examined explicitly, but we can also communicate sustainability to decision makers in a language they and everyone else understands: money.

In practice, initial reaction to this approach varies across the spectrum. Interestingly, people with strong or radical views on each side of the environmental debate tend to react the same way to the concept of using money to measure *everything*. Deep ecologists (those who believe passionately that most, if not all, development should cease) reject the very idea that a price can be put on nature—it is in fact *priceless*. They fear that any attempt to place a monetary value on ecological or social issues inherently demeans those things and could never accurately reflect their true worth. Equally, right-wing industrialists, businesspeople, and their political supporters disdain the idea of putting dollar values on "intangibles" such as ecosystems, rare species, and even sometimes on social issues. Monetizing these things would be dangerous because it might reveal the sometimes large costs that they are currently not asked to pay. The fear from both ends of the spectrum is perhaps rooted in the fact that monetization may challenge preconceptions and help force an accurate and dispassionate reckoning of the real situation. In the middle, however, are the vast majority of people, who desire a reasonable compromise between their own aspirations for prosperity, those of their neighbors and the people with whom they share the planet, and the beautiful, life-giving and utterly irreplaceable planet on which we all depend for our very existence.

THE RESPONSE OF BUSINESS

Business can deal with the issue of sustainability by examining a hierarchy of possible responses, as illustrated in Figure 1.8. At the most basic level, companies can react on a philanthropic level, as many already do, by implementing overarching CR policies and engaging with local communities and stakeholders. Moving one step further, businesses can manage and protect their reputation in the eyes of stakeholders by taking actions that allow them to be perceived as part of the solution and not the problem. Going a level up, companies can develop specific strategies designed to adapt to the changes that are coming, including carbon pricing and resource scarcity. At the pinnacle of the hierarchy is a truly strategic approach by which businesses seek to redefine themselves to take advantage of the considerable opportunities associated with sustainability, and seek to achieve competitive advantage by positioning themselves not only to withstand the shocks the future will bring, but also to provide the goods, services, ideas, and commodities that the world will need to achieve a truly sustainable society.

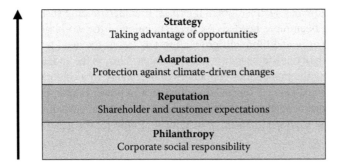

FIGURE 1.8 Hierarchy of corporate responses to environmental and social issues associated with sustainability and climate change.

OBJECTIVES AND STRUCTURE OF THE BOOK

How do we maintain and protect the benefits of industrialization, for the good of all, and work toward finding the harmonious, sustainable equilibrium we need? How do we make the idea of sustainability work, in real time, on real projects, so that industry and business can do their part in meeting this challenge? This book offers a pragmatic, tested, and highly transparent way for businesses, industry, government organizations, and policy makers to make better, more economic, and sustainable decisions that will stand the test of time. The approach is based on rationally assessing all of the key factors that go into a decision, starting with the traditionally dominant financial costs and benefits, but also including the equally important environmental and social impacts and benefits that the policy or project may entail.

This book is intended for a broad audience, including industry, government, and business decision makers; engineers involved in project design and technology selection; environmental and social technical disciplines; as well as policy making, government, and academic circles. However, the book is not written primarily for experts, and a preexisting knowledge of industrial practice or economic theory is not assumed.

Chapter 2 provides a broad overview of some of the mega-issues of the twenty-first century with which industry will need to cope to remain successful and sustainable over the long term. It examines briefly the state of the global environment, particularly with respect to the role that industry has had to play in bringing us to where we are, and the context in which industry finds itself at the start of the twenty-first century. Chapter 3 describes a rational and objective approach for quantifying sustainability and improving environmental decision making through the explicit monetization of environmental and social impacts and opportunities over a wide range of possible future conditions over the full life cycle of the project. This process is introduced as the *environmental and economic sustainability assessment* (EESA).

Chapters 4 through 7 provide a range of examples developed from real-world situations based on the author's own experience, illustrating how the full environmental-social life-cycle economic analysis embodied in EESA can reveal socially optimal sustainable choices and directions. These chapters cover four key issues: water, climate change, energy, and waste. The final chapter discusses the implications of this approach on the future of sustainability and decision making in industry.

Never before has such a rational approach to understanding sustainability in decision making been more needed. As the population of the world grows, the need for affordable energy and food will grow almost as fast as the need to preserve and protect the health of our life-sustaining environment. This, in the end, is what this book is about—facing the tremendous challenges of the twenty-first century. In this exponential era, we must balance our need for industry to power human prosperity with the equally important need to preserve and protect the natural environment of the planet. Only then can we hope to ensure a sustainable future for the generations to come.

NOTES

1. United Nations. 2007. *World Population Prospects.* United Nations, New York.
2. National Oceanographic and Atmospheric Administration (NOAA). 2008. *Arctic Report Card 2008.* U.S. NOAA. Washington, DC. http://www.arctic.noaa.gov.
3. Intergovernmental Panel on Climate Change (IPCC). 2007. *Fourth Assessment Report. The Physical Science Basis.* Cambridge University Press, Cambridge, UK.
4. Hardisty, P.E. 2008. Analyzing the role of decision-making economics in the climate change era. *Management of Environmental Quality*, 20(2) 205–218.
5. Cohen, J.E. 1995. *How Many People Can the Earth Support?* Norton, New York.
6. Martin, J. 2006. *The Meaning of the 21st Century.* Eden Project Books, London.
7. National Academy of Sciences. 2008. *Scientific Assessment of the Effects of Global Change on the United States.* Committee on Environment and Natural Resources, NAS National Science and Technology Council, Washington, DC.
8. Worldwatch Institute. 2008. *State of the World 2008: Ideas and Opportunities for Sustainable Economies.* Earthscan, London.
9. United Nations. 2008. *Global Income and Wealth Statistics.* United Nations, New York.
10. Sachs, J. 2008. *Common Wealth: Economics for a Crowded Planet.* Allen Lane, London.
11. Department for Environment, Food, and Rural Affairs (DEFRA). 2003. *The Scientific Case for Setting a Long-Term Emissions Reduction Target.* United Kingdom DEFRA, London. http://www.defra.gov.uk/environment/climatechange/pubs.
12. Intergovernmental Panel on Climate Change (IPCC), *Fourth Assessment Report. The Physical Science Basis.* Cambridge University Press, Cambridge, UK.
13. Intergovernmental Panel on Climate Change (IPCC). 2007. *Fourth Assessment Report. Mitigation of Climate Change.* Cambridge University Press, Cambridge, UK.
14. Commonwealth Scientific and Industrial Research Organization (CSIRO), Australian Bureau of Meteorology. 2007. *Climate Change in Australia.* Technical report 2007. CSIRO, Canberra, Australia.
15. Diamond, J. 2004. *Collapse: How Societies Choose to Fail or Survive.* Allen Lane, London.
16. McNeill, J. 2000. *Something New Under the Sun: An Environmental History of the 20th Century.* Allen Lane, Penguin Press, London.
17. Spratt, D., and Sutton, P. 2008. *Climate Code Red: The Case for a Sustainability Emergency.* Friends of the Earth, Fitzroy, Australia.
18. http://www.equatorprinciples.org.
19. Public Law 107–204-Sarbanes–Oxeley Act of 2002. Congressional Record Vol. 148(2002).
20. Pearce, D., and Warford, J. 1981. *World Without End.* World Bank Press, Washington, DC.
21. Costanza, R., d'Arge, R., de Groot, R., Farber, S., Grasso, M., Hannon, B., Limburg, K., Naeem, S., O'Neill, R.V., Parvelo, J., Raskin, R.G,. Sutton, P., and van der Belt, M. 1997. The value of the world's ecosystem services. *Nature*, 387: 253–260.
22. Hardisty, P.E., and Ozdemiroglu, E. 2005. *The Economics of Groundwater Protection and Remediation.* CRC Press, Boca Raton, FL.

2 Sustainability in the Twenty-First Century

A SHORT HISTORY OF SUSTAINABILITY

A 40-YEAR JOURNEY

The modern concept of sustainability has emerged over a journey of 40 years, during which individuals, governments, and nongovernmental organizations (NGOs) around the world have experienced a gradual awakening to the importance of the environment and our increasing impacts on the natural world. This journey has been punctuated by a number of seminal events that have shaped and in many ways defined the outlook of two generations. A timeline of some of these events is provided in Figure 2.1, starting in 1960 and extending to the present day.

SILENT SPRING

In 1962, Rachel Carson's book *Silent Spring* appeared in bookstores across America and quickly gained worldwide popularity.[1] The book described the increasingly noticeable and worrying disappearance of songbirds in the United States and traced the cause to the then-widespread use of DDT for the control of mosquitoes. DDT is an endocrine-disrupting pesticide, now banned in most Western nations but still used extensively in the rest of the world; it causes reproductive damage in birds, among other effects. The book, and the public outcry that resulted, brought environmental consciousness into the home of the average citizen for the first time. Here was a clear indication that while nature had always been considered vast and limitless, something we could not tame or harm, our technology and numbers were now such that we could cause real and lasting damage to the environment. This book is one of the symbols of the environmental movement that emerged from the 1960s, a time of public discussion about and increasing awareness of the damage we were inflicting on the natural world. It was one of the events that helped start a global environmental awakening.

REGULATIONS WITH POWER

In 1969, the U.S. government formed the United States Environmental Protection Agency, the USEPA. Driven by the surge in public concern for the environment, this new agency was established to protect and preserve the natural assets of the country and was afforded in law significant (and at the time revolutionary) powers to monitor the health of the waters, atmosphere, land, and biodiversity of the nation. This was the first time in history that any government had established a body uniquely dedicated to protecting the environment at this scale. In addition, that agency has

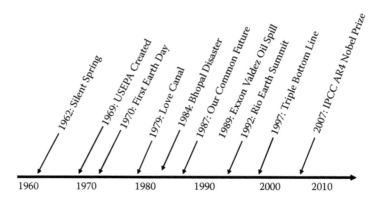

FIGURE 2.1   A timeline of global environmental and social awakening, 1960 to 2007.

been given legal and regulatory powers to enforce compliance with standards and, if necessary, impose penalties and undertake legal action against those who caused damage. Today almost every government in the world has a similar agency or department, and many are modeled on the USEPA.

EARTH DAY

The world's first Earth Day was held in 1970. Decades later, we are still celebrating the only planet known to sustain complex life, or indeed (at the time of writing) life of any kind. The growing popularity of Earth Day is another potent symbol of our growing awareness of the beauty and fragility of this jewel of life set in an incomprehensibly immense universe, with only the stars and dead planets for company. It is our only home, unique and irreplaceable, and it is vital that we care for all that makes it special, and that we realize that without its delicate perfectly evolved biosphere we could not exist.

INDUSTRY AWAKENS

During the 1960s and 1970s, as the public began to take notice of the increasing signs of the environmental impacts of our way of life, from loss of biodiversity, to city smog, to the contamination of rivers and lakes, industry was also jolted awake to the dangers of environmental liability. As regulatory sophistication and power grew, pollution was increasingly being scrutinized, and industry began to see tangible evidence that the days of uncontrolled and unmonitored discharge of waste and effluents were beginning to end.

One example is the Love Canal incident in the United States. Over decades, the Hooker Chemical Company had used an old canal within the Niagara River basin system in the northeastern United States to dispose of waste chemicals. In 1953, the company covered over the site and sold it for $1. In the late 1950s, over a hundred homes were built on the site. In the 1970s, residents began complaining of health problems, including an unusual frequency of birth defects. Then, after a particularly heavy rainfall event, organic wastes began appearing in the community. Much of the waste was highly toxic and carcinogenic, and as the steel drums in which it was held rusted away, large volumes of organic liquids were released into the groundwater. Some of it migrated through shallow aquifers and eventually found its way into the river system. Many people became sick from exposure to the contaminants, sparking an investigation that eventually resulted in several major firms, including Occidental Chemical Company (the owners of Hooker), being made to pay for the cleanup costs and damages. The offending companies were widely criticized by the public and in the media. Millions of dollars have been spent so far on remediation, but subsurface contamination is often difficult to remove completely, and money continues to be spent on the problem.[2]

This, and many other similar incidents, served to crystallize industry's understanding that they also had something to lose if the environment was not protected. Cleanup costs, fines, legal fees, and negative publicity all have a direct impact on the bottom line, and companies quickly found that the damage to their corporate reputations also made doing business more difficult.

THE EMERGENCE OF AN IDEA

The modern concept of sustainability was first expressed in 1974 in a coherent fashion by the World Council of Churches[3] and later expanded to the notion of *sustainable development* (SD) by the International Union for the Conservation of Nature and Natural Resources (IUCN).[4]

However, in the environmental and development disciplines, sustainability as a concept is commonly linked to the Brundtland Commission and its 1987 report, *Our Common Future*.[5] At the heart of their recommendations was the balancing of the competing demands of economic development and environmental protection, adopting the term *sustainable development* from the IUCN. The oft-quoted formal definition of sustainability produced by the Brundtland Commission was development that "meets the needs of the present without compromising the ability of future generations to meet their needs."

The Brundtland report explored the path that was required to achieve this vision. In particular, the commission concluded that the concept of sustainability implied limits: "not absolute limits, but limitations imposed by the present state of technology and social organisation on environmental resources, and by the ability of the biosphere to absorb the effects of human activities." However, the report made it clear that these limits could be overcome through technological advancement and perhaps most importantly through social changes. Explicit in the discussion was the concept that poverty should no longer be considered inevitable. Indeed, the report argued that poverty and environmental degradation tend to reinforce each other in a vicious self-sustaining cycle. So from the outset, sustainability has been firmly expressed in terms of social justice. The other key aspect of the Bruntland report's position was that the future itself must be safeguarded for people and the environment—that a sustainable society must allow "sustained human progress not just in a few places for a few years, but for the entire planet into the distant future." As such, the concept was firmly rooted in the notion of equity within and between generations. As Dresner stated, "Concern about sustainability must be based on moral obligations towards future generations—not just personal self-interest."[6]

Thus, from the outset, sustainability has been expressed fundamentally as an ethical construct in which the moral imperative of self-evident truths must impose itself on our economic and decision-making systems to create a new and better world.

ENVIRONMENTAL ECONOMICS

In 1980, David Pearce published one of his first works on the new science of ecological economics. Over the following years, he crystallized the concepts that many economists and environmental thinkers had been developing: the idea that the environment actually provides services to us that are of great value and so can and should be measured and expressed in monetary terms.[7] This idea is encapsulated in the term *externality*, which essentially refers to the value of something that is usually held to be external to our conventional market-based financial and economic analysis—and thus is not captured by our traditional accounting methods but nonetheless exists and has real value. Pearce argued that if we could include these values in our overall assessment, our view of what is economic and what is not would change, and the environment would thus be better protected. If a good or commodity has no value, we do not treat it as valuable or scarce, and we use it indiscriminately. At first, this notion of "putting a price on nature" was seen as anathema by conservation groups and environmentalists, and the whole notion of environmental economics languished for decades as the preserve of academics and theoreticians.

BHOPAL

In December 1984, a leak of 42 tonnes of poisonous methyl isocyanate gas at a Union Carbide agrochemical plant in India caused the deaths of over two thousand people in the nearby village of Bhopal.[8] It is estimated that up to 25,000 people have died since from diseases related to gas exposure that day. Anyone who was alive at the time will remember the horrific photos of dazed and blinded villagers streaming from their homes and the heartbreaking lines of corpses. The story made the front page of *Time* magazine worldwide. Industry was provided with a horrendous lesson on the importance of running their operations in a way that protects society and the environment and the consequences of failing to do so. The importance of corporate social responsibility (CSR), which has become a watchword of good corporate governance in the twenty-first century, was tragically crystallized at Bhopal.

EXXON VALDEZ

When the supertanker *Exxon Valdez* ran aground off the coast of Alaska in March 1989, it released over 40 million liters of crude oil into one of the most sensitive ecosystems in the world. The slick covered over 28,000 km^2, and the world watched aghast as the oil washed up onto the pristine coastal wilderness, killing thousands of birds and marine mammals.[9] Exxon poured over US$2 billion into the cleanup, which was also supported by a huge outpouring of volunteer and government effort. In the end, the cleanup costs were estimated at over US$3 billion. Exxon was pilloried in the press, was fined US$1.15 billion in the courts,[10] and was assessed another

US\$4.5 billion in punitive damages.[11] This was one of the biggest indications yet that industry and business needed to pay attention to their environmental performance, and that a poor record could directly affect profits and shareholder value. Interestingly, the huge expenditure on the cleanup (billions spent on goods and services in the local economy) provided a major boost to the gross domestic product (GDP) figure for the state of Alaska that year. Since GDP did not, and still does not, reflect the value of damage caused to the environment, the *Exxon Valdez* incident appears in the economic record as a positive event. More information on the numerous shortcomings of GDP as a measure of real wealth in society are discussed in Chapter 3.

The Fight to Save the Ozone

In the 1980s, thanks in part to work by James Lovelock, later to become the father of Gaia theory, the world awoke to a new threat. Widespread use of a new family of man-made chemicals, the chlorofluorocarbons (CFCs), was causing the natural ozone present in the outer atmosphere to disappear. The effect was most apparent at the South Pole, where a large hole in the ozone layer had been detected. Ozone protects the Earth against the withering intensity of the ultraviolet radiation hurled at us from the sun; without it, life as we know it could not exist. It was a problem that we could not see with the naked eye. We had to trust in the measurements and projections of scientists, and what is more, it was everyone's problem. It did not matter where the CFCs were emitted, they damaged the same ozone layer that protected everyone.

In response, the governments of the world got together in 1987 in Montreal and developed a protocol for rapidly phasing out the use of the ozone-destroying chemicals and replacing them with less-harmful substitutes. The Montreal meeting showed that there were indeed environmental issues, caused by our industrial activities, that were global in nature—that we could in fact affect not only a nearby river or village but also the whole of the planet and everyone on it. The Montreal protocol also demonstrated clearly that the world could indeed act quickly, and in a united way, to deal with a significant global environmental problem. This understanding will become increasingly important as time passes.

Sustainable Development Is Born

Relatively rapidly after the release of *Our Common Future*, governments, NGOs, international agencies, and corporations began adopting the new term *sustainable development* (SD), each shaping the definition to better suit its unique perspectives and needs. In 1992, still basking in the euphoria generated by the end of the Cold War, the world assembled in Rio de Janeiro for the Rio Earth Summit (United Nations Conference on Environment and Development). All of the money that had hitherto been spent on weapons and maintaining mutually assured destruction capability could perhaps now be diverted to making the world a better place. In the enthusiasm of the summit, a host of visionary treaties and protocols was developed and signed, including agreements on forest principles, biodiversity, and perhaps most importantly, climate change (the U.N. Framework Convention on Climate Change, UNFCCC). The idea of sustainability had finally become mainstream, and people, organizations, and governments from every part of the globe rallied around the idea that we had to act urgently to shift our behavior to a more environmentally and socially friendly platform. Sustainability was to replace doctrinal conflict as the new order of the world.

But, the notion of SD as defined by Bruntland and others, and as subsequently adapted and developed by various organizations around the world, has also been widely criticized. Particular criticism has focused on the fact that SD appears to offer all things to all people and can be readily interpreted to mean that we can have as much of each as we want, both development *and* sustainability, without sacrifices or trade-offs.[12] In some ways, this has explained the wide support for SD; it has provided a useful and popular message that offers something for everyone. Many environmentalists feel that it has simply been used as a screen to legitimize development.

Strong versus Weak Sustainability

Sustainability is fundamentally about risk management. Depletion of the natural resources of the planet is clearly fraught with risks to current and future generations. But, history has repeatedly shown that human ingenuity has been able to solve problems and, through the application of technology, develop solutions to problems previously believed to be intractable. The proponents of "strong sustainability" argue that we cannot be sure that technological solutions will be found to the problems of resource depletion and the degradation of the natural world in the future; therefore, we must prevent any net depletion of natural capital, even, theoretically, depletion of nonecospheric natural capital such as mineral resources and hydrocarbon reserves. The "precautionary principle" is widely used to express the need to mitigate against risks that are believed to be too catastrophic to contemplate, such as dangerous climate change and widespread ecosystem loss.

In contrast, supporters of "weak sustainability," including many economists, confidently predict that humankind will continue to find new ways to substitute technological progress for natural resources. They argue that overspending on protecting ourselves and the environment against difficult-to-assess future risks is inappropriate because it does not allow for the ability of future generations to find new, as-yet-undiscovered, ways to deal with the problems. Depletion of ecological natural capital within the biosphere can occur but must be balanced.

The Triple Bottom Line

In 1997, John Elkington coined the idea of the "triple bottom line."[13] The traditional view had been (and still largely is) that companies exist to generate profit for shareholders and nothing more—they manage the bottom line and ensure that it is positive. The triple bottom line concept advanced the idea that as companies go about their business, they must, in this new more environmentally and socially aware and regulated world, also pay attention to how they interact with environment and society. In fact, there are now three bottom lines: environmental, social, and financial. This concept is now widely quoted in corporate literature and increasingly appears as part of corporate responsibility (CR) statements and policies for companies. However, as discussed next, while the concept may be laudable, it has had little real effect so far on overall environmental and social performance worldwide.

THE NOBEL PEACE PRIZE

The 2007 Nobel Peace Prize was jointly awarded to the members of the Intergovernmental Panel on Climate Change (IPCC), for their Fourth Assessment Report, and Al Gore, former vice-president of the United States, for his work on bringing the issue of climate change to widespread attention around the world. As discussed here, climate change is one of the (but not the only) most pressing sustainability issues that we face today. The award of a Nobel Prize for work on an issue of sustainability reflects how far humanity has come in its understanding and acceptance of the issue.

Interestingly, the award has paradoxically also focused the politicization of sustainability. Many, particularly in the United States, have come to see sustainability as the preserve of the liberal, left-wing part of society. This is a shame. As this book will show, balanced and rational sustainability, objectively applied, is a good for business, good for profits, and good for society.

FROM CONCEPT TO CORE PRINCIPLE

As shown, the concept of sustainability has been around for a long time. It is almost a generation since the Brundtland Commission called on the world to start balancing the competing demands of economic development and environmental protection and to undertake development that "meets the needs of the present without compromising the ability of future generations to meet their needs." Since then, the concept of SD has been adopted, coopted, and at times abused by literally thousands of governments and organizations around the globe.

The definition of sustainability, in the modern context, remains elusive. Depending on one's professional background, employment, education, and political orientation, the term can have many definitions. Environmentalists tend to focus their view on the protection of natural habitat, species, and overall environmental quality. International development organizations and the governments of less-developed countries concentrate on poverty alleviation. Developers and industrial corporations talk about how they will build infrastructure, exploit resources, and produce products in a way that can continue over longer periods of time. This spectrum of definitions and understandings is perhaps one of the failures of sustainability—that there appears to be no uniformly consistent definition that everyone understands and can clearly articulate.

Yet, despite this, there seems to have developed, over the last 40 years of the modern sustainability movement, a basic understanding and agreement on at least two of the central themes of sustainability: First, the health of our natural environment is vitally important to our own well-being and prosperity; second, we should conduct ourselves in a way that allows our children and grandchildren to live at least as well as we have. These are perceived and expressed largely as ethical and moral concepts, part of a new definition of right and wrong that seems to have bridged cultures and national boundaries to emerge as what is perhaps among the few universal human truths.

THE CHALLENGES OF THE TWENTY-FIRST CENTURY

Over the last 40 years, then, the people of the world have been on a journey of environmental and social awakening, culminating in the realization that the world is not infinite, that we are now so numerous and have at our disposal such power and technology, that we can significantly affect our environment, often to the point of causing real and lasting damage. This realization has led to the widespread internalization of the concept of sustainability, in its traditional Bruntlandian sense, as a fundamental core principle in the minds and moralities of much of Earth's citizenry. Indeed, most governments, major international corporations, NGOs, and other organizations have now formally recognized sustainability as a core value and guiding principle. The evidence is there in public policy, government rhetoric, corporate sustainability policies and reports, and the burgeoning of thousands of environmental organizations, societies, and think tanks around the world, from Albania to Australia.

Over this same period of fundamental awakening and realization, therefore, we should expect to have witnessed a commensurate and real change in our overall behavior toward the resources, environment, and people of this planet. Given the broad-based support and almost universal acceptance of sustainability's principles and goals, it would be logical to anticipate a broad improvement in some, if not all, of the key metrics with which we measure sustainability. Unfortunately, this has not been the case.

An Overview of Global Trends

A number of international, authoritative surveys of the state of the planet have been completed by organizations such as the United Nations.[14] These exhaustive in-depth reviews have covered the widest spectrum of sustainability indicators and issues, from ocean health to poverty, from biodiversity to health and sanitation. The picture they paint is uniformly one of accelerating degradation of the natural capital of the world, of forest and biodiversity loss, significantly depleted marine resources, growing atmospheric pollution, declining and polluted water resources, and increasing numbers of people living in poverty.[15] These trends are clearly underpinned by the fundamentals of a rapidly growing world population, the legitimate aspirations of 2 billion of those people to rise above their current level of poverty, and an expanding world economy. More people, with greater demands, put increasing stress on the natural environment that provides the food, water, and raw materials necessary for that prosperity. Now, overlaid on this, there is clear evidence that climate change is beginning to affect the way the weather systems of the planet operate. Predictions are that these changes will only exacerbate the decline of our already weakened natural environment. Taken together, these overlapping and mutually reinforcing effects are now increasingly being called simply "global change."[16] This new terminology signals a broad-based scientific realization that the real problem facing us right now is a sum of the compounding effects of simultaneous overexploitation of resources, unchecked emissions of pollutants, and anthropogenic climate change.

Our Changing Relationship with the Planet

In his book *The Meaning of the 21st Century*, James Martin describes the unique coming together of circumstances that makes this new century unlike any ever before experienced by humankind.[17] The basic tenets of our relationship with the environment and the planet have for the first time in human history shifted fundamentally. Our ability to harness energy and use increasingly powerful technology has allowed us to tip the balance in our epochal struggle to subdue nature. Human hands are now reshaping the planet to a degree such that untouched natural habitat is now rare and becoming rarer. We are no longer a small band trying to survive in a hostile environment; we are now many, and the population of the world is rising inexorably. The U.N. Population Project and the U.S. Census Bureau both put current world population at above 6 billion, and both estimate that by 2050 we will reach close to 10 billion.[18] The current rate of growth, about 1.3% per year, means the equivalent of a city the size of Calgary (Canada), Perth (Australia), or Aberdeen (Scotland) is added to the planet every week. Most of the 4 billion new souls who must be clothed, fed, and housed on the planet over the next 40 years will not be living in the United States, Europe, Canada, or Australia, but in the poorest countries, where vast numbers already live in conditions that can only be described as shockingly miserable. Of this new 4 billion, it is estimated that more than 60% will be born and will grow up (for many more who are born will not grow up but will die of malnutrition, preventable disease, and violence) in the vast and growing urban slums that sheathe many of the major southern cities of the world.

More People, Less to Go Around

The current equation is essentially quite simple: more people, less planet to go around. The most direct effect of the large and rising world population is that humans are now for the first time in history using more of everything that the world gives us (topsoil to grow food, fish and animals to eat, water to drink and grow food, wood to use and burn) than the biosphere of the planet produces and renews each year.[6] In effect, we are now mining our natural capital.

The platform of prosperity that has served us so well for the last 20,000 years is now rapidly being stripped down. Between a third and a half of the original forest cover of the planet has been cleared.[7,8] Topsoil, centuries in its accumulation and the basis for modern agriculture, is being lost at alarming rates, with almost one-quarter lost according to recent studies.[9] The oceans of the planet are in deep trouble. It is estimated that over 90% of the edible fish stocks of the world have either been destroyed or are being fully exploited,[10] and overall ocean health is threatened by pollution, mismanagement, overfishing, and climate change.[11] Biodiversity, the intricate web of marine and terrestrial life that makes the world a living place, is also being seriously eroded by the combined forces of deforestation, land clearing, widespread chemical pollution, overexploitation by hunting and harvesting, and climate change.

The sections that follow provide more detailed information on the state of current research into each of these important areas: energy, food and poverty, biodiversity, the oceans, water, and climate change. However, in many ways, it is the combination of all of these trends—a large and rising world population demanding increased affluence and

FIGURE 2.2 Providing for the 3.5 billion additional people who will arrive on earth over the next 40 years, most in developing countries, will put significant new pressures on an already-stressed world ecosystem. An African child's ecological footprint is only a fraction of that of a child born in the developed world.

(rightfully) an end to poverty; the deterioration of air, soil, and water quality in much of the world; massive loss of biodiversity on land and in the sea; and arching over all of this, the prospect of serious and irreversible climate change—that provides the essential context for this book and the challenge facing us all in the twenty-first century.

PROVIDING A DECENT STANDARD OF LIVING FOR 10 BILLION PEOPLE

With more people on the planet, all demanding a standard of living similar to that of the developed nations, industry must expand to provide the necessary goods and services. How can industry continue to provide the essential energy, food, and products the world needs without contributing to further unacceptable damage to the planet? As discussed in the following sections, current methods—the business-as-usual approach—have provided prosperity for the current 6.5 billion people at too high an environmental and social price. The current path is unsustainable. The next 3.5 billion's aspirations cannot be met using the same methods, systems, and decision making without significant risk of even more intense damage (Figure 2.2).

THE FOSSIL FUEL INDUSTRY: AN EXAMPLE

The ability of the fossil fuel industry to provide large quantities of relatively inexpensive energy to the world has been one of the chief contributors to the amazing levels of prosperity and achievement that have been experienced over the last century. Oil, gas, and coal today provide about 80% of the world's energy. However, these industries also directly or indirectly contribute to many of the major negative environmental and social trends listed and discussed in more detail in this chapter.

As demand for energy has grown, exploration for hydrocarbons has moved to more and more remote areas. In turn, this has led to the opening of many previously inaccessible areas. Driving roads and rail lines into remote wildlands and habitats provides increased public access, which can lead to exploitation and damage of the natural environment. Disruption of indigenous local communities can also follow as new migrants displace traditional societies, clear land, and bring in unfamiliar technology and ways. As newfound resources are developed and exploited, new impacts arise. Fossil fuel production can be water intensive, and particularly in arid areas already susceptible to water stress mining operations can have significant impacts on local water supplies. Water quality may also be affected from a variety of discharges and wastes.

Offshore developments often involve dredging of sensitive habitats and the discharge of various types of effluent, including large volumes of produced formation water, into the marine environment. Production, refining, and shipping of fossil fuels consume significant quantities of energy in their own right, creating air emissions of all kinds, including greenhouse gases (GHGs).

Fossil fuels, particularly coal, are abundant and relatively cheap to extract and process. There is a massive existing stock of capital invested in the global fossil fuel infrastructure that will continue to operate and expand over the decades to come. Realistically, the world will not and cannot walk away from these fuels. The challenge will be to find ways to continue to find, extract, and use oil and coal in a way that is far less damaging to the Earth and its people.

The technologies and systems exist now to dramatically reduce the impacts on biodiversity during exploration, improve end-of-life reclamation of mines, reduce waste, improve water and energy efficiency at all stages of the production life cycle, and significantly if not completely cut the emissions of GHGs that result from the combustion of fossil fuels. However, despite some notable successes, many of the things that can be done to improve sustainability are not being implemented—and the reason is cost.

Improving sustainability costs money (investment in technology and systems requires investment). But even when these changes are actually cost negative overall, if payback periods are too long, or if internal rate of return (IRR) hurdles are too high, they will not be undertaken. This is the "NPV (net present value)-IRR trap," explored in more detail in Chapter 3. The result is that, despite our awakening, industry remains trapped in business-as-usual by the forces of traditional financial and economic thinking.

THE METRICS OF SUSTAINABILITY

Over the last several decades, governments at all levels, NGOs, and even private corporations have begun to steadily track a wide range of key indicators of sustainability. Data are now being collected that allow us to measure the amount of water we are using, pollution loads to atmosphere, carbon emissions, planetary biodiversity, fish stocks, and a huge variety of other factors. The following sections provide a brief overview of a few of the key "megatrends" of the twenty-first century, specifically in the areas of food and poverty, water, biodiversity, the oceans, and climate change. The discussion is not meant to be exhaustive by any means but rather to provide context to the evolving picture of sustainability. Examining some of this information allows us to gauge, in an overall sense, how well the concept of sustainability has served us.

FOOD AND POVERTY

One of the most immediate and basic indicators of overall progress is the access to good food and freedom from the ravages of poverty. Over the last 40 years, the world's population has risen from 2 billion to over 6.5 billion. The rate of population growth peaked at about 2% per annum in 1970 and has now dropped to about 1.3% per annum. But because we are so many, we are still on course to add 70 to 75 million to the population each year well past 2020.[19] During the last 40 years, the number of undernourished has risen steadily, so that today one in seven human beings goes without enough food, and every sixth child dies of hunger or hunger-related causes.[20] In fact, the number of people going hungry is increasing every year, rising by 75 million in 2007 and 40 million in 2008.[21]

The Hidden Costs of Food

So, while the "green revolution" of the 1960s and 1970s dramatically increased food production, largely through the application of energy and fertilizers in agriculture, more needs to be done to keep pace with the growing population. Globally, food production is more dependent than ever on large inputs of fossil energy, both for running mechanized farming systems and for creating the synthetic nitrogen-based fertilizers (made from natural gas) on which 40% of world's food production relies (this figure is expected to rise to 60% by 2050).[22] This also means that embedded in the calories the world consumes is an increasing carbon emission footprint.

Food production also accounts for 90% of total freshwater use on the planet, and in many parts of the world freshwater resources are already stretched to the limit—aquifers are being depleted, and many major river systems, such as the Colorado River in the United States, the Murray-Darling system in Australia, and the Yellow River in China, are so overused that they routinely now fail to reach the sea. The natural ecosystems that these river systems support are at the bottom of the water allocation list. In many parts of the world, this situation is exasperated by heavy use of chemical fertilizers, which pollute surface and groundwater resources. In addition, loss of fertile cropland to salinization and soil erosion continually reduces our food-growing potential. The damage to freshwater systems worldwide has been inestimable. There is a considerable hidden cost to modern food production, one that is growing and is still rarely reflected in market prices.

Feeding a Growing Population

These combined effects mean that while overall global food production continues to rise, the rate of growth has slowed considerably since the green revolution, and in some parts of the world has actually peaked and is in decline (Table 2.1). Major questions now exist around our ability to sustainably feed the current 6.5 billion people on

TABLE 2.1
Food Metrics

Factor and Location	Baseline and Year	Current	Causes and Implications
Grain: China[25]	392 million tonnes (mt) in 1998	322 mt in 2006	China is now the major importer of grain on the world market.
Total arable land devoted to cereal production[26]	7.2 million km² in 1980	6.6 million km² in 2002	Arable land is being lost to salinization, soil erosion, and encroachment of urban areas.
Rice: Australia[27]	1.64 mt in 2001	0.019 mt in 2008	Severe drought in Australia dramatically reduces rice production.
Global wheat production[28]	683 mt in 2007—record global production	667 mt in 2009	Global wheat production has increased almost linearly year after year since the 1960s, from 220 mt in 1960 to the 2007 record.[29]
Atlantic cod: Canada[30]	800,000 tonnes in 1968	0 tonnes—resource destroyed—1993	Chronic overfishing, lack of regulation, warnings of impending crash were not heeded for economic reasons. Tens of thousands of jobs were permanently lost, along with permanent loss of resource.
Sturgeon fishery: Caspian Sea[31]	20,000 tonnes in 1998	1,400 tonnes in 2002	Overfishing, pollutions of rivers from industrial sources, destruction of nesting grounds from oil development. Fishery revenues and jobs lost.

Earth, let alone the additional 3.5 billion or so who will be added to the world's population over the next 40 years.[23] Figure 2.3 shows that the area of arable land under cultivation has risen less than 10% since 1950 (there is a finite supply of arable land), and the area devoted to cereal production has actually dropped significantly, due to land degradation and urbanization, since a peak in the mid-1980s.[24]

Poverty: Progress and Setbacks

Poverty and hunger often go hand in hand. In 1998, over 3 billion people, half the world's population, lived on less than US$2/day. More than 1.3 billion lived on less than US$1/day.[32] By 2005, half the population was living on less than US$2.50/day (3.2 billion people), and the number of people living on less than US$1/day had dropped to 0.9 billion. Globally, poverty fell by about 10% between 1981 and 2005, but almost all of that improvement was in China, where over 600 million people were removed from poverty over that period. There was little real progress on poverty reduction over the same period in the rest of the world.[33]

This level of poverty worldwide seems to contradict global economic indicators of success. Global average per capita GDP rose from US$5,927 in 1987 to US$8,162 in 2004, but the vast majority of this increase in personal wealth was experienced by

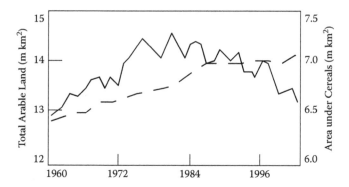

FIGURE 2.3 Area of arable land (solid line) and area used for cereal crops. (After data from United Nations Environment Programme [UNEP], *Global Environmental Outlook 4: Environment for Development [GEO-4]*, UNEP, Nairobi, Kenya, 2007.)

relatively few people. In fact, the latest United Nations data showed that wealth and income are concentrated as never before in very few hands. In 1968, people living in rich countries were on average thirty times better off than those living in countries where the world's poorest 20% live. By 1996, they were sixty-one times better off, and by 1998 they were eighty-two times better off. This growing disparity between rich and poor is further illustrated by the global statistics on distribution of wealth. In 2000, the richest 1% owned 40% of all the wealth on the planet, and the top 2% owned more than half of everything, leaving the poorest half of the population controlling less than 1% of the wealth.[34] By 2005, wealth was even more concentrated, with a full quarter of the riches of the world owned by 0.13% of the population. Over the last 40 years, the rich have become much richer, and the numbers of poor have increased. A truly sustainable world is one in which all people are free from hunger, disease, and the ravages of poverty.

The last 40 years of talking about sustainability seem to have had little effect. Business as usual continues to concentrate wealth, and international aid programs have been inadequate to materially improve the lives of hundreds of millions of the poorest people in the world, who continue to suffer with no end in sight.

WATER

An Unevenly Distributed Renewable Resource

Of all the water on the planet, only 3% is fresh, and of that the majority is locked away as snow and ice at the poles. Less than a third of the freshwater on Earth is actually available to support the ecosphere, flowing in lakes and rivers, falling as rain, and filtering slowly through underground rocks as groundwater. The sun acts as a huge desalination plant, evaporating freshwater from the seas and driving it back to Earth as rain in a continuous hydrologic cycle.

Freshwater is a renewable resource, but a finite one, and unevenly distributed. Some parts of the world, like Canada and Finland, are blessed with huge water

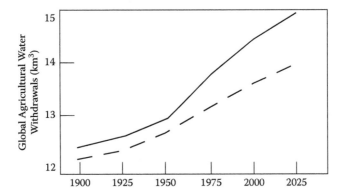

FIGURE 2.4   Total global agricultural water withdrawals 1900–2025 (after United Nations Environment Programme [UNEP], *Global Environmental Outlook 4: Environment for Development [GEO-4]*, UNEP, Nairobi, Kenya, 2007). Solid line is total withdrawals for agriculture, and dashed line is consumptive use. The difference between the two lines represents water that is wasted or returned to the system.

surpluses. Others, like sub-Saharan Africa and the Middle East, experience chronic shortages. We are also using more water every year as populations grow and as people's lives become more water intensive. In 1950, the world used about 1,360 km^3 of freshwater. By 2000, we were using over 5,190 km^3, almost a fourfold increase.[35] Figures 2.4 and 2.5 show past and future expected water withdrawals for agriculture and industrial use, respectively.[36]

Moving Water

In ancient times, civilizations were born and flourished in places where water was plentiful and available. The city of Sana'a in Yemen was founded in a wide valley surrounded by mountains of porous and permeable volcanic rocks, where perennial

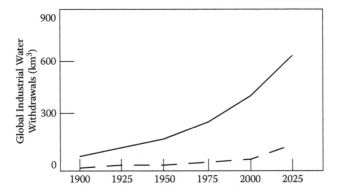

FIGURE 2.5   Total global industrial water withdrawals 1900–2025 (after United Nations Environment Programme [UNEP], *Global Environmental Outlook 4: Environment for Development [GEO-4]*, UNEP, Nairobi, Kenya, 2007). Solid line is total withdrawals for industry, and dashed line is consumptive use. The difference between the two lines represents water that is wasted or returned to the system.

springs bubbled up pure and sweet from the ground (but no longer—overpumping has dried up the springs, and the shallow aquifer is now badly contaminated with sewage). But as populations expanded and the needs of agriculture grew, water had to be harvested and moved to where it was needed. From Roman aqueducts, to Victorian distribution networks, to the major dam and interbasin transfer schemes of recent times, capturing and moving water has been a human preoccupation on the grandest of scales. There are over 45,000 large dams currently in existence in 140 countries, and China alone is planning another hundred.[37] Today, over 60% of the world's largest river systems have been significantly reengineered, dammed, or canalized to move water to the places we want it.[38]

Increasing Water Stress

Worldwide, the number of people who lack access to safe, clean drinking water exceeds 1.1 billion and is increasing every day. Over 2.6 billion people lack access to basic sanitation services. If present trends continue, according to the United Nations Environment Programme (UNEP), 1.8 billion people will be living in conditions of absolute water scarcity by 2025, and two-thirds of humanity would be subject to water stress.[39]

The main problem with water, according to the *Economist*, is not scarcity, but "man's extravagantly wasteful misuse of it."[40] In its survey of the water of the world, the *Economist* argues that water has been ill-governed and hugely underpriced. Cheap, or indeed even free, water encourages waste and misallocation and ignores the huge costs of the dams, reservoirs, pipelines, and pumping systems needed to deliver it. It also leads to using water for the wrong things, in the wrong places.

Irrigation of water-intensive crops, using inefficient methods, is one of the worst offenders. The small Mediterranean island of Cyprus has some of the best rainfall and water use databases in the world, stretching back to the middle of the nineteenth century. Despite some recent years of average rainfall, Cyprus is in the midst of a pro-tracted downward shift in average rainfall that began almost 30 years ago and is a direct consequence of climate change.[41] And, while it is raining less in places like Cyprus and Australia, we are using a lot more water. Again, echoing the trends in many other countries, agriculture accounts for almost 70% of water withdrawn in Cyprus, much of it for unlicensed small private farms and the growing of thirsty crops such as bananas. Per capita water use in Cyprus continues to grow. Perhaps the most outrageous example of water misuse is the cultivation of wheat in Saudi Arabia using water from oil-fired desalination plants. In that climate, it takes 1,000 tonnes of water to produce 1 tonne of wheat, making the real cost of Saudi wheat about one hundred times the world price.

The explosion in worldwide construction of desalination plants is a direct response to growing water scarcity and growing demands. Table 2.2 shows the growth in both the number and the capacity of desalination plants worldwide.[42] The vast majority of the more than 12,300 plants now operating globally (60% of which are in the Middle East, with increasing numbers in the United States, Australia, and Europe) are pow-ered by energy based on fossil fuels and thus represent a significant anthropogenic climate change feedback loop: Climate change-driven drying trends mean that we need more water, so we respond by burning oil, gas, or coal to make water, adding yet more CO_2 to the atmosphere, reinforcing the effects of climate change.

TABLE 2.2
Worldwide Desalination[43]

Year	Number of Facilities	Total Global Capacity (millions m³/day)
1985	4,600	10
1993	7,500	30
2007	12,300	47

Water Is Life

In our preoccupation with our own needs, it is easy forget that water is also vital to all other forms of life on Earth. And unlike us, our fellow creatures cannot shape their world; they depend entirely on finely balanced ecosystems, to which they have adapted over millennia. In our thirst for more and more water, we risk leaving nothing left for Mother Nature. Our attempts to harvest water and move it to where we need or want it most have vastly reshaped the natural hydrology and ecology of huge parts of the planet. The Colorado River in the United States, one of the largest in the world, has been dammed and diverted and its water allocated and siphoned off to the extent that not a drop now flows into the Sea of Cortez. The Yellow River in China and the Ganges in India suffer the same fate. The Everglades in Florida, one of the largest freshwater ecosystems in the world, supporting countless endangered and endemic species, has been engineered almost to the point of extinction, largely to feed the demands of agriculture.[44] The Aral Sea in Central Asia has been virtually destroyed since the 1950s, when Soviet engineers, working under a similar slogan to the one adopted in Cyprus ("water which is allowed to enter the sea is wasted"), started to divert the two rivers that fed the Aral. The plan was to use the water to grow cotton, a notoriously "thirsty" crop, in the desert-like plains of Kazakhstan. What resulted was one of the worst environmental disasters ever. The Aral Sea has shrunk to nearly half its area, losing over 70% of its volume. Its water has become so salty that all the fish have died, and the dried out seabed has turned into a dust bowl. The salt flats that have been exposed by the retreating sea are now the source of more than 77 million tonnes of salt and chemical-laden dust whipped up by the winds and carried back onto the cotton fields, reducing soil fertility and lowering yields (Figure 2.6).[45] A cruel irony indeed.

Water Pollution

Unfortunately, water has also become our favored medium for moving waste. Where once water had an almost spiritual power, giving life and sustaining communities, it has now become a carrier for every imaginable form of refuse, from sewage to toxic chemicals. Scarce and valuable freshwater is being polluted at an alarming rate, putting even greater stress on the supplies that remain. In many parts of the world, agriculture is a major contributor to water pollution as well as the major water user. In Europe, contamination of aquifers by nitrogen compounds from fertilizers is widespread and chronic and has been identified as a

FIGURE 2.6 The Aral Sea from the air, 2001. The expressions of successive retreating coastlines appear as a series of concentric lines formed as the sea shrank year by year. White areas are exposed salt flats.

major challenge by the European Environment Agency. Worldwide, between 1970 and 1995 nitrogen flows to river systems increased by a third.[46] Increasingly, persistent and often toxic herbicides and pesticides are showing up in groundwater supplies, the result of careless and uncontrolled use by farmers and households. A wide variety of industries, from mining to steel making to computer chip manufacturing to wood preserving, use and dispose of chemicals that can and in some cases do contaminate groundwater.[47] In China, water scarcity exacerbated by pollution in 2004 was estimated to have cost the country over 147 billion yuan, or 1% of GDP. And, while there have been some notable improvements, such as the overall reductions in DDT levels in rivers in China and Russia between 1988 and 1994, the overall global trend is that availability of freshwater resources continues to decline as the combination of excessive withdrawal, pollution, and the growing effects of climate change takes its toll.[48]

Water and Industry in the Middle East

An example of the important relationship between industry and water is the oil and gas sector in the Middle East. While blessed with huge reserves of oil and gas, the Middle East is also water poor. It is one of the driest regions on Earth, where 5% of the world's population have access to less than 1% of the planet's freshwater resources.[49] But far from being a "desert," it is a region of surprising ecological wealth and diversity. From the Nile Delta to the Iraqi marshlands, the

Middle East boasts a unique but fragile natural heritage that depends on naturally occurring freshwater. As populations and economic activity have expanded, driven by a thriving petroleum industry, so have the stresses placed on this most fundamental of resources.

In a major 2002 study, the UNEP identified key environmental issues facing the Middle East.[50] Freshwater leads the list. Water resources in the Middle East are under stress due to overexploitation and pollution. In Sana'a, Yemen, for example, near-surface aquifers, recharged by rainfall and used sustainably for centuries, have become polluted, forcing exploitation of deeper fossil water (recharged only slowly over hundreds or thousands of years). These deeper groundwater resources have in turn become overexploited, resulting in falling water levels, increasing costs, and within the next decade, aquifer exhaustion.[51] In Qatar, all available natural sources of freshwater are currently being exploited, and many have been damaged through intrusion of seawater caused by overpumping.

Depletion of natural freshwater supplies can lead to significant impacts on local ecosystems and biodiversity. The Azraq basin in Jordan is a clear case in point. Once a rich and thriving inland wetland system, home to hundreds of unique species, systematic large-scale pumping of groundwater for irrigation has dried up the springs that fed the wetlands, causing wholesale loss of habitat on a scale visible clearly on satellite photos.[52] The Mesopotamian marshes in Iraq covered over 20,000 km^2 in the 1950s. Decades of draining and water diversions reduced their area to less than 400 km^2 by 2000, with some slight recovery occurring since the end of the second Gulf War.[53] Similar losses have occurred throughout the region.[54]

In a way, the Middle East is a microcosm of the challenges facing industry worldwide. The petroleum industry of the region operates within a context of a rich and diverse but already stressed environment where scarce water resources are at the core of the issue. Every part of the industry's development cycle, from exploratory drilling to the development of production infrastructure, onshore and offshore, through to construction and operation of refineries and terminals, has the potential to have an impact on the environment. Water resources, already overstressed, are especially open to impact. Contamination of aquifers by crude oil or refined hydrocarbons can render water unfit for use for decades, and restoration is extremely expensive.[55] Major terrestrial oil spills throughout Kuwait and Iraq continue to threaten groundwater. Introduction of produced water into the subsurface can have similar effects and, considering the potentially large volumes involved, can be widespread. Ongoing operations generate effluents and by-products, including produced water, requiring treatment and disposal.

Global Trends

The most recent compilations of global data on freshwater resources and use are unanimous: The world is using more freshwater, and supplies are dwindling.[56] The water we do have is being increasingly contaminated, and climate change is predicted to exacerbate water scarcity worldwide.[57] The business-as-usual ways of managing water threaten our ability to feed a growing population, could affect the future prosperity of the world, and will put an even heavier burden on the poorest people of the world.

BIODIVERSITY

Ecosystem Services

From a purely selfish, anthropocentric perspective, we should care about biodiversity for one reason: The natural ecosystems of the Earth provide the essential elements, climate, food, and resources on which all life, and thus our own survival, depends. The ecosystems of the world have evolved over millions of years to provide the ideal conditions for human life, and because they have done so, we have flourished. Ecosystems provide a huge range of services to humanity, including

- support services, such as nutrient cycling, soil formation, and the massive primary productivity on which all of life depends
- provision services, such as food, fuel, freshwater, and a huge range of materials on which our economy depends
- regulating services, including climate, flood, and erosion control; water purification; and pollination, without which modern agriculture would not be possible
- cultural services, including aesthetic (beautiful landscapes), spiritual, recreational, and educational opportunities, without which the entire tourism and holiday industry would not exist (Figure 2.7)

FIGURE 2.7   An alpine ecosystem in British Columbia, Canada. This system provides primary nutrient cycling and water purification and storage (both as groundwater and in the glaciers); acts as a store of valuable biodiversity; sequesters and stores carbon; and is a major generator of tourism revenue.

The other side of the coin is that biodiversity loss affects human beings directly in many ways, including

- increased food insecurity
- increased energy insecurity
- health impacts
- declining water quality and quantity
- erosion of cultural heritage

The modern awakening to the plight of the environment of the world found expression in the damage that pollution, land clearing, and overexploitation of resources was causing to the creatures with which we share this planet. Over the last 40 years, a huge amount of data gathering and research by professionals, academics, NGOs, and private citizens has revealed a shocking story: We are literally eating away at the global treasure of biodiversity. Perhaps in no other area is the unsustainable character of humankind's current path more visible and frightening. There is no part of the biosphere that we have not affected, no country or region that has been left untouched by our activities.

The Living Planet Is Ill

The Living Plant Index was created to act as a measure of the overall health of the planet's ecosystems and biodiversity.[58] The 2008 global index was based on 4,642 populations of 1,686 species covering all of the major biomes around the world. Figure 2.8 shows the startling decline of 28% in overall biodiversity since the 1970 baseline year.[59] Table 2.3 shows the data broken down into some of the various

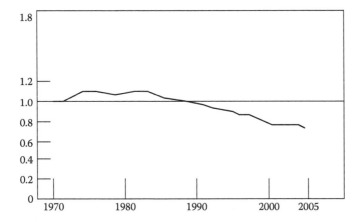

FIGURE 2.8   Living Planet Index (after WWF, *Living Planet Report, 2008,* World Wide Fund for Nature, Gland, Switzerland, 2008). An index score of 1.0 represents baseline conditions in 1970.

TABLE 2.3
Living Planet Index Subindices

Index	Change 1970–2005 (%)
Global	−28
Tropical	−51
Terrestrial	−33
Marine	−14
Freshwater	−35
Dryland ecosystems	−44
Grassland ecosystems	−36
Birds	−20

component indices, including freshwater species, forest species, and marine species. The plight of tropical ecosystems, increasingly under threat from logging and land clearing for agriculture (including notably palm oil and biofuel crops), is particularly worrisome. Over the last 40 years, tropical biodiversity has declined by more than half. If we continue to use up our tropical resources at this rate, there will be virtually no intact ecosystems girding the equator by 2050.

The five major threats to biodiversity that are responsible for the majority of the decline over the past four decades are[60]

1. Habitat loss, fragmentation and change, especially due to clearing for agriculture
2. Overexploitation of species, especially from hunting and fishing
3. Pollution of air, soil, water, and the marine environment
4. The threats from nonindigenous invasive species
5. Climate change (discussed in more detail in a separate section of this chapter)

Using More of Everything than the Earth Produces

If we were few, and lived within our means, the threats discussed could be absorbed by the Earth's ecosystems. But as we have moved into the twenty-first century, neither condition applies. We are many, and we all want more; in the wealthy industrialized countries, we have taken more. In fact, the total ecological footprint of people living in the developed Organization for Economic Cooperation and Development (OECD) countries increased by 76% between 1961 and 2005, including a ninefold increase in our emissions of GHGs.[61] As shown in Figure 2.9, sometime in the mid-1980s humanity started to use more of annual productivity of the Earth every year than the planet actually provides. We had moved into ecological deficit. In 2005, our way of life required the equivalent of 1.2 planets to sustain, and if we continue with business as usual, we will need 2.5 planets by 2050. We are living on environmental debt, borrowing from the future at an increasingly unsustainable rate. We are mining into our children's birthright, permanently damaging the health of the planet, and in the process compromising its ability to sustain us.

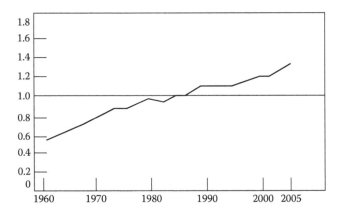

FIGURE 2.9 World's biocapacity and global human footprint (after WWF, *Living Planet Report, 2008,* World Wide Fund for Nature, Gland, Switzerland, 2008). World's biocapacity is 1.0. An index above 1.0 means that we are in deficit—we are using more every year than the earth produces.

The Plight of the Oceans

Special mention of the oceans of the world is deserved. The cradle of life on the planet, and the largest of our ecosystems, the seas and oceans of the world regulate our climate, help to feed us, and are a source of everlasting mystery and wonder for humankind. And yet, this ecosystem, once considered so limitless that we could never in our wildest imaginings have an effect on it, is in deep trouble. In particular, the fish stocks on which so much of humanity depends (2.6 billion people rely on fish for at least 20% of their annual protein intake) are being rapidly destroyed by a combination of overfishing, habitat destruction, pollution, and climate change.[62] Figure 2.10 shows the percentages of global fish stocks in various states of exploitation (from underexploited to crashed) between 1950 and 2003. Alarmingly, the

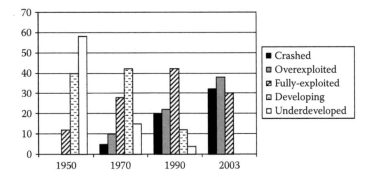

FIGURE 2.10 State of the world's fish stocks, 1950–2003 (after data from United Nations Environment Programme [UNEP], *Global Environmental Outlook 4: Environment for Development [GEO-4]*, UNEP, Nairobi, Kenya, 2007).

global data showed that 90% of the world's fish stocks are now fully exploited or overexploited, and 29% have crashed completely.

Yet, not only do we continue to fish out the remaining stocks at alarming rates, but governments around the world are actually subsidizing this effort. Every year, as much as $40 billion in subsidies help to replace old boats and launch newer, larger, more powerful factory-freezer vessels capable of staying on station for long periods of time.[63] Our ability to catch fish now outstrips the ability of the oceans to provide them. So, inevitably, stocks continue to dwindle. Bottom trawling is a particularly destructive practice. It literally rips up the ocean floor, pulling up everything in its path, including tonnes of so-called by-catch—all of the small fish and undesirable marine life that is caught with the valuable prawns and scallops. It is estimated that for every tonne of market-sold fish or seafood, 20 tonnes of by-catch are thrown back into the sea, dead and wasted. What is left is a swath of destruction on the ocean floor. Yet, even this highly unsustainable and destructive practice remains heavily subsidized: Over US$150 million a year in support payments from governments equates to as much as a quarter of the value of the total yield and is the only reason that the practice is profitable for operators. Without these subsidies, the bottom trawlers could not continue to operate.[64]

Perhaps the most poignant example of a thriving fishery being exploited to extinction was the Atlantic cod fishery on the Grand Banks of Newfoundland, Canada. At one time the most prolific fish resource on the planet, heavy fishing through the 1960s and 1970s put massive stress on the resource. The appearance of huge factory freezer ships from all over the world, trawling up and down the banks for weeks and months on end, added to the pressure. Despite repeated warnings through the 1980s from Canadian fishery scientists, the government was slow to impose restrictions and quotas, fearing the loss of jobs and economic activity that would result. Heavy exploitation continued, and then, "without warning" in 1993, the fishery crashed.[65] The cod were gone, and with them went thousands of jobs and a way of life that had continued for decades in Atlantic Canada. Given the enormous challenge of feeding humanity over the next four decades, the continued loss of our fisheries is of real concern.

No Pain, No Change

There is sufficient information available to everyone to understand the magnitude of the threat to biodiversity and to gauge how much damage we have already done. Treaties have been promulgated and ratified to protect the natural ecosystems of the planet, including the 1992 Convention on Biodiversity, which led the signatories in 2002 to commit to "significantly reducing the current rate of biodiversity loss at all levels."[66] The data provided here illustrate that this commitment has not led to action, and biodiversity loss, if anything, has *accelerated* since 2002. Altruistic declarations and treaties have had little effect on our collective behavior and none on planetary trends. As discussed in Chapter 3, the fact that most of these losses are not measured or valued in our market economies means that we see and feel no pain as our ecological assets vanish. As a consequence, we continue to draw down the world's remaining natural living capital with little regard to the longer-term consequences.

FIGURE 2.11 If current rates of ecosystem destruction are maintained for another 30 years, the children of today may witness a collapse of the living fabric of the planet, with unpredictable but surely negative consequences. And what of generations as yet unborn? Change is inevitable. The question is whether we will change for our own benefit or whether change will be forced on us to our detriment.

Even though we know that we should behave more sustainably, our economic and decision-making systems do not allow us to do so (Figure 2.11).

CLIMATE CHANGE

As discussed, climate change is only one part of an overall sustainability context that is becoming increasingly relevant for business. That wider context is underpinned by the fundamentals of a rapidly growing world population and declining global environmental health, as outlined in the previous sections. Climate change overlays and reinforces many of the other sustainability issues, which on their own would provide significant cause for concern. Climate change exacerbates them all. It makes our dependency on fossil fuels more precarious, inhibits our ability to grow food, increases poverty, and is a growing threat to the biodiversity of the planet. Increasing concentrations of carbon dioxide in the atmosphere mean more is absorbed into the oceans, threatening acidification of the marine environment, with potentially serious implications for marine biodiversity and productivity (feeding back into the issue of food).[67] Climate change, if unchecked, is expected to wreak havoc with the freshwater resources of the world. Unfortunately, climate change is happening now. It is not something that will only affect our children and grandchildren. There is clear scientific evidence that climate change is beginning to strongly affect the way the weather systems of the planet operate now.[68]

The Effects of Climate Change

The predicted effects of climate change, if emissions continue to grow unchecked over the next 30 years, are global in scale and pervasive in extent.[69] Rising sea levels threaten vast parts of the world with inundation (even a few centimeters of increase in sea level translate into several meters of inland migration of tide lines and major increases in the risk and damage associated with storm surges). Rising sea levels cause salination of coastal aquifers and destruction of coastal ecosystems. The resulting displacement of populations would place massive stresses on neighboring countries, and widespread elimination of low-lying agricultural lands would threaten our ability to feed ourselves. That in turn would lead to significant risks to global security, with the prospect of massive civil unrest and war predicted in a recent study by the Pentagon.[70]

And what of the ecosystems that are the life support systems of the planet? The latest science suggests that the majority of the planet's ecosystems and species cannot adapt quickly enough to the rate of warming predicted for the coming few decades under business-as-usual conditions (unchanged emissions growth).[71] That means there is the real risk that the ecosystems that generate the oxygen we breathe and support the biodiversity that binds the intricate web of life on Earth may start to unravel and disappear—essentially, a geologically abrupt and widespread change in the biosphere of the planet, leading to the extinction of as many as half of the species on the planet today.[72] The specter of abrupt dangerous climate change is now being examined as a matter of urgency.[73] The problem is that scientists have little way of knowing how quickly this could happen or how startling the changes could be.[74] The downside risks of climate change are almost certainly too frightening to allow.[75]

Public Opinion versus Science

The best-available science now clearly indicates that climate change is real, is happening, and is starting to impinge on our world. That same science is now unequivocal about the fact that we are the cause. Interestingly, however, public opinion still significantly lags behind the science. Opinions among senior business leaders often do not reflect the state of knowledge and the findings of the best-available science.[76] Many commentators in the public media have continued to attempt to discredit the climate change science,[77] claiming first that climate change was not happening, then that it was happening but it was natural and not our fault, and now that it may be partly our fault but it is too expensive to do anything about.[78] As Ross Garnaut, who led Australia's recent review of climate change economics and policy, stated: Many of the climate change skeptics presenting views in the public arena are in most cases simply not professionally qualified to challenge the science in the first place.[79]

The degree to which public opinion has shifted in recent years is striking, however. A BBC survey of 22,000 people in 150 countries found that 79% of respondents recognized climate change was a serious issue and wanted governments to take action.[80]

The demographics of the survey are also telling. Australian and Chinese responses to the survey were almost identical, among the highest levels of concern in any countries, with 95% of respondents recognizing the need for action. This puts a different perspective on the commonly voiced perspective that any action the rich developed countries take on climate change is of no use because their efforts would be dwarfed by the rapidly rising emissions of China and India, who purportedly "don't care." They clearly do care—after all, developing countries are predicted to be much more vulnerable and to be much harder hit by the effects of climate change.[81]

A New Sense of Urgency

A significant body of the latest research into the impacts of climate change is bringing a new sense of urgency to the issue. The warmest year in the Arctic on record in 2007 led to hugely accelerated melting of the polar ice cap,[82] and despite a slight recovery in 2008 and 2009, the trend of decline continues. The years between 2004 and 2009 have seen the lowest ice extents on record.[83] Summer Arctic ice cover in 2008 was the second lowest on record after 2007, 34% below the long-term 1979–2000 average.[84] Current estimates suggest that without a major reduction in the rate of warming, the Arctic may be ice free in summer well before 2030.[85] Polar bear survival aside, the implications of such rapid northern melting are significant; the flip in albedo from reflective white to heat-absorbing dark will accelerate the thawing of the permafrost, releasing significant quantities of methane into the atmosphere. Were this to happen, the potential releases of GHGs from thawing tundra could be so large that they dwarf the contribution we make from the burning of fossil fuels.[86]

Governments worldwide are being made aware that the Earth system appears now to be responding much more rapidly to the effects of warming than previously predicted. We may have far less time to tackle the issue than we thought even a couple of years ago. People around the world are starting to realize that, in effect, we must act now—30 years of denial, debate, and inaction mean that we have run out of time.[87] Climate change carries a large procrastination penalty: The longer we wait, the more difficult it will be to slow and reverse the effects of warming and the more expensive mitigation becomes for everyone.[88]

A Climate Change Risk Assessment

Risk assessment is widely practiced in industry and is now part of standard operating procedure in most companies. The process attempts to identify any and all possible risks associated with a project or activity and then assesses them based on the probability of occurrence and the magnitude of the expected effect (Figure 2.12).[89] Risks with very high probability and low impact are deemed unacceptable and are mitigated against. Risks with catastrophic effect (which could put the company out of business or result in fatalities) and even very low likelihood are also typically deemed unacceptable and are mitigated against.

Considering climate change from the risk assessment perspective is instructive. Even if we heed the skeptics' calls and assign climate change a low probability of occurrence of say 25% (the Fourth Assessment Report of the IPCC states with 95% confidence that climate change is real and that we are causing it),[90] the predicted effects, discussed in this chapter, are clearly in the "catastrophic" category. Any

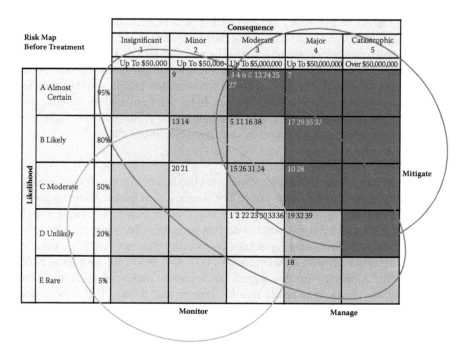

FIGURE 2.12 Risk matrix for assessing risks in industrial project design and execution, comparing the consequences and likelihood of risks. Catastrophic consequences combined with even moderate likelihood mean risks should be mitigated against. (WorleyParsons, 2008, Risk Assessment Methodology, EMSv3. Worley Parsons, Sydney, Australia.)

corporate or engineering risk assessment would then classify the risk of climate change as unacceptable and would result in immediate and comprehensive mitigation. It is interesting to consider why the planet as a whole is not deemed worthy of the same risk assessment practices used in our businesses.

An Unsustainable Course

Now, at the start of the twenty-first century, with over 6 billion people on the planet, armed with powerful technology, and nurturing ever-increasing expectations of wealth and prosperity, there is clear evidence from across the scientific and social spectrum that the prosperity has come at a price. In every part of the world, developed and developing, the process of development and industrialization has resulted in significant damage to the natural environment that not only provides all of the raw materials that underpin this growing human prosperity, but also sustains life itself. As discussed, the figures are alarming. Worldwide, over half of primary forest cover has been lost, and with this serious loss of habitat, biodiversity globally has been significantly impoverished. In fact, current estimates suggest that species loss is now more rapid than at any time in the last 75 million years—we are living through the sixth major extinction in the history of the Earth.[91] The oceans, the cradle of life on Earth, are also suffering. The vast majority of fish stocks worldwide are fully exploited, close to collapse, or near exhaustion. We continue to lose valuable arable land from topsoil erosion and salinization due to unsustainable farming practices, affecting our ability to grow enough food. And now, climate change is starting to compound the already significant accumulated physical and chemical impacts on the environment. In fact, most major environmental organizations worldwide, the vast majority of national academies of science, and a stunning consensus of the peer-reviewed literature confirm that we have now reached a new era in human history—one in which we are the dominant natural force on the planet and in which we are outstripping the ability of the living ecosystems of the planet to provide the resources we are taking and assimilate the wastes we are producing.

Despite Local Successes, Accelerating in the Wrong Direction

Among the global trends, there have been many local successes, instances where environmental issues have been tackled and quality has improved. Examples include improvements in surface water quality in Western Europe;[92] cleaning up the Great Lakes in North America; reductions in particulates in air since 1980 in Japan, Canada, and Germany;[93] and a stabilization of elephant populations in the southern cone of Africa.

In a highly popular book by Danish political scientist Bjorn Lomborg, many such local improvements were cited as proof that in fact the global environmental picture was actually improving, and that environmentalists were guilty of unjustified alarmism.[94] What Lomborg consistently failed to mention is that the environmental improvements cited, without exception, have not simply occurred by themselves—the environment has not somehow fought back on its own. These improvements have been the direct result of targeted effort and expenditure, usually supported or initiated by specific environmental legislation or regulation, often driven by the warnings of environmental professionals and concerned citizens. For example, between 1990 and 2003, nitrogen oxide (NOx) emissions in Europe dropped by a third. This progress was largely the result of new regulations requiring the fitting of abatement technology to emitting facilities.[95] Environmental improvement occurred precisely because society was no longer willing to accept the damage from acid rain that had plagued large parts of Europe for years.

The positive message that can be taken from these examples of environmental improvement is that if concerned citizens push government and industry to take action and money is spent to deploy the appropriate remedies, the environment can recover—things can be made to improve. The obverse is also true: When we do not take action, if money is not spent, when concerned citizens and action groups do not speak out, where regulation is not put in place, business as usual will drive a continuing deterioration of the environment, with all of the resulting consequences to society. Under unremitting pressure, the environment cannot, and will not, simply improve on its own. But, it can recover, as long as critical thresholds are not violated, if we can reduce the pressures we put on it. Unfortunately, there has been too little action over the past four decades, and a real balance between economy, environment, and society remains but a distant hope.

The importance of this balance, and the urgency with which we must find it, are brought into imposing relief by two facts. The first is that, by every measure, we are now accelerating ever more rapidly on our current unsustainable course. The global population continues to grow. Fossil fuel consumption, the biggest current contributor to the GHG emissions that drive climate change, increased 35% from 2006 to 2007, releasing over 55 billion tonnes of carbon dioxide and other insulating gases into the atmosphere.[96] Deforestation, another major GHG contributor, continues to accelerate as forests are cleared for lumber and to make way for agriculture (including ironically, bio-fuels). Yet, despite the now very clear evidence of the damage that these emissions cause, we continue to rip up the remaining forests, burn ever more gas in ever more cars, and build new coal-fired power stations using old inefficient technology. Despite the warnings of the IPCC and scientists around the world, and even after decades of mainstream discussions about the need for action, we continue to obfuscate, procrastinate, and deny the need for change.

TIME TO CHANGE DIRECTION

We are now many more on the planet than we were two centuries ago, and Earth is no longer the limitless universe it once seemed to be. At every level, in every part of the world, the magnitude of our impact on the planet can be clearly seen. We can no longer continue to exploit the natural world at a pace that exceeds its ability to regenerate. This is not only because key resources will be exhausted, but also because as we eliminate the natural habitats that these resources embody, we erode the fundamental ecosystems that maintain life on Earth. The resources that are the inputs to our industrial system are also the engines that provide us all with oxygen and clean air to breathe and water to drink, which regulate and maintain a benign climate that has allowed our species to thrive for 10,000 years, and create the beautiful homelands and waters that inspire us and give us our sense of place and spirit.

We now live in a time when we must consciously strike a harmonious balance between these two fundamental uses for the planet's bounty: that which keeps us alive and that which provides material prosperity. Our ability to sustain and expand human prosperity relies on finding this balance. Our aspirations for achieving personal wealth, for finally finding an end to poverty, for providing safe, comfortable, and meaningful lives for our children and more distant progeny, are all at risk if we fail to find this equilibrium. So far, this goal has eluded us. The reason lies not in some failure of the human spirit, but rather with a fundamental inadequacy at the heart of our modern system.

WHY SUSTAINABILITY HAS NOT WORKED

OVERVIEW

Over the past two or three decades, a preponderance of governments, major international organizations, and businesses have adopted some form of sustainability policy in response to the growing public acceptance of these fundamental themes of sustainability as a core value. However, as discussed, over the same period, by every meaningful measure, we are living in a less-sustainable way than ever before. By every metric, then, we cannot but conclude that sustainability as a guiding concept of human behavior has failed. It has failed because it has not had a discernable effect on our behavior or on our fundamental relationship with our planet. To understand why, we need first to look back in history.

A TIME OF PLENTY

When the Europeans launched humankind's industrial journey in the eighteenth century, the world was a vast, limitless place to be tamed and bent to human will. Earth provided everything we needed: water, forests, minerals, wild animals, arable land. We had only to take what we wanted—the only costs were the labor to convert the resources into products and the technology to do it better and faster. The moral, philosophical, and economic systems of the time reflected this view. In particular, our systems for measuring wealth and productivity focused on the benefits that flowed to us and what we had to pay to get them. Natural resources existed to be exploited and harvested, and wastes were simply discharged into the seemingly limitless environment without consequence. No matter how much of the planet's bounty we took, there was always more. No matter how much waste we discharged, nature was always able to process it. We were few, and the world seemed inexhaustible.

THE INDUSTRIAL REVOLUTION CHANGES THE RULES

By the nineteenth century, industrialization and technology were making us much more efficient at exploiting resources and were increasing the quantity and potency of the wastes we produced. By the 1850s, Europe had cleared over 75% of its forests,[97] and many higher mammals such as bears and wolves had disappeared. Waterways in the new megacities of the time, such as London, Paris, and New York, were so polluted that drinking the water was hazardous to health, and even swimming and fishing were next to impossible. Whales had almost been exterminated from their ancestral feeding and calving grounds in the Southern Ocean. In North America, the great herds of bison, once estimated in the tens of millions, had been wiped out to tame the native tribes and make room for agriculture. Clearly, there were limits to what nature could absorb, and they were staring to be felt.

THE TWENTIETH-CENTURY PROSPERITY EXPLOSION

As we moved into the twentieth century, our ability to substitute one lost or declining resource for another or to apply technology to synthesize natural products allowed unabated expansion and development. Even taking time out for periods of intense slaughter during the world wars of 1914–1918 and 1939–1945 (and several smaller episodes), the twentieth century was a period of unparalleled growth in human prosperity, characterized by a surge in fossil fuel use and rapid development in industrial, transport, and communications technology. Per capita GDP increased more than fourfold over the course of the century, from about US$1,200 to over $5,100 (in 1990 dollars).[98] Fossil fuel use went from about 2,000 million tonnes of oil equivalent (mtoe) per year in 1900 to over 11,730 mtoe per year in 2006.[99] Cheap and plentiful energy provided the basis for unprecedented economic expansion and wealth creation.[100] And for the first time in history, wealth, health, and prosperity were no longer the preserve of the elite but were becoming available to a wider cross section of society, especially in the Western industrialized nations. So, in short order, the Western industrial model, epitomized by Europe and the United States and underpinned by a philosophy and financial system rooted in the eighteenth century, became the reference point and aspiration for the rest of world.

THE ENERGY–CLIMATE PROBLEM

Burning all this fossil fuel was also pouring billions of tonnes of GHGs into the atmosphere every year. By the 1970s, scientists were already warning that this could lead to dangerous climate change. The IPCC issued its second climate change assessment report in 1999 and its third in 2001. But in an eerie parallel with the Atlantic cod fishery in Newfoundland, the warnings of the scientists were played down and ignored by industry and governments on the basis that acting to halt climate change would damage our economy. So, for the last 30 years, despite growing knowledge of the problem and how to solve it, nothing has changed.

Figure 2.13 shows energy use in the global economy for 1987 and 2004, broken down by energy type.[101] Two things are evident: First, we are using a lot more energy. Second, there has been no perceptible change in our energy mix. Despite overwhelming scientific evidence of the perils of climate change, our dependence on fossil fuels has increased, not decreased, over the last three decades. In 2006, annual GHG emissions rose to a record of 44 gt (billions of tonnes) of CO_2 equivalent (CO_2e). Not only was business as usual here to stay in the energy sector, but the dawn of a new century has seen us accelerating away from a sustainable path.

If we do not fundamentally change the current course, world energy use will increase 45% by 2030 to 17,000 mtoe/year, and fossil fuels will make up 80% of that total. This will cause GHG emissions to rise to over 60 gt/year by 2030 and will cause

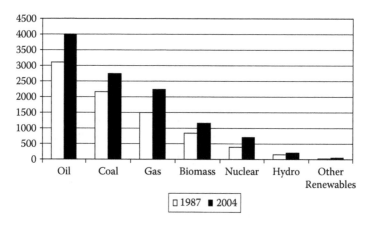

FIGURE 2.13 World's energy mix, 1987 and 2004. (Based on data from United Nations Environment Programme [UNEP], *Global Environmental Outlook 4: Environment for Development [GEO-4]*, UNEP, Nairobi, Kenya, 2007, and *World Energy Outlook*, International Energy Agency [IEA], Paris, France, 2006).

concentrations of GHGs in the atmosphere to double by the end of the twenty-first century, leading to an increase in global average temperatures of 7°C or higher.[102] Anything above 2°C of warming is widely held to entail significant risk of unpredictable violent and dangerous climate change.[103] The business-as-usual approach to energy will irreversibly change life on Earth as we know it.[104]

The International Energy Agency (IEA) and the OECD stated in 2008 that "preventing catastrophe and irreversible damage to the global climate ultimately requires a major decarbonisation of the world energy sources," and "the world's energy system is at a cross-roads. Current global trends in energy supply and consumption are patently unsustainable—environmentally, economically and socially. This can and must be altered. There is still time to change the road we are on." They also concluded their 2008 World Energy Outlook with the statements: "The consequences for the global climate of policy inaction are shocking," and "Time is running out and the time to act is now."[105] And yet, for all the warnings, there is still no evidence that the kind of fundamental shift we know we need to make has started or is anywhere on the horizon.

TANGIBLE IMPACT OF SUSTAINABILITY

As we have moved into the twenty-first century, it has become clear that our widespread awakening to the basic tenets of sustainability has not been matched by meaningful action. It is clear that sustainability as a concept has largely failed to have a perceptible impact on our behavior, our decision making, or the health of the planet we inhabit. Even though the majority understand that our future well-being and prosperity require that we protect our environment and halt the ravages of climate change, we continue with business as usual. How could such a seemingly powerful and self-evident set of principles, widely held, have achieved so little over the last generation? Why has sustainability not found its way into our decision-making systems, particularly in engineering design and project delivery?

EIGHTEENTH-CENTURY ECONOMICS

Economics for a Different Time

A large part of the reason that business as usual has not been affected by the principles of sustainability can be traced to the origins of our economic system, deep in the eighteenth century. Adam Smith's 1776 treatise *The Wealth of Nations* enshrined in modern economic thought the concept of the "invisible hand" of the competitive free market. The value of goods and services was linked explicitly to the labor input required to produce them.[106] At that time in history, we were few, and the world was vast. We could take whatever we required from the land and the seas, and there was always more. We could discharge our wastes with near impunity, such was the absorptive capacity of nature. So, the economic system that developed placed no value on any of these "external" assets.[107] The cost of water was the cost required to obtain it. The value of a forest was nil until it was harvested and sold as lumber. Ecosystems were ascribed no value in our accounts until someone converted them into products for human use. Two hundred years ago this worked because there was always enough of everything—always more water, wood, fuel, space, air, and fish.

The Tragedy of the Commons

Clearly, the world has changed considerably since Adam Smith's time. As discussed, more people with greater demands are putting tremendous pressure on natural resources and ecosystems worldwide. Costs for resources of all types, from oil to coal, from food to water, are rising rapidly now all over the world—a direct consequence of rising demand and dwindling supply.[108] But, our economic and decision-making systems have not kept up; they remain firmly rooted in centuries old ignorance of the value of the natural world that sustains us. Decisions at all levels are still based largely on NPV and IRR expectations that do not include any accounting for the value of natural assets damaged, lost, or used in the process.[109]

What results is a global realization of the "tragedy of the commons," a concept tracing its origins back to Aristotle but popularized by Garrett Hardin in 1968. The concept suggests that when there is free and unrestricted access to and demand for a finite common resource, such as the bounty of the oceans or the assimilative capacity of the atmosphere, that resource is structurally doomed to collapse through overexploitation.[110] In essence, without a price signal that limits demand and access, common assets are used until they disappear. Jared Diamond's book *Collapse* describes such an episode on Easter Island that essentially wiped out civilization there.[111] The planet as a whole is now hurtling toward a global version of the tragedy of the commons.

Without Price Signals Nothing Will Change

In a world where money is the universal measure of wealth, where national, corporate, and global success is gauged by GDP, profit, and growth, it is not surprising that we have used, abused, and failed to protect the things that have "no value." While we may personally believe that a coral reef, the tiger, or indeed even the future of the planet are important, we still have to go to work, earn a living, and do our jobs so we can make the money we need to support our families and ensure our children a future (the irony here is inescapable). So, we put aside personal views of how the world should be, and we do our jobs because that is where the money is. Without price signals in our economy that put real value, in monetary terms, on nature and society and all of the other things we care about, business as usual will be here to stay. Without a price on carbon, carbon emissions will not fall. Without a price on water that reflects its real value, water will be wasted and misallocated. Without a value being placed on coral reefs, they will continue to be destroyed. Altruism and wanting to "do the right thing" are simply not enough—our journey of the last 40 years has proved that.

Twentieth-Century Engineering

The other factor that stands in the way of a more sustainable world, and more sustainable industry in particular, is the tendency in engineering design and project delivery to stick with what has worked—business as usual. Engineers are inherently a conservative fraternity, necessarily guided by convention and codes of practice. This approach protects against substandard and dangerous designs and ensures quality in the delivery of engineering projects, from civil infrastructure to the construction of passenger aircraft. However, the vast majority of current standards were developed in the twentieth century, when energy prices were low and fundamentally stable, water was free, waste could be freely discharged (dilution was the solution to pollution), and the major emerging concerns of global change, including the need to dramatically and quickly reduce carbon emissions, did not exist. Engineering, like economics, needs to adapt to a new reality—a reality described by the sustainability imperative.[112] Project specifications need to begin to take wider energy, carbon, water, biodiversity, and community issues into explicit consideration. Changing the specifications allows engineers and designers to bring their creative talents to bear, to find better ways of doing just about everything.[113]

Twenty-First Century Economics

Better Decision Making Will Make Us More Sustainable

Moving toward greater sustainability requires that economics and engineering evolve to deal with the realities of the twenty-first century. The fundamental economic analysis that governs decision making throughout our economy needs to explicitly take environmental and social sustainability into account by providing real price signals that monetize impacts on those heretofore "external" assets and bringing them into decision-making accounting. By doing this, policy and project decisions can be designed from the outset to maximize profit, increase human welfare, and benefit a multitude of other competing environmental and social issues.[114] This will in turn unlock the ability of engineers, designers, operators, and managers to develop and run more sustainable operations.

That is essentially what the rest of this book is about—a decision-making support system that allows the complete real economics of decisions to be evaluated and compared on a like-for-like basis, across the full life cycle, including all of the relevant environmental, social, and economic considerations. By explicitly valuing all of the external benefits and costs and including them in an overall analysis with the more traditional financial costs and benefits, true social optima can be found and the trade-offs between various issues identified and rationally examined. By incorporating sensitivity analysis and explicitly recognizing that the value of all of the parameters we deal with, from energy prices to the value of carbon or water, are inherently variable over time, decisions that are robust over time can be identified and the reasons for their selection communicated in the universal language: money.

It Is Not Only about Cost

Part of the problem is that we continue to treat the economy and the environment as mutually exclusive and in conflict. We think that saving the planet will hurt the economy, and that saving the economy means we cannot save the planet. This is perilous thinking. In fact, the economy and the environment are inextricably linked; the economy cannot exist without a healthy robust environment. Inaction on climate change is widely justified by the high cost of taking action to eliminate carbon from our economies. Indeed, it will be expensive. As much as US\$4.1 trillion in investment over the next 20 years (about \$200 billion or 0.24% of global product each year) will be needed to stabilize atmospheric concentrations of GHGs at 550 ppm CO_2e, costing each person on Earth US\$17 each year to 2030. To reach a 450-ppm stabilization target will cost about $5.1 trillion from 2010 to 2030.[115] This is a lot of money. In comparison, world military spending in 2006 alone was US$1.2 trillion.[116]

However, in any other decision we make, cost is balanced against the value we expect to realize from the expenditure. As individuals, we do this every time we purchase something—a new car, a watch, even a meal—we compare the cost to the utility and satisfaction that we receive. If we feel we are getting more from the purchase than we have to pay, we proceed. Inherent in this is the concept of value; we will not always simply opt for the cheapest thing available. Sometimes, we feel that there is value in spending more—for just the right clothes, the higher-quality appliance. The decisions we make about the sustainability of our planet should be no different. Battling climate change to an uncomfortable stalemate will cost a lot of money. We hear about those costs all the time. But, what about the benefits of taking action?

It Is Also about Benefits

The biggest benefit of action on climate change is preventing the damage that will occur if we do not act. This damage is expected to be massive. Inaction on climate change will result in widespread damage to the global economy, costing the planet at least 5% and as much as 20% of global product (GP) every year, now and forever[117] (these figures do not include the value of many of the environmental ecosystems and species that would be irretrievably damaged by significantly altering the climate). This level of permanent economic damage would make the 2008–2009 global financial crisis seem trivial in comparison.[118]

So, for a cost of as much as 1% of GP per year, we avoid permanent losses of 5% to 20% of GP. And there will be other benefits as well. Limiting fossil fuel emissions will also result in significant overall air quality improvements. According to the OECD and IEA, meeting the 550-ppm stabilization target will also result in over US$7 trillion in energy cost savings to consumers alone— almost double the expected cost of the transition.[119] Moreover, studies by industry groups such as the U.S. Electricity Production Research Institute (EPRI) have shown that there is a significant economic procrastination penalty associated with inaction on climate change;[120] the longer we take to make the necessary changes, the more it will cost us overall. From a purely economic cost–benefit perspective, investing in preventing climate change and arresting global environmental degradation is a very good deal for everyone, perhaps the deal of the century. (Despite this, Lomborg, in 2001, maintained that it would be preferable to accept climate change and adapt to its consequences than spend the money required to prevent the worst of its impacts.)

We Can Do This Now

The kind of decision making that can lead to a more sustainable world can be done today. Corporations and industry can include "shadow costs" for a variety of externalities in their economic analysis now, and many already are. But, personal and corporate altruism is not enough to achieve the kind of sustainability we need. Creation of real market price signals for externalities requires government action and involvement. In Australia, for example, the proposed federal emissions trading scheme would start to put an effective price on carbon, which would then begin the process of internalizing the social cost of carbon. The United Kingdom has already put an official shadow cost on carbon.

Understanding and using both quasi-market prices and the larger social costs of external assets will allow businesses to make better decisions. They can then set specifications that take sustainability into account and make it a key project consideration (guided by the economics), unleashing the creative powers and talents of engineers and designers. If we can send men to the moon and probes to Saturn, then we can design and deliver projects that use energy more efficiently, reduce or eliminate carbon emissions, use less water, and have smaller impacts on biodiversity and local communities. We may even then find solutions to the wider problems of global change.

CONCLUSION

Sustainability is emerging as one of the pivotal issues of the twenty-first century. But, despite being widely accepted as a core value, it is poorly practiced. Even when decision makers intuitively know the right and best course of action, it is rarely taken, usually because the "economics" do not work. Typically, we fall back on what we know and what has worked in the past.

Every metric of planetary well-being is telling us that we can no longer continue on this course. The impacts of global change are being felt everywhere, by everyone, and are accelerating. The cause of our present predicament is a virous combination of outdated economic thinking, obsolete decision-making systems, and previous-century engineering practices and design codes, all exacerbated by a lack of political and social leadership and the entrenched resistance of a few powerful organizations. By explicitly valuing environmental and social sustainability and including these costs and benefits into our decision-making analysis, more sustainable, profitable, and robust alternatives can be identified. In many cases, these new alternatives will change traditional project specifications and will require engineering and project delivery that goes beyond business as usual.

How business and industry respond to these challenges, and the increasing regulatory, stakeholder, investor, and shareholder pressures that they bring, will be a key factor in achieving a sustainable society on the planet, and will play a key part in determining how successful individual companies and business sectors will be in the future. Governments and NGOs will also need to embrace new economic thinking that puts society and environment on an equal footing with traditional financial concerns.

The following chapters describe a practical and robust decision-making methodology for maximizing social net benefit and provide a host of detailed examples from across the world, in a variety of industry sectors, from water to petroleum to mining. The examples reveal that business as usual rarely provides the environmental, social, and economic optimum, but also that if we spend *too much* on environmental and social protection, we pass the point of diminishing marginal returns and do not maximize our opportunities for true sustainability. Experience also shows that sustainability, properly practiced, improves overall corporate performance, shareholder value, and profits.

The scientific community of the world is telling us, loudly and clearly, that we must act now to preserve the environment of the planet and secure the future for our children. There is no time left for procrastination. In his book *A History of Economics: The Past and Present*, the economist John Kenneth Galbraith wrote: "Few problems are difficult of solution. The difficulty, all but invariably, is in confronting them. We know what needs to be done; for reasons of inertia, pecuniary interest, passion or ignorance, we do not wish to say so." More sustainable decisions, and a more sustainable world, require that we confront the issues of global change by making them our pecuniary interest.

NOTES

1. Carson, R. 1962. *Silent Spring*. Houghton Mifflin, Boston.
2. United States Environmental Protection Agency. 1979. The Love Canal Tragedy. *USEPA Journal*. Jan 1979 http://www.epa.gov/history/lovecanal.
3. World Council of Churches. 1974. *Report of Ecumenical Study Conference on Science and Technology for Human Development*. World Council of Churches, Geneva.
4. International Union for the Conservation of Nature and Natural Resources (IUCN). 1980. *World Conservation Strategy: Living Resources Conservation for Sustainable Development*. IUCN, Gland, Switzerland.
5. World Commission on Environment and Development. 1987. *Our Common Future*. Oxford University Press, Oxford, UK.
6. Dresner, S. 2002. *The Principles of Sustainability*. Earthscan, London.
7. Pearce, D., A. Markandya, and E. Barbier., 1989. *Blueprint for a Green Economy*. Earthscan, London.
8. Eckerman, J. 2005. *The Bhopal Saga: Causes and Consequences of the World's Largest Industrial Disaster*. Orient Blackswan. Hyderabad, India.
9. National Oceanographic and Atmospheric Administration (NOAA). 1992. *Oil Spill Case Histories 1967–1991*. Report MHMRAD-92–11. NOAA, Washington, DC.
10. *ExxonValdez* Oil Spill Trustee Council. 2004. *20th Anniversary of the Exxon Valdez Oil Spill*. Exxon Valdez Oil Spill Trustee Council, Anchorage, Alaska.
11. Damages Cut Against Exxon in *Valdez* Case. 2008. *New York Times*, June 26.
12. Dresner, *Principles of Sustainability*.
13. Elkington, J. 1997. *Cannibals with Forks: The Triple Bottom Line of 21st Century Business*. Capstone, Oxford, UK.
14. United Nations Environment Programme (UNEP). 2002. *Global Environmental Outlook Report 2002—GEO-2*. United Nations, New York.
15. United Nations Environment Programme (UNEP). 2008. *Global Environmental Outlook Report 2008—GEO-4*. United Nations, New York.
16. U.S. Committee on Environment and Natural Resources and the U.S. National Science and Technology Council. 2008. *Scientific Assessment of the Effects of Global Change on the United States*, National Science and Technology Council, Washington, DC.
17. Martin, J. 2006. *The Meaning of the 21st Century: A Vital Blueprint for Ensuring Our Future*. Eden Project Books, London.
18. United Nations. 2002. *World Population Prospects*. United Nations, New York.
19. United Nations. 2007. *World Population Prospects*. United Nations, New York.
20. United Nations World Food Programme. 2009. World Hunger Series: Hunger and Markets. WFP and Earthscan, London, UK.
21. Ibid.
22. Food and Agriculture Organization (FAO). 2003. *World Agriculture: Towards 2030/2050*. United Nations Food and Agriculture Organization, Rome.
23. Ibid.
24. United Nations Environment Programme (UNEP). 2007. *Global Environmental Outlook 4: Environment for Development (GEO-4)*. UNEP, Nairobi, Kenya.

25. Martin, J. *The Meaning of the 21st Century*. Elden Project Books. London, UK.
26. United Nations Environment Programme (UNEP) 2007. *GEO*-4.
27. World Rice Trade Industry. 2009. Australia Rice Output Down by Almost 90%. Feb 20, 2008. http://www.oryza.com Asia-Pacific/Australia-Market/.
28. Australian Bureau of Agricultural and Resource Economics (ABARE). 2009. *Australian Commodities*, 16, 4, 703.
29. Economic Research Service, USDA. 1997. *International Agricultural Baseline Predictions to 2007*. U.S. Department of Agriculture, Washington, DC.
30. Sachs, J. 2007. *Common Wealth. Economics for a Crowded Planet*. Allen Lane, London.
31. Food and Agriculture Organization (FAO). 2006. *The State of the World's Fisheries and Aquaculture (SOFIA) 2004*. Food and Agriculture Organisation of the UN, Rome.
32. United Nations Development Programme (UNDP). 1999. *Wealth Distribution Statistics, 1998*. UNDP, New York, NY.
33. World Bank. 2008. *World Development Indicators 2008*. World Bank Press, Washington, DC.
34. Davies, J., S. Sandstrom, A. Shorrocks, and A. Wolff. 2006. *World Distribution of Household Wealth*. World Institute for Development Economic Research, United Nations University Press, Tokyo, Japan.
35. McNeill, J. 2000. *Something New Under the Sun: An Environmental History of the 20th Century*. Allen Lane, Penguin Press, London, UK.
36. UNEP, *Global Environmental Outlook 4*.
37. WWF. 2007. *The World's Top Ten Rivers at Risk*. World Wide Fund for Nature, Gland, Switzerland.
38. Nilsson, C., C.A. Reidy, M. Dynesius, and C. Revenga. 2005. Fragmentation and flow regulation of the world's large river systems. *Nature*, 305.
39. Ibid.
40. Priceless. World Water Survey. 2003. *The Economist*. http://www.economist.com.
41. Wheater, H. 2004. Personal communication.
42. Danoun, R. 2007. *Desalination Plants: Potential Impacts of Brine Discharge on Marine Life*. University of Sydney, Ocean Technology Group.
43. California Coastal Commission. 1993. *Seawater Desalination in California*. Sacramento, CA.
44. McNeill, *Something New. Under the Sun*. Allen Lane, Penguin Press, London, UK.
45. Ibid.
46. UNEP, *Global Environmental Outlook 4*.
47. Hardisty, P.E., and E. Ozdemiroglu. 2005. *The Economics of Groundwater Remediation and Protection*. CRC Press, Boca Raton, FL.
48. World Water Development Report (WWDR). 2006. *Water—A Shared Responsibility*. United Nations World Water Programme. UNESCO, Paris.
49. World Bank. 2002. *Middle East and North Africa Water Resources Management: Sector Overview and Development Context*. World Bank, Washington, D.C.
50. United Nations Environment Programme (UNEP). 2002. *Global Environmental Outlook 3: Past, Present and Future Perspectives (GEO-3)*. UNEP, Nairobi, Kenya.
51. High Water Council. 1992. *Water Resource Management Options in Sana'a Basin*. Final Report, Vol. IX. UNDP/DESD Project YEM/88/001. Technical Secretariat of the High Water Council, Sana'a, Republic of Yemen.
52. United Nations Commission on Sustainable Development. 1997. *Qatar Country Profile, Agenda 21*. United Nations, New York.
53. UNEP, *Global Environmental Outlook 4*.
54. United States Geological Survey. 1998. *Overview of Middle East Water Resources— Water Resources of Palestinian, Jordanian and Israeli Interest*. Middle East Data Bank Project. VSGS, Reston, VA.

55. Hardisty, P. E., and E. Ozdemiroglu. (2005). *The Economics of Groundwater Remediation and Protection,* CRC Press, Boca Raton, FL.
56. Worldwatch Institute. 2007. *State of the World, 2008.* Worldwatch Institute, Washington, DC.
57. Spratt, D., and P. Sutton. 2008. *Climate Code Red: The Case for a Sustainability Emergency.* Friends of the Earth, Fitzroy, Australia.
58. Loh, J., and S. Goldfinger. 2006. *Living Planet Index 2006.* World Wide Fund for Nature, Gland, Switzerland.
59. WWF. 2008. *Living Planet Report, 2008.* World Wide Fund for Nature, Gland, Switzerland.
60. Ibid.
61. Ibid.
62. UNEP, *Global Environmental Outlook 4.*
63. Worldwatch Institute. 2008. *State of the World, 2008: Ideas and Opportunities for Sustainable Economies.* Earthscan, London.
64. Sumalia, U.R., and D. Pauly. (eds). 2007. Catching more bait: A bottom-up reestimation of global fisheries subsidies (2nd version). *Fisheries Centre Research Reports,* 14(6):49–53. Fisheries Centre, University of British Columbia, Vancouver.
65. Roughgarden, J., and F. Smith. 1996. Why Fisheries Collapse and What to Do About It. Proceedings of the National Academy of Sciences of the United States of America, 93: 5078–5083.
66. Sachs, J. 2007. *Common Wealth. Economics for a Crowded Planet.* Allen Lane, London.
67. Intergovernmental Panel on Climate Change (IPCC). 2007. *Fourth Assessment Report. Impacts, Adaptation and Vulnerability.* Cambridge University Press, Cambridge, UK.
68. Intergovernmental Panel on Climate Change (IPCC). 2007. *Fourth Assessment Report. The Physical Science Basis.* Cambridge University Press, Cambridge, UK.
69. Inter Academy Council. 2007. *Lighting the Way: Towards a Sustainable Energy Future.* IAC Secretariat, Amsterdam, The Netherlands.
70. Schwartz, P., and D. Randall. 2003. *An Abrupt Climate Change Scenario and Its Implications for United States National Security.* United States Department of Defense, Washington, DC.
71. Worldwide Fund for Nature. 2004. *Extreme Weather; Does Nature Keep Up?* WWF, London.
72. Thomas, C.D., A. Cameron, R.E. Green, M. Bakkenes, L.J. Beaumont, Y.C. Collingham, B.F.N. Erasmus, M.F. de Siqueira, A. Grainger, L. Hannah, L. Hughes, B. Huntley, A.S. van Jaarsveld, G.F. Midgley, L. Miles, M.A. Ortega-Huerta, A. Townsend-Peterson, O.L. Phillips, and S.E. Williams. 2004. Extinction Risk from Climate Change. *Nature,* 427: 145–148.
73. MacCraken, M.C., F. Moore, and J.C. Topping Jr. 2008. *Sudden and Disruptive Climate Change: Exploring the Real Risks and How We Can Avoid Them.* Earthscan, London.
74. Flannery, T. 2005. *The Weathermakers.* Text Publishing, Melbourne.
75. Stern, N. 2006. *The Economics of Climate Change—The Stern Review.* Cambridge University Press, Cambridge, UK.
76. McKinsey, 2007. How Companies Think about Climate Change: A McKinsey Global Survey. *McKinsey Quarterly,* December, 1–5.
77. Boykoff, J., and M. Boykoff. 2004. *Journalistic Balance as Global Warming Bias: Creating Controversy Where Science Finds Consensus. Fairness and Accuracy in Reporting Fair.* Fairness & Accuracy in Reporting, Nov/Dec 2004. http://www.fair.org.
78. Ward, M. 2006. *Answering the Climate Change Sceptics.* Paper for World Wide Fund for Nature and Environmental Defense Society. WWF and EDF, Auckland, New Zealand.

79. Garnaut, R. 2008. Climate Change and Australian Economic Reform. Speech to the Melbourne Institute 2008 Economic and Social Outlook Conference, March, Melbourne.
80. BBC. 2007. BBC Climate Change Poll. http://www.bbc.com.
81. IPCC, *Fourth Assessment Report. Mitigation of Climate Change.*
82. National Snow and Ice Data Center, 2007. Arctic Sea Ice Shatters All Previous Record Lows. NSIDC Press Release, October 1, 2007. NSIDC, Boulder, CO.
83. Comiso, J.C., C.L. Parkinson, R. Gersten, and L. Stock. 2008. Accelerated Decline in the Arctic Sea Ice Cover. *Geophysics Research Letters*, 35, L01703.
84. U.S. National Oceanographic and Atmospheric Administration (NOAA). 2008. *Arctic Report Card 2008.* U.S. NOAA. Washington, DC. http://www.arctic.noaa.gov.
85. U.S. National Oceanographic and Atmospheric Administration (NOAA). 2009. *Ice-Free Arctic Summers Likely Sooner than Expected.* U.S. NOAA. Washington, DC. www.arctic.noaa.gov.
86. Cox, P.M., R.A. Betts, C.D. Jones, S.A. Spall, and I.J. Totlerdell, 2000. Acceleration of Global Warming due to Carbon-Cycle Feedbacks in a Coupled Climate Model. *Nature*, 408: 184–187.
87. Department for Environment, Food, and Rural Affairs (DEFRA). 2003. The Scientific Case for setting a long term emissions reduction target. United Kingdom DEFRA. London, UK. http://www.defra.gov.uk/environment/climatechange/pubs
88. Stern, N. 2006. *The Economics of Climate Change—The Stern Review.* Cambridge University Press, Cambridge, UK.
89. WorleyParsons. 2008. *Risk Assessment Methodology*, EMSv3. WorleyParsons, Sydney, Australia.
90. IPCC, *Fourth Assessment Report. The Physical Science Basis.*
91. WWF, *Living Planet Report, 2008.*
92. European Environment Agency. 1995. *Europe's Environment: The Dobris Assessment.* EEA, Copenhagen, Denmark.
93. Organization for Economic Cooperation and Development (OECD). 1999. *Environmental Data Compendium 1999.* OECD, Paris.
94. Lomborg, B. 2001. *The Skeptical Environmentalist: Measuring the Real State of the World.* Cambridge University Press, Cambridge, UK.
95. European Environment Agency. 2006. *Energy and Environment in the European Union.* Report 8/2006. EEA, Copenhagen, Denmark.
96. Organization for Economic Cooperation and Development/International Energy Agency. 2007. *World Energy Outlook.* Paris, France.
97. McNeill, J. 2000. *Something New Under the Sun.* Allen Lane, Penguin Press, London, UK.
98. Ibid.
99. Organization for Economic Cooperation and Development/International Energy Agency (OECD/IEA) 2008. *World Energy Outlook 2008.* OECD/IEA. Paris, France.
100. McNeill, J. 2000. *Something New Under the Sun.* Allen Lane, Penguin Press, London, UK.
101. Organization for Economic Cooperation and Development/International Energy Agency (OECD/IEA) *World Energy Outlook 2008.* Paris, France.
102. Ibid.
103. World Business Council on Sustainable Development (WBCSD). 2007. *Energy and Climate Change: Facts and Trends to 2050.* WBCSD. Geneva, Switzerland.
104. National Academy of Sciences. 2008. *Scientific Assessment of the Effects of Global Change on the United States.* Committee on Environment and Natural Resources, NAS National Science and Technology Council, Washington, DC.
105. Organization for Economic Cooperation and Development/International Energy Agency (OECD/IEA) *Global Energy Outlook 2008.* Paris, France.

58

106. Galbraith, K. 1987. *A History of Economics*. Penguin, London, UK.
107. Pearce, D. 1981. *World Without End*. World Bank, Washington, DC.
108. World Food Programme. 2008. *WFP and Global Food Price Rises*. United Nations Food and Agriculture Organization (FAO), Rome, Italy.
109. Worldwatch Institute, *State of the World, 2008: Ideas and Opportunities*.
110. Hardin, G. 1967. *The Tragedy of the Commons*.
111. Diamond, J. 2004. *Collapse: How Societies Choose to Fail or Survive*. Allen Lane, London, UK.
112. Hardisty, P.E. 2008. Improving Sustainability in Project Delivery. *J. Aust. Pipeliner*.
113. Hawken, P., A.B. Lovins, and L.H. Lovins, 2000. *Natural Capitalism*. Earthscan, London. 396 pp.
114. Pearce, D. and J. Warford. 1981. *World Without End*. World Bank Press, Washington, DC.
115. Organization for Economic Cooperation and Development/International Energy Agency (OECD/IEA) *Global Energy Outlook 2008*. Paris, France.
116. Stockholm International Peace Research Institute. 2007. *SIPRI Military Expenditure Database*. SIPRI, Stockholm, Sweden.
117. Stern, N. *Economics of Climate Change*. Cambridge University Press, Cambridge, UK.
118. Hardisty, P.E. 2009. The Environmental Consequences of the Financial Crisis: Crisis Which Crisis? *Middle East Economic Survey*, 51(49), pp. 22–29.
119. Organization for Economic Cooperation and Development/International Energy Agency (OECD/IEA) *Global Energy Outlook 2008*. Paris, France.
120. Electric Power Research Institute (EPRI). 2007. *The Power to Reduce CO2 Emissions*. EPRI, Palo Alto, CA.

3 Quantifying Sustainability for Improved Decision Making

BALANCING ENVIRONMENT, SOCIETY, AND ECONOMY

INTRODUCTION

The global crisis of sustainability is a direct result of an economic system that has not kept up with the realities of this new century. On a macroeconomic level, this fundamental problem is embedded in the way we measure national and global economic success: gross domestic product (GDP) and global product (GP), respectively, which represent the value of market-traded goods and services produced in an economy. As goes our macroeconomic view of the world, so goes the microeconomic view. Business and industry measure their success in the same fundamental way. By examining the flaws in GDP, we can see the essentials of a solution: decision making that balances environment, society, and economy.

THE PROBLEM WITH GDP

Gross domestic product (GDP) is the de facto measure of wealth, well-being, and progress in our society. GDP measures the total value of market-traded goods and services produced and consumed in the economy in a given year and therefore equates to the total income earned and spent by consumers in that same year. It is the measure that governments around the world use to assess the performance of their economies. When politicians and economists talk about "growth," they are referring to a net annual increase in GDP. If the rate of GDP growth should slow over the course of a year, we call it a *recession*. The global financial crisis (GFC) of 2008–2009 saw the largest contraction in economic activity (as measured by GDP) since the Great Depression and triggered a massive spending spree intended to "stimulate" the economy. In fact, as much was spent in a 2-year period, bailing out banks and insurance companies and stimulating the economy with public spending packages, as the Organization for Economic Cooperation and Development (OECD) estimates is needed to decarbonize the world's economy and prevent the worst effects of climate change (over US$4 trillion).[1] This shows just how highly we regard GDP—if it is not increasing year after year, there is quite literally panic in the streets.

Promoting Unsustainable Behavior

But for all the hysteria, GDP is actually a very poor measure of real wealth, well-being, and happiness in society. It assumes that only the things that are traded in the market economy contribute to well-being—none of the so-called externalities discussed in Chapter 2 are included. So, when the *Exxon Valdez* spilled millions of liters of oil onto the Alaskan coastline, the GDP of the state actually *increased* that year because of all of the money spent on the cleanup. The fact that tens of thousands of birds died and a pristine wilderness was marred for years was not reflected. In fact, GDP has several well-known flaws:[2]

- *GDP does not include the value of unpaid labor.* All of the time spent by people who raise children, care for sick and elderly family members, do housework, and volunteer in society is not included in GDP. This is perhaps part of the reason that governments in the West in recent years have been diligently pushing those who do unpaid work into the labor force; it is an easy way to get GDP statistics to increase—the same work is being done, the children are still being cared for, but now it is being done by a child-care worker in a day-care center, and the parent can enter the labor force.

- *GDP does not account for the external costs of economic activity.* None of the damage done to the environment or society by our economy is reflected in the statistics used to compile GDP. So, GDP growth can look very strong, as China's did in 2003, but the cost to China of that growth in terms of polluted rivers and lakes, loss of biodiversity and wildlands, species loss, and health impacts from chronic air pollution was estimated at almost 4% of GDP.[3] The value of environmental damage should be subtracted from GDP.

- *GDP counts expenditure used to protect ourselves against environmental damage as a positive.* So, when a child is hospitalized for respiratory illness caused by air pollution in China or Egypt, all of the money spent on the child's medical care is counted on the positive side of the GDP ledger. Other examples of defensive measures include provision of bottled water as a response to water pollution, increased holiday travel to unspoiled places to escape degraded environments, medical costs from environmentally triggered illnesses, and our multibillion-dollar-a-year efforts to remediate environmental damage worldwide.[4] All of these things boost demand for goods and services in the economy and so, perversely, boost GDP—we think we are better off. Such expenditure, like the cleanup costs for the *Valdez* spill, has become a sizable and growing part of our GDP and can be considered akin to digging a hole, filling it again, and then calling the effort productive work.

- *GDP does not account for depreciation of natural capital.* All of the natural resources on the planet are finite, some renewable, others not. Depletion of those stocks, on a planetary or national scale, means we are worse off. The value represented by that capital has been converted into energy, goods, or waste and is no longer available for future use. The coming generations have lost the option of using that resource later. But, as we use up our petroleum reserves, erode away our topsoil, or mine into our natural capital of biodiversity, these losses are not reflected in our calculus of well-being; all we see is the value of the goods and services produced from those raw materials.

Combined, these effects severely discredit GDP as a measure of real success and progress in society; in fact, our dedication to the GDP concept actually promotes unsustainable behavior. It does so first by ignoring the costs of unsustainable actions, second by giving credit to all of our efforts to repair environmental and social damage, third by not recognizing any of the valuable contributions that people make to society and the places they live through unpaid work, and finally by turning a blind eye to the mining of the natural capital of the Earth.

A More Sustainable Alternative: Net National Welfare

Many economists have long been promoting the idea of an adjusted measure of economic success: net national welfare (NNW).[5] In principle, NNW adjusts GDP upward for the value of nonmarket efforts and consumption and the value of capital services. Then, all of the costs of economic growth (both the defensive costs included in GDP and the externalities not included) are subtracted. Finally, the depreciation of natural and created capital is removed. The genuine progress indicator (GPI) provides a similar adjustment.[6] In 2004, the GDP of the United States was US$10.8 trillion; but the GPI was only US$4.4 trillion.[7] If this type of adjusted version of GDP were used, we would find that our notions of progress and growth would be greatly altered, and almost surely our behavior would change. At its core, this chapter is based on the same notion of adjusting our economics to account for externalities and capital depreciation, but on a microeconomic project decision-making level.

FROM MACRO TO MICRO

Our microeconomic project and policy decision-making systems consistently reflect the macroeconomic bias embedded in GDP. Project and business decisions are based on a narrow view of net present value (NPV), which also does not include the value of costs or benefits external to the project. Thus, the economic analysis that lies at the core of our business decision-making and project selection and prioritization systems places *no value* on the environment or society. All that matters is what revenue is produced and what market costs must be incurred to generate that return. Under these conditions, external issues are relegated to a secondary, qualitative position, as footnotes to the decision.

Furthermore, because these issues are quoted in a variety of nonmonetary units (liters of water, tonnes of carbon, number of bald eagles, hectares of coral reef, number of people affected by particulates in the atmosphere), trade-offs between the various issues are difficult to make, and their relative importance in terms of overall project economics is hard to understand. What this means, effectively, is that business-as-usual (BAU) decision making remains largely unaffected by the notion of sustainability. Traditional NPV-based economics continues to hold primacy, and environmental and social issues are not considered on an equal footing with the financial. Because our national and global economies are simply the sum of all of the individual project and policy decisions made by businesses, industries, and individuals every day, the net result is what we see now: a global deterioration of our environment and an ever-more-unsustainable world.[8]

How Industry Makes Decisions

Ask anyone in industry today what they mean by the term *economics*, and they will answer in terms of project NPV—the time-discounted profit that a project is expected to generate for the company. *Profit* is defined as the expected proceeds from sale of the commodity on the open market minus the capital expenditures (CAPEX) and operating expenditures (OPEX) over the project lifetime. This "economic" analysis is done purely from the perspective of the company (or proponent)—no issues external to the company, other than those regulated by government and applied to the project in terms of taxes, royalties, or penalties, are included. So, currently, damages to the environment that result from the project (an obvious example here is the emission of greenhouse gases [GHGs], which contribute to environmentally damaging climate change) are not included in the analysis.

What the private sector calls an economic analysis is actually a financial analysis. An economic analysis requires a complete evaluation of all of the costs and all of the benefits accruing to all segments of society as a result of the project.[9] A financial analysis is inherently skewed since potentially significant costs and benefits that the company does not feel or see directly are not included. Misconstruing the financial for the economic means that important social and environmental price signals are not available to decision makers. People cannot make rational decisions with incomplete information, and what has been missing for the last four decades, and arguably for the last two centuries, is the value (measured and expressed in the same units as the rest of the analysis—money) of the common environmental and social assets of the planet.[10]

ECONOMIC QUANTIFICATION OF SUSTAINABILITY

If a lack of transparent and complete accounting is at the heart of the sustainability problem, then it follows that a shift in behavior and significant improvements in sustainability can be triggered by changing the way we assess policies, projects, and investment. If sustainability can be defined more rigorously and usefully through full social and environmental economic analysis (in much the same way as GDP can be theoretically adjusted for its flaws through the computation of NNW or GPI), then it can be more fully reflected in our decisions.

This chapter therefore introduces the concept of the *environmental and economic sustainability assessment (EESA)*, a whole life-cycle, physically based economic analysis designed to integrate sustainability into decision making in a fully quantitative and objective way.

To do this in practice, we need to be able to do the following:

1. Set the objective of the assessment.
2. Determine a range of practical options for meeting that objective, spanning the full range of possible solutions from the most basic, through BAU, to options at the technological limit of feasibility.

3. Identify environmental and social assets at risk or impacted by a proposed project or policy.
4. Quantify, in physical terms, the damage done to those assets (environmental and social) as a result of each option that meets the project or policy objective, over a period of time that reflects the full natural life cycle of the project or policy (taking a longer-term perspective), and in a way that considers the full life cycle of the inputs and outputs of the project.
5. Quantify the benefits, in physical terms, that accrue to society and the environment as a result of the project or policy option (over the longer term and the full life cycle of the project).
6. Estimate monetary values for each of the external environmental and social assets affected by each option (which are not normally included in economic analysis conducted by industry).
7. For each impacted asset, apply monetary values to the physical estimates of impact and benefit.
8. Combine the external valuation of costs and benefits with the traditional financial costs and benefits of each option to produce a combined socioeconomic NPV estimate.
9. Compare the socioeconomic NPV to the financial NPV for each option to determine the relative importance of the external issues.
10. Using the same technique, compare a range of options designed to achieve an outcome, in terms of their overall socioeconomic and financial NPVs, to identify an economically optimum solution. In doing this, quantify, for each option, its impact (positive or negative) on each key parameter. This allows a quantitative examination of the trade-offs that exist between options.
11. Conduct sensitivity analysis to determine how the choice of optimal solution varies (or not) as key assumptions vary.
12. Select the option that provides maximum overall net benefit over the widest range of likely future conditions and values for key assets.

This is the basic procedure that lies at the heart of this book. EESA allows all of the implications of a project or policy decision to be balanced—the financial, environmental, and social—and so can help decision makers to see the full, real implications of their decisions. The following sections of this chapter describe the EESA methodology in more detail in practical terms. The following chapters provide a series of real examples from around the world where this approach has been applied and has led to fundamental changes in decision making.

AN ECONOMIC DEFINITION OF SUSTAINABILITY

What this approach provides is a useful, quantitative definition of sustainability that moves away from the more qualitative, emotional, and altruistic definitions without compromising the ideals they represent. By explicitly valuing the environment and society and including them in the overall economic analysis for industry and government projects, strategies, and policies, decisions that are optimal for all of society are revealed. And, if we also take the definition of sustainability at its most basic and literal (that what we do should continue to provide real benefits *over long periods of time*), we end up with what is essentially an economic definition of *sustainability*:

> If over the long term a proposition delivers more benefit than cost over its complete life cycle, when all environmental, social, and economic factors are taken into account, using a socially acceptable discount rate, then the proposition is sustainable.

If overall financial, social, and environmental costs are greater than benefits, then the project is unsustainable; society will not want to continue to fund and support it over the long term because it simply costs more than it is worth. Even when projects receive government subsidy, the information provided by EESA will eventually drive society to recognize that the expenditure of time, effort, energy, and materials is simply *not worth it*.

A Double-Edged Sword

Economists justify an expenditure based on the anticipated benefits resulting from that expenditure. If the benefits accruing to society (including the proponent) from a project exceed the costs of implementation, then the project is worth doing. In environmental terms, therefore, what industry spends on environmental protection should have some relation to the value of the benefits that result. If costs vastly exceed benefits, then society loses—the funds could have been spent in a way that would benefit society more. Conversely, by explicitly valuing natural resources such as freshwater and biodiversity, appropriate restoration and protection expenditure levels can be determined. In this way, common goods such as air and water are much less likely to be treated as worthless sinks to be damaged or exploited without cost.

The Externalities Can Be Worth a Lot

Application of economic tools to the environment is a relatively new approach, brought about by recent advancements in valuation of natural resources. The total annual value of all of the services provided by the world's biosphere (most of which depend directly on freshwater), previously considered as "free" common goods, has been estimated to be as much as US$33 trillion, more than the world's combined conventional annual GP.[11] So, when we prevent damage to a river or remediate a contaminated aquifer, we can calculate a resultant benefit to compare with the cost of the action.

This is the basis of regulatory impact assessment (RIA), now required within the European Union and the United States, by which the economic costs and benefits of a new policy or regulation are explicitly considered. This prevents the imposition of regulations or standards that unnecessarily harm industry (which has to pay the costs) without society experiencing sufficient gain.

AN ILLUSTRATION: THE NPV-INTERNAL RATE OF RETURN TRAP

Many firms require that process and equipment modification to achieve reductions in energy consumption or reuse waste heat meet financial hurdle rates that match, or in some cases actually exceed, those for new capital projects. In many instances, energy efficiency projects examined without carbon costs cannot provide internal rates of return (IRRs) that meet these hurdle rates and are therefore rejected. The result is that many environmentally worthwhile projects are rejected by industry because they are NPV negative—they are profitable (or cost negative, as discussed) but not profitable enough to meet IRR targets. These calculations almost always exclude any accounting for environmental or social externalities, which might make the overall economics look starkly different.[12] This "NPV-IRR trap" is one of the biggest barriers to improving sustainability in industry.

Example: Heat Recovery in the Petroleum Industry

An example from the petroleum industry in Canada, involving the recovery of waste heat from steam-assisted gravity drainage (SAGD) oil sands mining operations, illustrates the NPV-IRR trap dilemma.[13] In one such operation, steam is generated using power from the coal-fired electrical grid and pumped into the reservoir to enable oil recovery. The oil and the condensed steam from the reservoir are produced to the processing plant at 150°C to 180°C. Approximately one-quarter of the energy produced in the steam generator is recovered within the closed-loop cooling system of the plant as low-grade heat at approximately 100–120°C. Because the water must be discharged to a nearby river after treatment, it must be cooled to prevent environmental impact. This is accomplished with electrical grid-powered air coolers that send the heat into the atmosphere (Figure 3.1). Without consideration of the costs of carbon associated with the energy required for cooling (and heating), this BAU approach provides the highest financial returns.

However, a brief examination of options for this operation reveals that this low-level waste heat could be used to produce power utilizing an organic Rankine cycle (ORC) power generation system. For a capital investment of around US$9 million, 10 MW of waste heat can be used to generate up to 1 MW of electricity. The power can be fed back into the project, reducing power consumption from the grid. The IRR for this heat recovery project was estimated at about 8%, using conventional financial analysis, based on the current $80/MWh cost of power. The proposal was not implemented, however, because the rate-of-return hurdle for the company for assessing all projects was 11%. The project fell into the NPV-IRR trap: It was profitable but not profitable enough to appear as NPV positive within the company's project decision-making paradigm.

FIGURE 3.1　Air cooling system for wastewater from SAGD system, Alberta, Canada.

Examination of this same proposition in the context of a carbon-constrained future boosts the IRR considerably. Imputing a nominal cost of carbon (heretofore in the analysis considered to be an "externality") and considering even a modest rise in the socioenvironmental value of carbon over the 30-year life of the project has a profound impact on project IRR. The energy savings predicted from implementation of waste heat recovery would result in a reduction of approximately 8,000 tonnes of CO_2 equivalent (tCO_2e) per year in emissions from the coal-fired grid.[14] Using the current Alberta carbon tax level of US$10/$tCO_2e$ (which currently only applies to emissions over a set maximum annual amount and so would not be incurred on this project as a direct financial cost) over the 20-year life of the project, the NPV of the project rises to 9.5%. At the social cost of carbon (SCC; see separate discussion on SCC) of US$85/$tCO_2e$, the IRR of the heat recovery project is almost 15%. Explicit consideration of the carbon externality, and its prospect of gradual internalization, can lift the project out of the NPV-IRR trap and into NPV-positive territory. The heat recovery project is worth doing.

THE ENVIRONMENTAL AND ECONOMIC SUSTAINABILITY ASSESSMENT: EMBEDDING SUSTAINABILITY IN DECISION MAKING

Escaping the NPV-IRR trap, and creating more sustainable outcomes, requires that we make decisions in a way that effectively, rationally, and objectively balances environmental, social, and economic concerns. This section describes an overarching approach for achieving this balance, the EESA, which combines elements of several well-known systems, such as multicriteria analysis (MCA), life-cycle assessment (LCA), cost–benefit analysis (CBA), and sensitivity analysis, within a structured decision-making framework designed to be objective and quantitative. The EESA thus explicitly marries environmental, social, and economic considerations.

Over the last 10 years of applying the concepts discussed in this book, the author's experience has shown that each project or cluster of like projects needs to be considered on an individual basis. In practice, the circumstances of each proposition are so unique and widely variable that no "standard procedure" can be completely applicable. For this reason, this section focuses on the concepts of a workable assessment approach rather than on a detailed methodology. Application of the approach is illustrated through a number of examples provided in the following chapters.

APPROACH OVERVIEW

The key to a successful and balanced EESA is a structured process that includes setting a clear objective, identifying distinct options for analysis, applying constraints to each option, choosing the external assets to be valued, incorporating risk analysis (in whatever form is applicable), and clearly understanding the limitations of the analysis. The approach described here is flexible and can be used in a number of different ways. In broad terms, it can be used to help select the optimal environmental

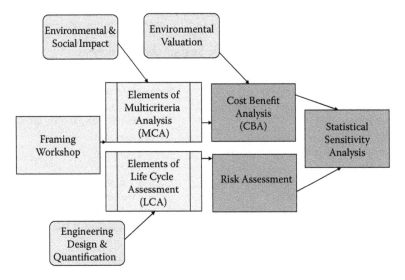

FIGURE 3.2 Main components of an environmental and economic sustainability assessment: an integrated social, environmental, and economic life-cycle decision-making methodology.

and sustainability objectives for a given project or the most economic project among a list of competing project options. Once an objective or project has been chosen, the approach can be used to determine the most sustainable way of actually reaching that goal or executing that project. Figure 3.2 shows the integrated EESA process, which includes elements of risk assessment, MCA, CBA, and sensitivity analysis and takes inputs from environmental, social, and health impact assessment (ESHIA), engineering design and cost estimating, and economic valuation to provide a complete decision-making support analysis with sustainability at its heart. This process is developed in more detail in Table 3.1. Note that sustainable decision making demands more than simple CBA. CBA plays a central role, but it is only one component of the overall EESA methodology. Each of the steps and their corresponding processes are discussed in more detail in the following sections.

FRAMING WORKSHOP

The framing workshop brings together the widest possible range of internal and, if possible, external stakeholders for a structured examination of the issues. As indicated in Table 3.1, this would have been preceded by an engagement process through which the key stakeholders would have been identified, contacted, and briefed on the proposed project and the intent of the EESA process. Ideally, the workshop participants will have available the results of some level of environmental and social impact analysis, or baseline assessment, as a reference point for the project. Facilitation by a third party can provide an important element of balance and perspective to the process. The main steps in the workshop are outlined in Table 3.1 and are discussed next.

TABLE 3.1

Key Steps in Performing an Economic Sustainability Analysis

Step	Process
Identify and consult with stakeholders	Preliminary evaluation, master plan
Determine environmental and social baseline and predicted impacts of project	Environmental and social and health impact assessment (ESHIA)/baseline assessment
Determine the objective of the assessment	Framing workshop
Determine level of assessment: whole project or marginal analysis	Framing workshop
Identify project options for achieving objective	Framing workshop
Identify constraints to each option	Framing workshop
Design, layout, conceptualize each option	Engineering design, master plan
Evaluate life-cycle implications of each option: inputs and outputs	Life-cycle assessment (LCA)
Catalogue and evaluate project risks, including external environmental and social risks, associated with each option	Project risk assessment (RA)
Quantify risks and predicted impacts in physical and temporal terms	Project RA and ESHIA
Estimate capital, operation, and energy costs for each option	Engineering cost estimation
Monetize risks and predicted impacts for each option	Environmental and social economic valuation
Apportion value of predicted impacts by option	CBA
Evaluate NPVs for each option	CBA
Conduct sensitivity analysis	Sensitivity analysis
Evaluate implications and inform decision making	Evaluation, reporting, and decision making
Communicate findings to stakeholders	Sensitivity analysis presentation

Determine the Objective and the Level of Assessment

When a wide range of stakeholders is brought together to discuss a particular issue, it is not uncommon to find that they initially have an equally wide range of views, opinions, and objectives in mind. This diversity of views and preconceptions is a powerful force and can be extremely useful when opening the sphere of consideration to encompass sustainability issues. However, it can also be an impediment to agreement on the scope and objective of the assessment. A key goal of the framing session is to identify a simple, clear statement of the objective of the assessment, preferably in a single concise sentence. An example of such an objective might be

> Identify the most economic and sustainable turbine technology for generating power at the facility.

This objective would lead the assessment team to assess various turbine design options that would be able to provide the required amount of power for the facility. Some might have higher capital costs, some better operation and maintenance (O&M) costs, some might produce more oxides of nitrogen (NOx) emissions than others, or produce more CO_2, or use more or less fuel. The assessment would then examine the trade-offs between all of these factors and determine the optimally economic and sustainable choice over the project life cycle.

But, it might be that the real underlying question for this organization is not simply which turbine technology to use, but whether turbines should be compared to other methods of supplying power to the site. This more strategic question might be expressed, for instance, as the objective:

> Determine the most economic and sustainable way of delivering the required amount of power to the facility over the long term while protecting against energy cost and security risks.

This is a fundamentally different question and would lead the assessment to consider a much wider range of alternatives, of which on-site use of turbines might be one, but which might also include renewable energy on site or purchase of power from the grid, for example. But again, the objective could be elevated to a yet more strategic level. Perhaps the underlying question for the organization is actually centered around the fundamental choice of which processes they are using at the facility and how much power they actually need: Could there be alternative processes that might allow them to use less power overall rather than simply searching for ways to provide a set amount of power? This process of elevating the perspective, from technology level, to examining project or portfolio implications, to examining options for strategy, or even considering options for overarching policy, is a key part of the framing workshop (Figure 3.3). By the end of the session, the participants

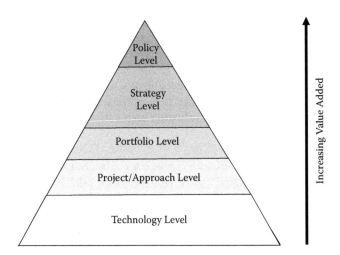

FIGURE 3.3 Schematic of the hierarchy of assessment. The higher the level of the assessment, the greater the potential is for the EESA to add value. Decisions made at a technology level affect only the sustainability of the particular project in question and only the part of the project dependent on that technology. In contrast, decisions made at a policy or strategy level can have widespread effects of improving sustainability throughout the organization.

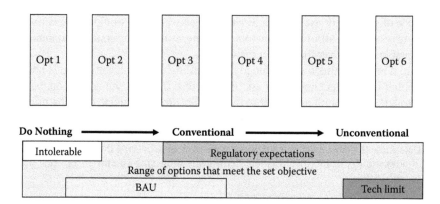

FIGURE 3.4 A wide range of options needs to be considered, including a reference case that represents "do nothing" or business as usual (BAU) and a technological limit option.

need to be happy that they have articulated a clear objective that satisfies the level of consideration that fundamentally underlies the problem. Of course, the higher up this hierarchy the assessment is conducted, the greater the potential value is. The difference between choosing a particular technology rather than another may only amount to a few hundreds of thousands of dollars over the life cycle of the project. Variations in project configuration may result in swings of millions of dollars, and variations in strategy or policy may imply differences of tens or hundreds of millions of dollars between options.

Identify Options for Achieving the Objective

Next, the stakeholder group identifies a range of options that can feasibly meet the stated objective. Ideally, the range of options to be considered should include the widest possible spectrum, from the cheapest available option, to BAU (which will often be the same thing), to a range of more conventional (but perhaps more costly) alternatives, to options that might exceed regulatory standards or community expectations, to those that might be considered to be at the technological limit of sustainability and efficiency (Figure 3.4). Evaluating such a broad range is important for a number of reasons:

1. *It provides a suitable baseline for comparison.* The cheapest (perhaps BAU) option provides a reference point against which other less-conventional options can be measured. The technological limit brackets the analysis on the high cost side, revealing how much better or worse options fare compared to the most expensive solution available. This helps the results to be examined in the context of the full possible range of what could be done to meet the objective and provides a sense of scale to the assessment.
2. *It provides for options that are robustly distinct.* If the options to be considered are too similar, with only minor variations, then the key trends within the environmental, social, and economic parts of the assessment are more difficult to see and interpret. Indeed, if the options are too similar

and do not cover a wide range, then the value of the assessment overall is diminished; there is simply not going to be as much difference between their performance, and a decision based on more conventional and less-rigorous decision-making systems would probably have sufficed. This is revealed in the examples provided in the following chapters.

3. *It allows for future changes in market, environmental, and social conditions.* Because the assessment will inherently be considering a longer time frame (the full asset life cycle), there is every likelihood that conditions may change over time. Indeed, the sensitivity analysis component of the assessment (discussed separately here) will explicitly consider a wide range of possible future conditions and values. Thus, options that may not be attractive today may become real contenders for selection under future conditions.

4. *It reinforces the objective nature of the assessment.* As discussed here and in the previous chapters, this assessment process is designed to provide objective decision making that fully integrates the three elements of sustainability. As such, all options that might conceivably meet the objective should be considered side by side. Screening out options too early based purely on our personal judgment or experience automatically biases the result, and whether we like to acknowledge it or not, such qualitative screening brings with it all of our in-built prejudices, preconceptions, and experience—most of which has been learned and developed in a BAU world, where sustainability has not played the role it perhaps should have. Eliminating this bias, and creating a truly objective assessment, requires that all of the stakeholders are challenged to consider options that they may even consider to be outlandish or crazy. There is no harm in this: If the option is crazy, the analysis will reveal it to be so. But, it may turn out to be worthy of serious consideration when all of the environmental, social, and economic factors are fully considered into the future.

Identify Assets to Be Included in the Assessment

The next step in the framing workshop procedure is to invite all stakeholders to identify the internal and external (environmental and social) assets that are important to the analysis and need to be considered in the analysis. It is perhaps here, above all, that the diversity of the stakeholders involved in the workshop plays its most important role.

The goal is to ensure that all of the economic, social, and environmental issues that could be relevant to the project are identified. Table 3.2 lists some of the key internal (private) and external (special) factors that should be considered in the assessment. Capital, O&M, and energy costs are almost always included. Typical external assets include the total economic value (TEV) of water saved or damaged and the value of air emissions such as carbon dioxide. The list that is adopted by the stakeholder group provides the basis for the valuation of costs and benefits that become inputs into the CBA portion of the assessment.

Table 3.2 immediately reveals that the possible implications of any decision include a much larger range of external factors than internal ones and reinforces the importance of including externalities in decision making if true socially and economically optimal decisions are to be made.

TABLE 3.2

Typical Internal (Private) and External (Social) Assets and Issued Considered

Internal (Private)	External (Environmental and Social)
CAPEX: capital expenditure	Total economic value (TEV) of freshwater resources created or eliminated, including through contamination, waste, or protection, be it surface or groundwater
Nonenergy OPEX (operation and maintenance costs) over the life cycle	Greenhouse gas emissions
Energy operational costs: over the asset life cycle	Ozone-damaging air emissions (chlorofluorocarbons)
Key process input costs: raw materials, water (particularly important when whole-project analysis is being undertaken)	Other air emissions, including NOx, SOx, VOCs, particulates, mercury, dioxins
Corporate reputation (applies to a purely financial analysis but not to a wider social assessment): very important to many companies and can be directly related to corporate environmental and social performance (see Chapter 1)	Terrestrial ecosystems and components damaged or protected, including the value of tropical forests, boreal forest, alpine ecosystems, dryland systems, native bushlands or grasslands, important habitat
Value of property owned by the proponent or company	Freshwater ecosystems damaged or protected, including wetlands, river systems, and lakes
Financial environmental liability held on the company's accounts, including the possibility of litigation or prosecution (civil and criminal) and the resulting penalties	Marine and coastal ecosystems, including beaches, coral reef systems, sea-grass beds, and mangrove swamps
Health and safety of the workers of the company insofar as this relates directly to impacts (positive or negative) on production or payments to workers	Terrestrial and avian species
Corporate revenues from the project	Marine species, especially commercially valuable fish stocks, whales and other cetaceans, and other iconic marine species
	Amenity and nuisance factors, including noise, visual blight, traffic, and odor
	Human health, either improved or impacted
	Property value, which reflects a combination of other values and can be used as a proxy for improvement or erosion in amenity
	Cultural or heritage assets, including aboriginal cultural assets, historical assets
	Social infrastructure improved or damaged, including roads, buildings, public utility infrastructure and assets
	Market-traded commodities or production of external parties, including agricultural production eliminated or improved and natural resource production impacted
	Recreational amenity value, both for nearby residents and visitors to the affected area

Identify Risks and Constraints

While the framing session is not intended to *solve* the problem, initial stakeholder input into the risks associated with the project and the constraints that apply to the project should be elucidated. For each option identified, the likely risks posed to each of the assets of interest can be discussed, their sense determined (positive, negative, neutral), and a preliminary view of required action developed through the use of a standard qualitative risk assessment matrix, as shown in Figure 2.12.

This process informs the physical quantification of the expected damage incurred or avoided, informed by the results of ESHIA and any previous risk analysis completed. In addition to the conventional risk assessment, which examines possible impacts of the project or proposition on environment and society, the risks posed by environment and society to the project should also be considered. This process, sometimes known as nonfinancial risk management (NFRM), is not widely considered but should form an integral part of the overall analysis.[15]

An example of this "flow back" of risk is the current situation in the Niger Delta, where years of oil production have negatively impacted the environment and the ability of traditional cultures to maintain their way of life.[16] As a result, some have taken up arms and now occasionally kidnap oil workers, capture and damage production platforms to demand ransom, and destroy oil field infrastructure. The initial impacts on the environment and society of the oil operations have created feedback risks to the operations themselves, costing money and time, and putting workers at risk.

The framing session participants should also identify critical constraints that may apply to the analysis. These could include physical, regulatory, temporal, legal, or other factors that would render a particular option infeasible. Applied to the developed list of options, certain alternatives may be eliminated from further consideration at this stage.

Agree on Planning Horizon for the Assessment

The final step is to determine the planning horizon and the financial parameters that the stakeholder group wants to use in the assessment. Inherently, this analysis is about improving the sustainability of decisions; therefore, a key goal is to lift the decision-making perspective up and out of the typical short-term payback-dominated view and into the longer term. This is illustrated in Figure 3.5. Theoretically, a generation is considered the minimum for social economic analysis, in part because the benefits provided by the environment, and the concerns of society, are longer term in nature. Short decision-making horizons of the type typically used in industry, which focus on payback periods that can be anywhere from a few months to 3 or 4 years, ignore the equilibrium-based long-term flows of services provided by ecosystems and do not support inherent concerns of society about providing for the well-being of future generations.

Good business decision making in the twenty-first century needs to consider investments over their full life cycle. Profits from investment can and should continue to flow even after payback has been achieved. Assets, if well-managed over the longer term and designed for sustainability, can provide stronger returns coupled with decreased liabilities and improved stakeholder relations, all of which lower cost and boost output, contributing to the bottom line over the longer term.

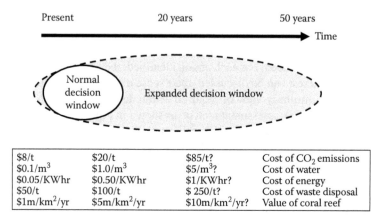

FIGURE 3.5   Using an expanded decision window, typically a generation or the expected life of the asset, provides an inherently more sustainable perspective. When coupled with a social discount rate, a longer-term perspective helps decision makers future-proof their projects and operations.

The future is, of course, inherently uncertain. As shown in Figure 3.5, the value of key commodities (such as water, energy, and a variety of natural resources) will almost certainly be subject to fluctuations over time. As the population and demand for resources grow, scarcity will likely drive long-term increases in resource costs. Designing for sustainability in the long term allows businesses, industry, and government organizations and utilities to explicitly take these types of fluctuations into account and manage the associated business risks. From this perspective, a useful planning horizon is the expected life of the asset being evaluated. For power stations, for instance, this could easily be 30 to 50 years. Refineries and petrochemical complexes routinely operate for 25 years or more, and major infrastructure has a life span of at least 20 years. Using guidance from government, a typical social economic assessment would use a planning horizon of 20 to 30 years.

Set the Life-Cycle Boundaries of the Assessment

The physical and life-cycle limits of the assessment need also to be decided. A key consideration is how much of the external life cycle of inputs and outputs to consider. Typically, in a more basic assessment, simplifying assumptions are made that deem that once flows have reached the point of exchange (bought or sold from another organization), the costs and benefits associated with past and future management of the thing are included in the price. This of course assumes that the other organizations are pricing in external issues properly (which much of the time they are not). More complex assessments, which require more time and effort, can extend the explicit examination of externalities further afield, spatially and temporally, and bring them into the assessment. At the limit, a cradle-to-grave analysis can be completed. The framing session participants need to reach consensus on where to set the physical and life-cycle limits of the assessment. This will depend on the complexity, scale, profile, and importance of the issue being investigated and on the time and data available for completing the work.

The team also needs to consider the perspective of the assessment. In some cases, it may be important to examine the project as a whole, including all of the revenues and costs from the entire operation. However, depending on the objective selected for the assessment, the analysis can be done on the margin—only those parts of the operation directly affected by the decision need be taken into account. This last approach is often favored because it is inherently simpler, less data intensive, and does not require the firm to disclose possibly sensitive commercial and financial data.

Identify Range of Discount Rates to Use

Another key choice is the discount rate to be used or, rather, the range of discount rates to be considered. For social economic analysis, discount rates are typically significantly lower than typical commercial rates and vary from about 2.5% to 6% per annum.[17] Lower rates are used in social analysis because higher discount rates effectively devalue the future; at a 10% discount rate, a benefit that arises in a hundred years is almost worthless in present value (PV) terms. An asset, such as the lumber contained in a tropical forest, that might be worth $100 million if cut and sold today is worth only $852,000, using a 10% discount rate, in PV terms, if cut and sold in 50 years. This creates a powerful economic impetus to cut the trees now.

So, higher discount rates mean large costs and benefits that occur in the future have very small impacts on decisions made today.[18] Some have called this "writing off the future."[19] The result is that decision making focuses on the present and the short term and largely ignores the future.

Another benchmark is the weighted cost of capital (WCC) for a particular organization, which is essentially their cost of borrowing money to implement projects. Rates of return on investment lower than this mean that the borrowing organization would lose money; it needs to make returns on investment that are at least equivalent to its WCC. Commercial rates, typically 10% but often as high as 30% or more, imply very short-term payback expectations and very short-term thinking. CAPEX dominates decision making in industry, while longer-term operating and energy cost considerations, and certainly longer-term external factors, have relatively less impact on decisions, even when access to capital is not a major concern.

Unsustainable behavior is structurally embedded in our fundamental decision-making processes for two reasons: first because we tend to make decisions based on very short planning horizons and second because our search for high rates of return inherently devalues the future. The two are connected, of course, and combine to drive individuals, companies, and even governments down a path of suboptimal, socially, and (quite often) privately unsustainable behavior. As discussed in Chapters 1 and 2, these things are so deeply entrenched in our business systems today that many feel powerless to make real change, even when they know that it is the right thing to do.*

---

* Ecological economists argue that people also make decisions based on ethical and moral grounds and not purely to maximize their personal utility, as orthodox economics would maintain. However, when making decisions on behalf of companies and organizations, which are inherently financial beings, personal ethics can readily be subsumed.

Framing Session Output

By the end of the framing session, the multistakeholder team should have developed consensus around a succinct outline that will serve to guide the assessment. This outline should include a clear statement of the objective of the analysis, a list of wide-ranging options capable of meeting the objective, and a complete catalogue of the key internal and external assets and issues that could be impacted by each option should it be implemented. An overarching constraints map along with a preliminary qualitative risk assessment provide the context for the more detailed analysis that is to follow. Finally, the group has agreed on the range of discount rates to be tested, the life-cycle boundaries to be used, and the planning horizon to be considered. If the facilitator has guided the group effectively, all stakeholders have contributed, and all stakeholders' concerns, issues, and favored solutions are represented. The assessment team must now shape this framework into a quantitative analysis that reveals the advantages, disadvantages, and trade-offs between each of the options in a rational and objective way.

PHYSICAL QUANTIFICATION OF OPTIONS

The next stage of the process is to physically describe and quantify, to the degree possible with the data available, each of the options. The process is based on physical reality, so the core of it lies in estimates of the volumes and mass flows of material inputs and outputs required to realize each option. This part of the assessment borrows heavily from LCA, which is used widely in industry, to evaluate these flows over the planning horizon. Table 3.3 provides examples of the types of data typically required for the physical quantification of each of the options across the asset life cycle and common sources of those data.

SOCIOENVIRONMENTAL ECONOMIC ANALYSIS

As discussed, the now well-established science of environmental and ecological economics has made it possible for a range of realistic monetary values to be placed on a wide spectrum of environmental and social assets. These values can be used within an economic analysis using the techniques of CBA. This section provides a brief overview of the CBA element of the overall EESA and a number of references for readers who wish to explore the subject in more detail.

FULL SOCIAL COST–BENEFIT ANALYSIS

An economic model for assessing the benefits of environmental and social protection was presented by Hardisty and Ozdemiroglu (2005).[20] This method explicitly describes and measures sustainability in economic terms by explicitly monetizing the widest range of external costs and benefits as possible and appropriate and

TABLE 3.3

Examples of Physical Data that Might Be Required for Each Option

Element	Source	Comments and Examples
Option location	Master plan	Facility could be located offshore on an island or reef, on the coastline at a remote location, adjacent to an existing industrial area, or as part of an expansion of an existing similar facility. Each location would have different implications in terms of distances from sources of inputs, expected damage to environmental assets, and impacts on communities.
Physical layout of option	Engineering design, master plan	Facility complex layout, providing footprint, location of lines, roads, associated infrastructure, and placement of all key components. Comparing options for layout of a new housing development could include leveling of land or maintaining natural topography to varying degrees, retention of natural habitat corridors to varying degrees, coastal setback differences, or locations of roads, sewerage systems, and parks.
Capital costs (CAPEX)	Engineering cost estimation	CAPEX for each option, depending on the items mentioned in previous steps. Often, organizations may have already developed CAPEX estimates for one or more of the options to be considered. This can form a starting point for developing offshoots or other related options for assessment. As discussed, options need to be considered on a fair and objective basis; this requires that assumptions made in developing cost estimates be used rigorously across all options being compared. The importance of this will become apparent in the sensitivity analysis section.
Nonenergy operation and maintenance costs (OPEX)	Engineering cost estimation	Nonenergy OPEX is a critical and often underappreciated part of an EESA. OPEX costs become more important in a longer-term social economic assessment, as discussed. This would include, for each option, estimates of all costs, for each year across the planning horizon, required to keep the facility operational. This would include, as needed, the costs of raw material inputs, labor, and periodic additional capital works required to keep the facility operational.
Energy use	Engineering cost estimation	Energy is often one of the most important factors in an environmental and economic sustainability assessment. Many facilities use large amounts of energy, and often the differences of energy use between options may be considerable. For example, water treatment options may include simple biological systems that use little or no energy or options that rely on energy-intensive processes such as reverse osmosis (RO). Data are provided in kilowatt hours, joules, barrels of oil, or cubic meters of gas, as appropriate for each time period across the asset life cycle.
Emissions to atmosphere	Engineering design or estimation (LCA)	Estimation of annual emissions of all key compounds that may impact human health, the environment, infrastructure, or economic productivity, including CO_2, methane, NOx, SOx, particulates, heavy metals such as mercury, and dioxins. Data are provided in units of mass for each year across the planning horizon.

(continued)

TABLE 3.3 (CONTINUED)

Examples of Physical Data that Might Be Required for Each Option

Element	Source	Comments and Examples
Other emissions	Engineering design or estimation (LCA)	Estimation of annual emissions to water and land, including wastewater discharges of all kinds, solid wastes, sludges, construction waste, and wastes generated during decommissioning, including identification of the ultimate receiving body or location and method of disposal. Data are provided in units of mass for each year across the planning horizon.
Environmental impact	ESHIA	Estimation of the physical damage caused to the environment (or damage avoided) that would occur if each option were implemented. The results of environmental baseline and impact studies of various kinds, including environmental investigations, are used to quantify the expected range of impacts. For installation of a new LNG facility on an offshore reef complex, options requiring dredging of the footprint will result in elimination of a certain number of hectares of reef. Generation of a spoil plume from the dredging operations would also result in additional coral mortality. Options that are not reef based would of course not incur these damages. Data are provided in appropriate physical units for the environmental asset in question, for each year throughout the planning horizon (e.g., hectares, numbers of individuals of a certain species affected, households affected).
Social and health impact	ESHIA	Estimation of the physical, economic, and other damage (or damage avoided) to people, communities, aboriginal cultures and communities created by implementing the option. Placing a new refinery complex in close proximity to an existing community may affect human health and amenity through air emissions, noise, aesthetic impact, traffic, and the like and may result in increased rates of illness, and property value impacts. This may also include economic impacts external to the project, including direct impacts on production from other facilities nearby, damage to agricultural production (loss of yields), and similar changes caused by implementation of particular options. Data are provided as numbers of individuals or households affected in specific ways or in physical quantities of production lost or gained in each year over the chosen planning horizon.

adding these to the conventional internal or private costs and benefits of a proposed project or action.

Net Benefits

CBA compares the costs of a project, action, or policy with the benefits that accrue to all of society from implementation of that project, action, or policy. To find net benefits, we deduct the flow of costs from the flow of benefits across the planning horizon. Thus, the present value of the net benefits (NPV) (benefits minus costs) of the selected project or action is given by

$$NPV = \sum_{0}^{t} \left[\frac{(B_p + B_x) - (C_p + C_x)}{(1+i)^t} \right]$$

where NPV is the total social NPV of project p summed annually from year 0 to year t; B_p and C_p are the flows of private (internal) benefits and costs of the project in each given year from 0 to t, respectively; B_x and C_x are the external benefits and costs of the project in each year from 0 to t, respectively; t is the planning horizon (expanded to cover the full project life cycle); and i is the applied discount rate.

Typically, businesses, industry, and even government bodies have considered only the private costs and private benefits of the proposition when undertaking their NPV analysis. What results is, strictly speaking, a *financial* analysis—it considers only the internal factors. Decisions based on financial analysis ignore all of the external environmental and social considerations, which can be so important if real sustainability is to be achieved. A true economic analysis brings in all of the impacts, positive and negative, that the rest of society experiences from the decision, over a longer period of time t, allowing sustainability to be quantified in monetary terms. If this can be done, what results is an NPV that measures the overall true economic, social, and environmental result of a proposition (Figure 3.6). Only decisions that provide an overall increase in net human welfare will be NPV positive. By examining the relative proportions of the NPV accruing to both the private and external stakeholders, an environmental, social, and economic optimum can be identified.

Objective setting must consider the benefits of achieving a given objective. The overall objective of any decision is assumed to be the maximization of human welfare over time. To compare the different benefit and cost streams over time, the process of

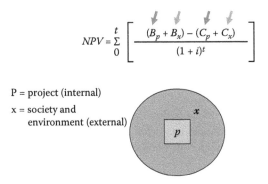

FIGURE 3.6   A social economic NPV analysis that includes both the internal and the external costs and benefits. The project (internal) is part of society and so is included, but the costs and benefits to the environment and society must also be included to determine a true overall optimum.

discounting is used, and amounts over time are expressed as PVs. Economic analysis recommends the decision with the maximum NPV (PV of net benefits, or benefits minus costs, over time) or the highest benefit cost ratio (BCR) (ratio of the PV of benefits to the PV of costs). Benefits of an action can be direct or can be expressed as the "damages avoided" by undertaking that action.

VALUATION OF BENEFITS

For the given equations to be used, both the costs and benefits of environmental and social protection need to be estimated in a common unit. Economic analysis uses money as this common unit based on what individuals are willing to pay (WTP) to avoid damage (or willing to accept [WTA] as compensation for damage) and what we would have to spend on the actions to protect or remediate environmental resources or assets. But, before economic value of damage can be estimated, the damage needs to be quantified in physical terms through examination of the expected impacts on identified receptors should the protective action not occur. The benefits of action are then equivalent to the avoided damage.

Economic benefits accrue due to the protection of the value of the environment or natural resources (as an input to production or consumption [direct use value], its role in the functioning of ecosystems [indirect use value], or perhaps its potential future uses [option value]). In the case of water, for instance (a key consideration in most assessments), people may value water and be willing to pay for its protection, not only because of their own direct use of the resource (direct and indirect use values), but also because of its benefits to others (altruistic value), for future generations (bequest value), and for its own sake (existence value). The sum of these different types of economic benefits or values is referred to as TEV in the economic literature.[21] The rest of this section illustrates what these different types of benefit may mean in the context of an environmental and economic sustainability assessment and how they can be estimated using economic valuation techniques.

Private Benefits

If the analysis is undertaken from the perspective of the problem holder, only the costs and benefits that accrue to the proponent are considered. This approach uses market prices of costs and benefits that accrue to the project proponent, which include subsidies and taxes. Private discount rates are used, which are determined by the cost of capital or rates of return expectations from alternative investments in the private sector. Private discount rates are generally higher than social discount rates. Financial analysis does not deal with environmental or other external social impacts. Any of the internal assets listed in Table 3.2 which are positively affected by the project option under consideration would become internal benefits in the assessment.

External Benefits

A full economic analysis looks at those costs and benefits that accrue to society as a whole. This includes costs and benefits to the project proponent as well as those to the rest of the society. The latter are commonly known as *external* costs and benefits (as they are external to the transactions in the market and hence not included in market prices) as long as they are not compensated by or paid to the proponent. This definition of costs and benefits requires them to be measured differently from a financial (private) analysis. The prices for marketed goods and services that are affected should no longer be market prices but real or shadow prices. Shadow prices are estimated by subtracting (or adding) the subsidy and tax elements from (to) market prices. Subsidies and taxes are referred to as *transfer payments*—their payment does not cause a net change in benefits to society as a whole but simply transfers funds from one party to another within society.[22] External assets of the type listed in Table 3.2 that are positively affected by the implementation of the project option under consideration become external benefits.

Internal Costs

The internal cost part of a typical financial analysis has been well-studied in the engineering and accounting worlds, and the capital (CAPEX) and operating (OPEX) costs of undertaking just about any type of project or policy can be readily and sometimes quite accurately determined by private and public sector organizations. Any of the internal assets or issues listed in Table 3.2 that are impacted negatively can be treated as internal costs. In practice, these are dominated by CAPEX and OPEX.

It is interesting to note, however, that in some complex, highly "siloed" organizations, project financial analysis can be carried out with such a strong focus on the project that the costs and benefits to the rest of the same organization (beyond the specific project) are ignored. This can sometimes mean that, from a practical perspective, damage done to the rest of the organization by the project is treated as an externality and not included in the project's financial analysis. An example of this type of behavior could be the single-minded maximization of financial NPV in project design and execution that leads to a rupturing of relationships with community and regulatory bodies. The company uses its political and economic muscle to push through the project because the short-term profits are considered to be worthwhile by the project leadership. However, if that behavior subsequently means that other projects are more difficult and costly to develop and approve because the lack of trust within the community, or because the regulatory approvals process is now longer, or if approvals and licenses to operate are withheld completely in favor of rivals, the company as a whole has suffered, perhaps greatly. Expanding the analysis to look at wider risks, and the flow-back nonfinancial risks (which eventually become financial), can be very useful for firms even if they do not wish to explicitly include wider environmental and social externalities.

It is also important to ensure that all costs are included and measured on the same basis. Sometimes, firms ignore the real value of resources they use and do not register them as internal costs. This can lead to distortion of the true sustainability picture. An example is the natural gas industry, where it remains common practice to treat natural gas burned to power production and refining facilities as "free." As expected, this leads to financial analysis that is skewed toward lower energy efficiency within production facilities, certainly lower than if gas was ascribed to a going market value. The EESA seeks to redress these distortions by treating the value of resources and assets consistently, allowing objective and rational comparison of options.

External Costs

In the process of undertaking a project, it is sometimes possible that environmental and social impacts are produced by those actions despite best attempts at mitigation. The economic value of these impacts should be included in the overall assessment. The costs of dealing with these effects (as a lower-bound estimate) or the value of the damages that they cause, which are not borne by the problem holder, are termed *external costs*.[23]

External costs Cx can be divided into two categories: planned or process-related external costs that cannot be mitigated against Cx_p and unplanned or inadvertent external costs Cx_{up}, such that

$$Cx = Cx_p + (p \cdot Cx_{up})$$

where p is the probability that the unplanned external cost will occur. The risk assessment process is used to determine the range of likely probabilities of these occurrences.[24]

VALUING THE ENVIRONMENT AND SOCIETY

To conduct the type of economic sustainability assessment proposed here and bring all of the various factors and trade-offs at play into the analysis, social and environmental externalities must be valued in dollar terms or "monetized." This is the preserve of the discipline of environmental or ecological economics.* The science began to develop in the 1970s, with work done by the World Bank and other development agencies and a new breed of economists, including Herman Daly and others.[25] Professor David Pearce of University College London was recently recognized as a pioneer in this area, and a number of his works lay out the case for not only how monetization can be done but also how the information can be used in CBA.[26]

This section provides a brief overview of some of the common methods of valuation, by way of introductory background, and summarizes a range of valuation studies that can be used in support of decision making. The focus is on practical valuations for some of the key areas that commonly confront decision makers in industry, business, and government, including water, biodiversity, key ecosystem types, air emissions of different types, and public amenity.

---

* Economists distinguish between environmental economics, which is usually considered to be within the realm of orthodox economic thought, and ecological economics, which espouses a far more radical approach to the protection of the environment. Environmental economics tends to examine the environment within the context of a traditional quantitative economic approach; that is, the environment is included in the economy through the valuation of environmental assets, but the fundamental notion of never-ending economic growth is maintained. Ecological economics espouses the concept that the entire economy must be seen within the context of the physical and ecological limits of the planet, and as such there exists a point at which economic growth can no longer occur without a net loss to society as a whole. Many ecological economists now believe that we are already in a position of "overshoot" and are arguing for a steady-state economy.

Overview

As discussed in Chapter 2, people clearly value both the environment that sustains us and their own cultural, aesthetic, and physical well-being. That the natural environment has value to humankind is self-evident. Without air we cannot breathe, without water we die of thirst in a few days, without food we starve. All of these basic necessities of life are provided to us by a healthily functioning ecosystem that has taken millions of years to evolve. Beyond mere survival, we also value, to differing degrees, the beauty of an alpine meadow, the majestic sweep of a pristine coastline, or simply the peace and quiet of a walk through our favorite park or woodland.

By definition, anything that is economic improves human welfare, on balance. If it does not improve human welfare, it is uneconomic, regardless of what a particular segment of society (such as a company) experiences. Thus, if the environment or society benefits from a project or is damaged by a decision, then to reflect the overall economic position that results from that decision, the values of these externalities must be taken into account.

More than three decades ago, economists began estimating the value of public goods and natural environmental assets. Initially, much of the work centered on developing countries and the value of damage incurred during major developments such as large-scale forestry projects and the construction of dams. While such megaprojects created jobs and value, they also sometimes resulted in serious environmental and social impacts, such as soil erosion, population displacement, and loss of biodiversity. To gauge whether the country was really better off as a result of these projects, economists began trying to estimate the value of these damages, which could then be set off against the benefits. The results were sometimes startling.

Over the last few decades, the science of environmental economics has burgeoned, and substantial work has been done not only to develop valuation techniques for externalities but also to estimate values for a wide range of environmental and social assets, from coral reefs to cultural heritage. It is not the purpose of this book to review in any detail the science of environmental valuation. Rather, the focus is on using the valuation data and guidance that has been developed over the past several years within a decision-making support analysis that allows business and industry to consider environmental and social "externalities" alongside the traditional financial aspects.

Valuation Techniques

Economists place ranges of values on specific environmental and social assets using a variety of methods. These techniques are summarized in this section. The next sections discuss the valuation of specific key assets of interest and provide indicative ranges of values that can be used within a practical decision-making framework. Example areas covered include GHG emissions, other atmospheric emissions, water, biodiversity, and social amenity.

Actual Market Techniques

When the good itself is priced on the open market as a salable commodity, an estimate of its overall value can be readily determined. For example, groundwater sold as drinking water has a price per unit volume, and land is bought and sold and has a specific value depending on location, zoning, and market conditions. One of the easiest ways of examining the economic impact of a project or the legacy of a project (such as site contamination) is to consider its effects on property value. This provides a direct, easy-to-measure, and robust reflection of people's real attitudes toward the impacts. Shadow prices should be used whenever possible to maintain rigor.

Surrogate Market Techniques

If a market good or service is influenced by an externality that itself is not reflected in a market, then this so-called surrogate market can be used to estimate value. For example, water might be used to irrigate crops that are sold at market prices. The crop market in this example is a surrogate market for water, and a proportion of the economic value of the yield is representative of the value of water as an input. This approach is especially useful when a good such as water is dramatically underpriced in the economy (provided for free or subsidized). Defensive measures, discussed in Chapter 2, are costs incurred to protect society from some form of environmental or social damage. So, for instance, returning to the *Exxon Valdez* example, the expenditures that society chose to make to clean up the coastline and protect the ecosystems at risk provide an indirect but still market-based estimate for the value of those assets; they act as a surrogate for the value of ecosystem.

Hypothetical Market Techniques

Economists can also create hypothetical markets via structured questionnaires that elicit individuals' willingness to pay (WTP) to secure a particular outcome. Alternatively, people may be WTA compensation to avoid or tolerate a loss or to forgo a beneficial outcome. Among these stated preference techniques are contingent valuation (CV) and choice modeling. These techniques are widely used but are also subject to uncertainty. For instance, questionnaires reflect only what a sample of the population claims they would be willing to pay or accept but do not actually require that the people pay in reality. Research has shown that people are much more likely to state WTP than actually to follow through with real payments. There is a significant body of research on hypothetical market techniques that can be explored by those interested.[27]

GREENHOUSE GAS EMISSIONS

When assessing the sustainability of any design or project, one of the key emerging considerations is the potential contribution to climate change. Emissions of GHGs, still unpriced in most parts of the world, are rapidly becoming one of the key metrics reported by companies in their annual sustainability reports.[28] This is underpinned by a strong consensus of recent science, which has concluded that human impact on the climate system of the Earth is now readily apparent and are certain to worsen over time.[29] As discussed in Chapter 2, from an overall sustainability point of view, climate change is an overarching issue. Other efforts to improve sustainability, protect habitat, promote environmental cleanup, and impact reduction are arguably worth relatively little if the entire fabric of the current climate regime of the Earth and the ecology it supports are threatened. The widespread and fairly recent recognition of GHG emissions as a key element of overall sustainability is particularly relevant to energy-intensive business sectors. The implications for decision making in these sectors are even more significant if the economic implications of GHG emissions are considered in an overall economic context.

The Carbon Markets

Carbon pricing, in one form or another, is quickly becoming commonplace. It represents the efforts of governments and society to create a market value for carbon, which will provide a price signal to the economy to drive change and reduce emissions. In Europe, the flourishing carbon market was worth over US$24 billion in 2007, trading over 1 billion tCO_2e. The EU ETS (Emissions Trading Scheme) long-term phase 2 average price has fluctuated between about US$20 and $25/$tCO_2e$ until the GFC of 2008–2009, which sent European carbon prices into freefall for a period of months, along with many other commodities traded on world markets. The Clean Development Mechanism (CDM), established under the Kyoto accord, traded over 500 million tCO_2e in 2006, worth over US$15 billion.[30] Other trading schemes, voluntary and regulated, are starting to appear around the world.

In Alberta, the oil- and gas-producing province of Canada and home of the Athabasca tar sands megareserves, the government has announced a CDN$12/$tCO_2e$ tax on GHG emissions exceeding specific reduction targets. The voluntary Chicago Climate Exchange has grown year after year since its inception, and the Montreal Stock Exchange is now in the process of developing a similar voluntary market in Canada. Also propelling carbon prices are carbon reduction and mandatory renewable energy targets being set by various governments worldwide and at the national, state, and local levels. The European Union has announced a new goal of cutting emissions by 30% from 1990 levels by 2020. The United States is moving rapidly to bring in a nationwide carbon reduction program. Full engagement of the largest economy of the world will have a resounding effect on the way the rest of the planet approaches carbon regulation in the coming decades. Australian state and federal governments are now proposing similar schemes. All of these measures will increasingly impose a real market price for carbon and a tangible cost to organizations that emit GHGs.

The Social Cost of Carbon

But, there is a fundamental difference between market-based (in the case of cap-and-trade schemes) or tax-based carbon prices and the real value of the damage caused by the emission of carbon into the atmosphere. The SCC reflects the value of the damage caused by each additional tonne of GHG put into the atmosphere. Carbon markets or taxes only reflect the cost that governments have imposed on emitters, and this cost represents only a fraction of the true value of the damage. The SCC is directly related to the total amount of GHG in the atmosphere, so the longer it takes to stabilize concentrations of GHG, the higher the eventual SCC will be.[31]

The Stern review examined the economic implications for society of the predicted effects of climate change. On a macroeconomic level, Stern estimated that the cost of taking action to stabilize GHG levels at below 550 ppm CO_2e, which gives us an even chance of avoiding warming greater than about 2 degrees Celcius (2C) on average, will cost about 1% of global GDP. However, the cost of not acting to control emissions and continuing on a BAU emissions trajectory will cost the global economy between 5% and 20% of GP now and forever. Stern concluded that combating climate change is the pro-growth strategy.[32]

A report on the economics of climate change prepared by the Australian government called for even more pronounced and urgent action.[33] The OECD and IEA (International Energy Agency) went even further in their most recent world energy review, suggesting that the costs of successful mitigation are lower, and the benefits higher, than estimated by Stern.[34]

However, Stern did not address specifically how these far-reaching findings affect businesses, investment decisions, and business planning. Climate change poses risks, uncertainties, and opportunities for business as society and governments increasingly demand action to regulate and reduce GHG emissions. Whether this takes the form of mandated carbon reduction targets and associated market structures or some form of explicit carbon tax, the economic costs and benefits of actions taken by businesses to reduce emissions need to be carefully considered as the marginal cost of carbon (now on the order of US\$5 to \$25/tCO$_2$e) climbs toward the social cost, which Stern estimated at US\$85/tCO$_2$e, assuming BAU emissions trajectories. However, Stern was also explicit in his acknowledgment of the uncertainty and variability around the SCC. Some studies have put the SCC as low as \$5.5/tCO$_2$e[35] and others as high as \$500/tCO$_2$e.[36]

The U.K. government identified a shadow price for carbon (SPC), which can be used on the margin to assess individual project decisions within the United Kingdom.[37] The SPC is based on the realization that a single nation cannot in isolation determine the trajectory of global emissions and thus the SCC. A more immediate value for carbon is required that reflects the current climate change goals and commitments of the U.K. government. The SPC serves this purpose. Based on a global stabilization of atmospheric CO$_2$ concentrations at 550 ppm, Stern calculated an implied SCC of US\$30/tCO$_2$e. The Department for Food, Environment, and Rural Affairs (DEFRA) set the U.K. 550-ppm stabilization SPC at about US\$50/tCO$_2$e, increasing at 2% per annum from 2007, and recommended that, for project decisions, the U.K. Treasury's standard social discount rate of 3.5% be applied.

Table 3.4 provides a range of current market-based prices for carbon and estimates of the SCC. This inherent variability in the value of many externalities is potentially a significant barrier to using any monetary value-based method of analysis. However, as will be discussed in the "Sensitivity Analysis" section of this chapter, it can also be used to great advantage in the decision-making context.

Whatever one's perspective on the uncertainties surrounding climate change, managing GHG emissions and understanding the economics of achieving sustainability objectives will become increasingly important for business and industry as time passes. Many companies are already establishing their own internal emissions reduction targets. Some, such as BP, have set up their own internal emissions trading schemes, which have been highly effective at reducing emissions in their operations and saving large amounts of money. With a focus on design and process efficiency, significant emissions reductions can be achieved at relatively low cost, in many cases

TABLE 3.4
Estimates of the Value of Atmospheric Carbon Emissions (2009)

Estimate	Value (US$/t CO_2e)	Notes
Stern review, SCC under business-as-usual emissions trajectory	85	Central estimate based on review of available studies.
Stern review, SCC under 450- to 550-ppm stabilization trajectory	25–30	Increasing at 2% per year.
IPCC,[38] SCC	12	Average of peer-reviewed estimates of SCC reviewed by IPCC based on studies available in 2005 (range was US$–3 to $95/tCO_2$e).
Hope and Newberry (2006),[39] SCC	18	Using data from IPCC and using higher discount rates than Stern; range of $3.5–50 tCO_2$e; value expected to rise at 2.5% per year.
Hope and Newberry (2006), SCC,	24.5	Using a climate sensitivity distribution mean of 3C from Stainforth et al. (2005).[40]
UK DEFRA shadow price for carbon,[41] SCC estimate	54	£27 in 2010, increasing at 2% per year, to reflect the increased damage expected as GHG concentrations in the atmosphere increase over time.
European ETS, market price (August 2009 average)	21.45	Prices have fluctuated considerably over time and were depressed, like all commodities. by the global financial crisis of 2008–2009.
Alberta carbon tax	10	Imposed on emitters greater than 100,000 tonnes per annum (tpa) (Can$15).
Norwegian carbon tax	60	In place since 1991, carbon taxes have limited GHG emissions growth in Norway to 15% over that time, despite overall economic growth of 70% over the same period.
Australian ETS, tax and market price to follow	7.5	Initial flat-rate carbon tax in 2010 of AUD$10 (legislation pending approval), to be replaced by a cap-and-trade scheme in following years.

actually reducing overall costs to operators, and improving profitability. A series of examples focusing on applying EESA to GHG management is provided in Chapter 5.

AIR POLLUTION

While the current focus worldwide is increasingly on GHG emissions, there is a wide variety of other emissions to atmosphere that can cause significant environmental and social damage and for which robust working valuation estimates are available. In general, atmospheric contaminants can be classified according to their persistence in the atmosphere and their geographic range. As shown in Figure 3.7, some emissions, such as volatile organic compounds (VOCs), are local and fall back out of the atmosphere within only a few kilometers of their source. Others, such as the ozone-destroying chloroflurocarbons and the GHGs, are both highly persistent

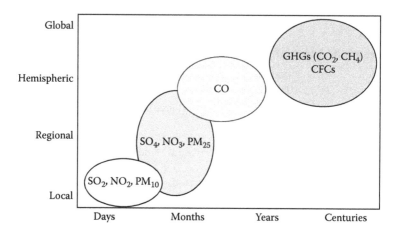

FIGURE 3.7   Residence time in the atmosphere of various types of industrial air pollutants. (After United Nations Environment Programme [UNEP], *Global Environmental Outlook 4: Environment for Development [GEO-4],* UNEP, Nairobi, Kenya, 2007.)

(the half-life of CO_2 in the atmosphere is estimated at between 19 and 49 years, with about 30 years a reasonable median estimate)[42] and are global in impact. Depending on the nature of their impact, the damage they cause, their persistence and extent of impact, the values of atmospheric contaminants can vary considerably. Table 3.5 shows a range of external cost estimates for a range of common air emissions.

Many of these estimates explicitly include the value of social externalities. So, for instance, the European Environment Agency estimate for the value of VOC emissions is dominated by the human health component.[49] VOC emissions, if inhaled, can cause cancer and respiratory illness, requiring hospitalization and costing the economy in terms of lost productivity. While these costs can be measured and priced within the economy, they are typically considered to be external to the project because they fall on society as a whole and are not borne directly by the emitter (and thus not passed on to the user). Environmental regulations, taxes, and penalties act to internalize some of these costs by forcing emitters to pay for damages or spend money on cleanup measures. While it is not the intent of this book to explore in detail the science of economic valuation of externalities, decision-making systems of the type described here must take into account the sources of the available valuation estimates, the assumptions embedded within, and what the various estimates comprise and ignore. This is discussed in more detail in this chapter.

As with GHGs, the values shown in Table 3.5 vary over a wide range for individual pollutants and among pollutants. Some, such as airborne mercury, for instance, carry high external costs due to their acute human and ecological toxicity. Others, such as NOx, are by-products of hydrocarbon combustion and can cause acidification of surface water, acid rain, and damage to agriculture and infrastructure and so carry market-based values that are an order of magnitude and higher than for CO_2e. Atmospheric emission values can also be categorized according to their longevity in the atmosphere and the scale of their impact. In general, pollutants that are short lived

TABLE 3.5

A Sample of Estimates of the External Costs of Common Air Pollutants

Pollutant	Value Range (US$/t)	Source and Notes
NH₃	13,750–38,750	Amalgamation of external costs from 25 E.U. countries, including damage to health, agriculture, infrastructure, and water resources; European Union Clean Air for Europe (CAFE) 2005 study[43]
NOx	5,500–15,000	European Union CAFE (2005) study
NOx	1,800–23,500	External costs of motor vehicle emissions in urban areas in the United States (in 1991 U.S. dollars)[44]
NOx	500–2,200	U.S. northeast NOx trading area market value spread, 2000–2009; July 2009 average monthly price US$802/t[45]
SO₂	7,000–20,000	European Union CAFE (2005) study
SOx	175–1,600	U.S. SOx market value range, 2000–2009; June 2009 spot price US$71.58/t[46]
SO₂	13,188	Mortality and morbidity effects and crop damage, ExternE study, Europe (1999)[47]
PM2.5	32,500–93,750	European Union CAFE (2005) study
PM10	1,600–32,800	External costs of motor vehicle emissions in urban areas in the United States (in 1991 U.S. dollars)
PM10	1,295–14,800	Total cost report American Institute of Chemical Engineering (AICE; 1998 U.S. dollars)[48]
VOCs	500–2,000	External costs of motor vehicle emissions in urban areas in the United States (in 1991 U.S. dollars)
VOCs	1,188–3,500	European Union CAFE (2005) study
Cadmium	26,125	Mortality and morbidity effects and crop damage, ExternE study, Europe (1999)
Arsenic	213,000	Mortality and morbidity effects and crop damage, ExternE study, Europe (1999)
Lead	926–8,934	Total cost report AICE (1998 U.S. dollars)
Mercury	3,300–9,900	Total cost report AICE (1998 U.S. dollars)
Dioxins	40,700,000	Mortality and morbidity effects and crop damage, ExternE study, Europe (1999)

in the atmosphere (such as nitrogen oxides [NOx] and sulfur oxides [SOx]) will have localized impacts, will generally create more acute damage, and will thus tend to generate greater damage costs per unit of mass. Longer-lived pollutants, such as the GHGs, are global in their scale of impact and create more chronic long-term damage. As discussed here and shown in the examples that follow, examination of the trade-offs between options that create more or less of various pollutants can sometimes yield surprising results.

WATER

Water has tremendous value. We all know that without it we cannot live and we cannot grow the food we need to eat. The world's ecological web that maintains all life on Earth depends on freshwater. Water is a key input to almost every industrial and economic activity. As discussed in Chapters 1 and 2, past failures to assign water its true economic value have led to wasteful and damaging use of the resource.[50] One of the most profound developments of recent decades is the realization that water is in fact an economic good, supporting almost every economic activity, and needs to be treated as such.

The Total Economic Value of Water

The concept of TEV can be applied to a wide range of public goods that provide a range of services to mankind. TEV is the sum of the value provided from a variety of uses (direct and indirect) and from other ways in which a good provides value to us without it being used or consumed (nonuse values). In the case of water, TEV is typically described as consisting of the following components:[51]

- *Use value*, which comprises
  - Direct use value:
    - Consumptive use: value derived from the direct consumption of water for potable use, agriculture, or industrial use. Market prices for water in the economy provide a direct market value for this component of TEV.
    - Nonconsumptive use: value derived from the use of water for recreation, boating, angling, swimming, and the like.
  - Indirect use value: value derived from indirect use of the resource, notably in regulating services such as nutrient cycling, flood management, waste assimilation, and the whole variety of ecosystem services that freshwater provides.
  - Option value: future direct and indirect use values. We may not be using a certain freshwater body today, but we are willing to pay to preserve the option of using it at some time in the future.

- *Nonuse value*, which comprises
  - Existence value: the value that people place on the knowledge that a resource or asset simply exists and will continue to exist in the future.
  - Altruistic value: the value that people place on the knowledge that a resource can be used by the current generation.
- Bequest value: the value that people place on the knowledge that a resource will be preserved for use by future generations.

TEV can also be determined using an ecosystem services approach, which examines value first from the perspective of the intermediate services provided by the water resource (such as water storage) and then the resulting final services (such as habitat provision), creating benefits to humankind, which can then be expressed as use and nonuse values.[52] Selection of the valuation approach and methodology requires significant expertise and experience. Valuation studies need also to consider the economic jurisdiction of the study (the number of people affected and to what degree). Use of TEV estimates in economic analysis using benefits transfer may also require adjustment as appropriate to reflect distance decay (change in value with distance from resource).[53] When apportioning TEV within a CBA, different options may realize different components of TEV depending on the use and the final state of the water. Some workers have found it useful to consider a stable ratio of the relative contributions of the three main categories of TEV as the total value of TEV of water is varied within a sensitivity analysis; typical relative contributions to TEV are 60% for use value and 40% for nonuse value.[54]

Water Value Estimates
In some parts of the world, where freshwater is plentiful, water is sold for as little as US$0.10/m^3. The cost of desalinating seawater, a necessity in more and more countries, especially in the Middle East, is now greater than US$3.00/m^3 (and this does not include valuation of the secondary costs of atmospheric emissions produced by this energy-intensive process). These prices generally reflect only the direct use value component of TEV and do not reflect the very substantial ecological value that water delivers to humankind. Nevertheless, these market prices reveal that water indeed has value, and when water resources are damaged, through overuse or pollution, someone somewhere suffers economic damage.

Table 3.6 provides a variety of estimates for the value of water from around the world. Most are market use value estimates, but the considerable range nevertheless provides an indication of how water scarcity can affect water valuation. Estimates from Australia and other water-stressed parts of the world tend to be greater than from parts of the world where water is more plentiful. Market-based prices provide a threshold value for TEV; the total value will certainly be higher when indirect and nonuse value components are added. Replacement cost (cost of desalination) provides another estimate of the value of water; society is willing to pay this much to generate freshwater from saline or brackish sources.

BIODIVERSITY

Overview
While air emissions and even water can and are now more commonly valued and priced within markets, or by surrogate markets, valuation of biodiversity is significantly more complex and uncertain. Ecosystems provide myriad essential services to humanity, as discussed in Chapter 2. These can be summarized in the following categories, based on the lexicography of the United Nations Millennium Ecosystem Assessment (MA) framework:[61]

- Provisioning services: products obtained from ecosystems, including crops, fish, fruit, pharmaceutical products, medicines, genetic materials and ornamental resources

TABLE 3.6
Various Estimates of the Value of Water

Estimate	Value (2009 U.S. dollars)	Source and Notes
South Australia: long run marginal cost of water, direct use component	$2.00/m^3	South Australia is a particularly dry part of the continent and was experiencing severe water stress after the longest protracted drought ever recorded (2009 data).[55]
Replacement cost of water: cost of desalinating water, direct use	$1.60 to $5.00/m^3	Estimates derived from survey of various desalination operations worldwide, 2001 study.[56]
Shadow price of water in manufacturing in Canada: direct use value estimate	$0.04/m^3	Mean value.[57]
Average cost of urban water in Australia: direct use	$1.20/m^3	Based on AUD$230/200 m^3, based on 2002 data.[58]
Direct nonconsumptive use component: swimming	$57.57/hh*/yr	WTP for water quality improvements, freshwater lake, Canada, per household per year.[59]
Direct nonconsumptive use component: boating	$33.13/hh/yr	WTP for water quality improvements, freshwater lake, Canada, per household, per year.
Direct nonconsumptive use component: fishing	$15.40/hh/yr	WTP for water quality improvements, freshwater lake, Canada, per household, per year.
Direct use: improved water quality	$225/hh/yr	Mean WTP for improvements in drinking water quality in terms of reduced risk of cancer and microbial effects.[60]

* hh = household

- Regulating services: benefits from the regulation of ecosystems, including air quality improvement and maintenance (e.g., trees both remove and store CO_2 and produce oxygen as well as remove a wide variety of other air pollutants)
- Cultural services: nonmaterial benefits to people from the inherent value of natural ecosystems as places of spiritual and religious value, aesthetic value, cultural heritage, and tourism (huge direct economic value is generated each year from tourism associated with the beautiful and spectacular landscapes of the world)
- Supporting services: the services that support and regenerate the other ecosystems themselves, including soil formation and retention, primary production, nutrient and water cycling, and provision of habitat

Some examples of valuation studies for various types of ecosystems and ecosystem services is provided in Table 3.7. New ecosystem valuation studies are becoming available on a regular basis. Interested readers should consult the economic literature to access the broad and ever-increasing volume of information in this area.

TABLE 3.7
Value Estimates for Various Ecosystem Services

Estimate	Value (2009 U.S. dollars)	Source and Notes
Coastal wetlands	$5,517/ha/yr	1999 data, based on costs of land preservation[62]
Inland wetlands	$2,577/ha/yr	Value of wetlands from Thailand, based on services provided to 366 villages[63]
Wetlands	$6,671/ha/yr	Flood control value (1981 study)[64]
Major wetlands	$9,358/ha	Proxy value: revealed cost, based on Florida State government purchase of 187,000 ha of Everglades habitat from U.S. Sugar Corporation for US$1.75 billion for restoration.
Forests: indirect use value	$34/person/yr	U.K. study of the value of recreation, biodiversity, aesthetics, and carbon sequestration of forests in Britain for approximately 2.8 million ha of forest–ecosystem services (2003)[65]
Native bushland	$0.45/ha//hh*/yr	WTP for preservation of natural bushland in Australia (2004 study)[66]
Temperate boreal forest	$541/ha/yr	Use values from nutrient cycling and recreation[67]
Forest	$171–$567/ha/yr	Value of preserving forests in United States (preventing deforestation) over and above raw land value[68]
Tropical forest	$3,624/ha/yr	Use values from erosion and climate control, raw materials provision[69]
Forest	$25/ha/yr	Based on WTP survey in South Africa in 2001, Western Cape Region[70]
Urban trees	$124/tree/yr	Value of urban trees in Sacramento, California, from reduced energy for cooling, air quality improvement, wildlife habitat, shade (1998 study)[71]
Urban trees	$175/tree/yr	Value of Adelaide street trees, from reduced energy for cooling, air and water quality improvement, aesthetic value (2002 study)[72]
Grasslands	$416/ha/yr	Indirect use value from waste treatment and assimilation and genetic resources[73]
Marine sea grass	$165/ha/yr	South Australian sea grasses[74]
Coral reef	$10,000/ha/yr	United Nations Environment Programme (UNEP) median estimate (2007), indirect use value, nonuse value[75]
Mangrove forest	$15,000/ha/yr	UNEP median estimate (2007), indirect use value, nonuse value[76]
Hardwood forest	$2,720/ha/yr	Value of carbon sequestration and desalination in Australia, based on a carbon value of $20/tCO$_2$e (2005)[77]
Individual species	$7/species, one off	WTP for each additional species protected in New South Wales, Australia (2004 study)[78]

* hh = household

The types of valuation data shown in Table 3.7 can be considered with a framework of examining the impact pathway of the change produced by the project or policy:[79]

1. The project or policy is considered within the context of current conditions, which requires an environmental baseline to be established.
2. Identify and qualify the potential impacts produced by the project or policy on the ecosystem.
3. Quantify the impacts of the project or policy through an EIA (environmental impact assessment) or an environmental assessment of some kind on specific ecosystem services.
4. Assess the effects on human welfare.
5. Value the changes in ecosystem services.

Values determined can then be used within the CBA part of an overall economic sustainability assessment discussed in the "Application of Social Cost-Benefit Analysis" section of this chapter.

SOCIAL EXTERNALITIES

Social costs are the value of damages that fall on society as a result of actions otherwise undertaken to create value in the economy or well-being. As discussed, many estimates of the value of specific pollutants carry within them explicit measurement of social external damage as well as perhaps wider economic and environmental costs. In some cases, improved decision making will also require that other social costs be examined and included, such as the impacts on visual amenity of a new development or increases in odor, noise, traffic, or other disamenities. Equally, some projects or project options by their very nature will create external social benefits, which also should be monetized and included to reveal the full social, environmental, and economic picture. Table 3.8 shows a range of social costs and benefits that may play a role in decision making and which may be monetized.

These various social values may be estimated through a variety of methods. One of the most readily available estimates of the value that society places on these issues is provided through the property markets. Hedonic pricing methods examine how changes in various factors affect property prices (among other things). This provides empirical market-based data on people's WTP to avoid negative impacts (such as noise, odor, visual blight) or to gain access to desirable outcomes (i.e., a lovely view, a quiet neighborhood, or separation from a smelly or contaminated site). Care must be taken to avoid double counting, however, as many of these values may be incorporated into other measures, for instance, in ecosystem values measured through CV studies. Table 3.9 provides some examples of studies on the valuation of various social impacts.

USING VALUATION DATA

Using the various estimates for the full real value of environmental and social assets in the CBA portion of an environmental and economic sustainability assessment is not straightforward. Figure 3.8 shows a procedure for considering valuation of

TABLE 3.8

Typical Social Costs and Benefits

Social Externality	Cost	Benefit
Health impacts from air emissions	Project creates health impacts that require hospitalization, purchase of medications, and defensive expenditures. Individuals and society experience lost production and earnings.	Project reduces emissions to create a benefit to society, which can be offset against the cost of the defensive or remedial measures.
Health impacts from emissions to water and land	Project creates health impacts from ingestion of contaminated water or soil or from the products that are grown with or within those media. Results in increased health costs, lost productivity, and defensive expenditures.	Project reduces emissions to create a benefit to society, which can be offset against the cost of the defensive or remedial measure.
Displacement and inconvenience	Project requires displacement of people from their homes or deprives them of access to traditional hunting or spiritual lands.	Project restores access to hitherto damaged or inaccessible lands or protects cultural heritage.
Loss of recreational value	Project eliminates the opportunity for recreation or reduces the quality of recreational experience for people. An external cost is imposed on the people who would otherwise have used and enjoyed that asset.	Prevention of damage or remediation of damage that restores a valued recreational asset can be considered a benefit.
Noise	If the project produces increased noise levels, temporarily or permanently, it imposes an inconvenience and loss of amenity to members of the public affected. At high levels, hearing damage can occur, and the resulting medical costs are an additional external cost.	Projects that eliminate or buffer noise can consider the increase in public amenity that results as a benefit of the action.
Odor	If the project or policy produces unpleasant odors in the nearby community, the people affected will suffer a loss of amenity. This may be reflected strongly in property values, for instance.	Projects that eliminate odor will produce a benefit for residents and visitors.
Aesthetic value	If the project or policy spoils the view enjoyed by residents or creates a perceived deterioration in the overall aesthetic quality of an area, people suffer a loss of amenity. May be strongly reflected in property and recreational values.	Project or policies that improve the aesthetic quality of an area, eliminate aesthetic blight, or produce amenity benefits for those either permanently or temporarily affected.

TABLE 3.9

Value Estimates for Various Social Impacts or Improvements

Estimate	Value	Notes
Disamenity from living in proximity to a landfill site	2.5–5% of property value	This study assumed that disamenity was reflected in property prices and was driven by health concerns, visual blight, noise, and odor disamenity.[80]
Amenity from living close to an attractive river	2.1–3.3% of property value	Benefit expressed in terms of increased property value within 300 m of the river[81] as a one-time increase, which drops to 0.2% to 0.7% 600 m or farther from the river.

externalities for use in CBA.[82] Significant experience is required for each of the steps shown, especially application of the benefits transfer methods, by which available valuation studies are adapted for use on a specific project for which no primary evaluation data are available or can be reasonably obtained. Benefits transfer can either be made using direct transfer (the simplest method) or using functional transfer, a more sophisticated approach in which valuation data are scaled and adjusted to better reflect the characteristics of the users group, the affected population, income, and the nature of the good itself.[83] One of the key concerns with monetizing externalities is that the range of values available in the literature for any given asset or issue is typically large. Managing this uncertainty within a decision-making context is discussed in detail in the next section.

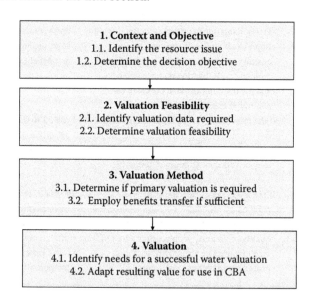

FIGURE 3.8 Process for applying valuation information into cost–benefit analysis. (After Canadian Council of Ministers for the Environment [CCME], *Water Valuation Guidance: Development of a National Water Valuation Guidance Document,* report by WorleyParsons and Eftec, CCME Water Agenda Development Committee, Ottawa, 2009.)

APPLYING THE ENVIRONMENTAL AND ECONOMIC SUSTAINABILITY ASSESSMENT

This section describes how the economic sustainability assessment is applied to decision making following the steps outlined in Table 3.1. Each of the elements listed is combined to move first from the framing session, through to data collection and development of the options in physical terms, and then into impact assessment and quantification. At this point, monetization and CBA are used to bring all factors into a common unit of measure so that the interrelationships inherent in the physical descriptions of the models and their impacts can be expressed as trade-offs. Then, once the CBA is complete, the comparison of the options in terms of their overall economic suitability and sustainability can be made.

At this stage, the discussion again moves away from money per se and focuses on how well each option performs against the others over a wide range of conditions and values. What is being sought is an option (or options) that meets all of the specified objectives under a wide range of likely future conditions and is consistently superior to the other courses of action available. At this stage of decision making, what is important is not the NPV of the option, but whether it outperforms other possible options under a wide range of conditions.

Money serves as a common unit of measure and basis of comparison, but by the end of the EESA money has become largely irrelevant. As described in the following discussion, the inclusion of external environmental and social costs and benefits into the CBA often reveals a very different economic picture than a traditional financial analysis (internal costs and benefits only being considered) would provide. Frequently, this means that a different option would be selected. Of course, communicating the relevance of the externalities to the private sector and moving toward social imposition of these aspects of a total economic analysis are challenges that need to be addressed. Both are discussed next.

APPLICATION OF SOCIAL COST–BENEFIT ANALYSIS IN DECISION MAKING

The framing workshop session has been described in detail. What results is a list of options that are to be compared, all of which can meet the stated objective. The session has also identified the key internal and external issues and assets that are important to the assessment and its stakeholders and has set the platform for monetization of those assets and the physical quantification of each of the options across its life cycle. The results of ESHIA and other environmental studies, along with engineering design input, are combined to provide the necessary data.

Apportionment of Costs and Benefits

Physical quantification of all of the environmental and social externalities, as discussed, provides the data with which the costs and benefits accruing to each option are apportioned over time. The benefits apportionment matrix (BAM) was discussed in detail by Hardisty and Ozdemiroglu (2005),[84] and so is not covered here in detail. The essential point is that the direction and magnitude of each environmental and social effect produced by the implementation of each option are recorded in the applicable physical unit within the BAM. So, for instance, emissions of GHGs are referenced into CO_2 equivalents, and the annual mass of emissions is recorded over the life cycle, for each option, as a negative (they represent a cost, or a disbenefit, to society). If in a given year for a specific option the proponent undertakes actions that absorb carbon from the atmosphere and produce a net negative emission in that year, the mass of carbon absorbed from the atmosphere is recorded as a positive (a benefit). Equally, the analysis team must decide early where to set the economic datum for the assessment. This is particularly important in terms of preventing double counting of costs and benefits.

For example, a firm has a current GHG emissions trajectory under BAU conditions over the next 20 years, and the team is examining various options for reducing emissions. Options can be assessed by treating all emissions reductions over the future predicted trajectory as benefits (positives), or it can simply count the total emissions in any given year as costs (negative). But, it cannot do both or mix these approaches across options. Baseline assumptions must be rigorously applied across all options for a proper and informative analysis.

In the full economic analysis, the prices for marketed goods and services that are affected should no longer be market prices, but real or shadow prices. Shadow prices are estimated by subtracting (or adding) the subsidy and tax elements from (to) market prices. Subsidies and taxes are *transfer payments* (no value is being created; money is simply being transferred among parties), and as such their payment does not cause a net change to real social welfare. Other transfer payments that come out of the economic analysis include litigation expenses, fines, and the value of bad publicity to the company. External benefits will also, importantly, include uncompensated environmental and health effects. Many of these "nonmarket" effects (there is no conventional market that trades these commodities and generates a market price) can still be estimated by using the monetary valuation techniques discussed above.

Externalities Change Perception of Optimality

At this stage, the options to be evaluated have been physically described and costed (CAPEX and OPEX), and in the case of a full-project analysis, the revenues resulting from each option have also been estimated. Figure 3.9 shows a hypothetical example of four options, each with estimated total PV internal costs and PV internal benefits (revenues). This results in a traditional financial analysis (benefits minus costs), as shown in Figure 3.10, and selection of the option that provides maximum financial NPV (option 2). But, if we consider the same four options (they could be designed to reach any objective, say the construction and operation of a liquefied natural gas [LNG] facility or perhaps the implementation of a site remediation project), with the same internal costs and benefits, but now also including the associated external environmental and social costs and benefits (as shown in Figure 3.11), a very different picture emerges. Figure 3.12 shows the overall social economic NPVs for each option, revealing that the economic optimum has now shifted to option 3, which is the only option that provides more overall benefits than costs as far as society is concerned.

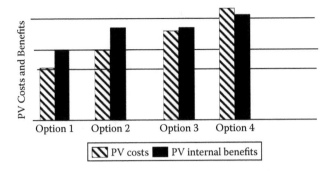

FIGURE 3.9 Hypothetical comparison of the present value (PV) costs and benefits of four project options from a purely financial (internal) perspective.

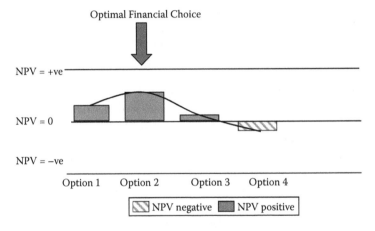

FIGURE 3.10 The options from Figure 3.10 resolved into net present values. The financial optimum, option 2, maximizes NPV and is chosen by the firm.

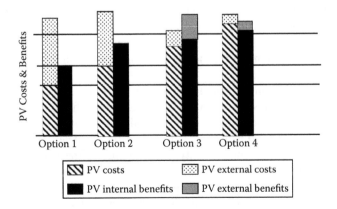

FIGURE 3.11 Adding PV external social and environmental costs and benefits to the same four options considered in Figures 3.9 and 3.10.

FIGURE 3.12 The full life-cycle total economic NPVs for the four options shown in Figure 3.11. Note that the economic optimum has shifted from option 2 to option 3, the only option that provides more overall benefit to society as a whole than the costs incurred.

Private Sector Organizations and External Costs

Of course, from the perspective of a private organization, external costs may seem irrelevant. Corporate decision makers may understand what external costs represent, but if companies do not have to pay those costs, there will be no effect on the bottom line. The question may simply become: "Who cares?" So why should private sector organizations pay attention to external costs?

First, knowledge of the magnitude and scale of the external damage that results when projects or policy decisions are implemented is in itself valuable to decision makers. Quantifying external damage in monetary terms, which hitherto may only have been communicated to decision makers in arcane technical terms (kilograms of a particular chemical emitted to atmosphere or increased mortality rates within a specific ecosystem), provides people involved in the process with a direct sense of the value of the damage being caused by their decision in the unit of measure that they use to evaluate their firm's performance (and their own). In practice, this alone can be extremely valuable and can sometime cause decision makers to shift their decisions away from damage that seems large in comparison to the total expected private profits of the project.

Second, private organizations may be interested in the explicit monetization of the damage they cause because it gives a direct measure of the potential future liability associated with that damage. Environmental and pollution law in many jurisdictions, and most OECD countries, is based on the "polluter pays" principle. In many instances, environmental liability can be applied retroactively (a firm may be found liable for damage being caused today as a result of actions taken in the past, even decades ago). Understanding the monetary scale of this liability will become increasingly important to firms around the world. Simply, the monetary value of the damage becomes a direct measure of the potential liability that the company may face in the future as a result of the damage they have caused.

Finally, companies need to understand that society implicitly uses regulation to internalize the damage associated with industrial activity. So, while a company may be aware that there is a significant external cost to their actions and may currently not face any regulation that makes that external cost "real" for them, in a cash sense, what they can be sure of is that over time, if society no longer wishes to accept that cost, it will find ways to make the company pay for it. What was external becomes a "real" internal cost to the company, and the penalty will spur a change in behavior.

Current efforts in various parts of the world to restrict GHG emissions to the atmosphere are an example. Either putting a price on carbon (a carbon tax) or limiting emissions under a cap-and-trade scheme effectively price carbon, so that what was once a free emission now becomes a real cost to the emitter. This provides a price signal that drives the emitter to reduce emissions and avoid costs. So, by monetizing a current externality, the assessment provides decision makers with a view of what future financial costs might be, particularly if the planning horizon is set at 20 years or more, as guidance suggests.

An Environmental, Social, and Economic Optimum

Figure 3.13 shows the net benefits of four hypothetical options for achieving the same objective. The shape of the "curve" created by connecting the absolute values of the NPVs is typical. If options are ordered from least expensive (lowest CAPEX) on the far left to most expensive (highest CAPEX) on the right, and if the NPVs represent the full life-cycle environmental, social, and economic costs and benefits, then the cheapest options (which will typically embody BAU lowest-cost options) rarely represent the optimum for all of society. By spending more money, a social optimum that balances profit for the proponent with a greater level of environmental and social protection can be found. In fact, such an optimum *always* exists, but rarely is the effort made to determine where it lies. It must be found.

However, once the optimum is passed, additional spending on environmental and social protection produces less and less benefit per dollar spent. The point of

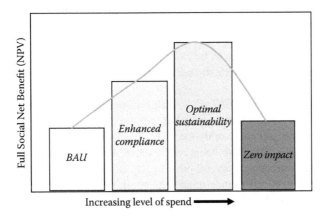

FIGURE 3.13 Finding the social optimum among a set of options that achieve the same objective.

diminishing marginal returns has been reached and passed. Typically, the most expensive options, which might provide for the highest level of environmental and social protection but at high cost, are also rarely the optimal choice. At either end of this spectrum—not spending enough on making a project sustainable or overspending on sustainability—society as a whole is worse off. Underspending results in real damage to society and the environment, which could be removed or mitigated with a net overall improvement in sustainability, and overspending means that money could be best used elsewhere to bring another project to optimality. At this point in humankind's development, when there is so much to be done to improve sustainability at all levels, with limited resources at our disposal to effect the needed changes, finding optimum solutions and being as efficient as possible in all that we do have never been more important.

FULL ENVIRONMENTAL, SOCIAL, AND ECONOMIC LIFE-CYCLE MODELING

Calculation Software

Much of the modeling and analysis presented in the following chapters was completed using the WorleyParsons EcoNomics™ DELTΔ© tool set,[85] which simultaneously combines risk assessment, traditional financial analysis, and full social economic analysis to produce a single outcome, quantified in monetary terms. The process can be used at any stage of a project to assist in choosing the most viable project options, both from the traditional internal perspective and from the wider external societal perspective. The external issues that can be included in this model are GHG emissions, water resource use and protection, loss of ecological diversity, and a wide range of other environmental and social issues. Implicit in this process is the recognition that, over time, external impacts are progressively internalized through regulation and thus will eventually have an impact on the traditional financial bottom line.

DELTΔ© is capable of a traditional purely financial assessment (NPV from the perspective of the company alone), an enhanced financial assessment, which includes risks to and from the project, or a full social economic assessment with or without risk. DELTΔ© also allows probabilistic analysis of the sensitivity of the economic results to all variables. This allows a wide range of possible future scenarios for future value trends to be examined, robustly sustainable (NPV-positive) solutions identified, and determinedly unsustainable options (NPV negative) rejected.

The approach used by DELTΔ© is based on the methodology described in this book and follows conceptual approaches espoused and approved by a number of government organizations worldwide. The fundamental calculations performed by EcoNomics DELTΔ© have been verified and benchmarked against the internal financial analysis models of several other organizations to ensure accuracy and consistency. DELTΔ© has also undergone rigorous in-house testing and benchmarking against other publically available financial analysis tools; in addition, it has been externally validated and verified for accuracy by a third-party independent validation firm.

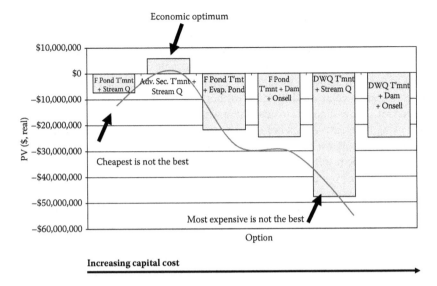

FIGURE 3.14 The economically sustainable choice is rarely the lowest-cost option and is often not the most expensive. Balancing environmental, social, and economic considerations often requires an optimal choice, which neither underspends nor overspends on environmental and social protection.

Optimizing Decision Making

Decision makers need to understand the life-cycle risks and opportunities associated with a range of options that might meet their objectives. As shown in Figure 3.14, typical base case analysis yields a picture of the total environmental, social, and economic NPV for a range of options under fixed conditions selected to represent either the current most likely values of all parameters or the expected median values over the life-cycle considered. An optimum choice can usually be determined on this basis. However, as has been discussed, all of the key inputs to such an analysis are subject to variability, particularly over a longer planning horizon of a generation or more. For instance, in Table 3.6, the TEV of water may be almost zero or may be as high as $5/m^3. NOx emissions may vary in value from $1,800 to over $23,000/t and carbon emissions from zero to the SCC. Note that while the value of externalities can be highly variable, the value of market-traded goods and services that also go into an economic analysis, such as oil, electrical power, steel, or labor, are just as variable. One only has to examine the recent historical price of a barrel of oil (see Chapter 1) to see just how variable the prices of these so-called tangible goods can be. Either way, the effect of this variability on the analysis, and more importantly on the decision to be made, must be explored.

Sensitivity Analysis

In its simplest form, the sensitivity analysis can take the form of a comparison between the calculated full social NPVs for various options using values for key parameters that are varied above and below the base case. Key input parameters such as the value of carbon, NOx, water, or the value of property and the blight factor applied, for instance, can be individually varied to high and low values and the option NPV results compared to see if the conclusions reached using base case values change. In addition, the analysis can be run by assigning all parameters their highest possible values and their lowest possible values to investigate the extremes of behavior. The essential is to determine whether conclusions about optimal choices made using base case values hold over a wider range of likely future values. If conclusions are robust over a wider range of possible values, then a greater degree of confidence in the result is obtained.

In conducting even this basic form of sensitivity analysis, the focus moves rapidly away from the absolute value of the NPV for each option and toward the ranking of options in relative terms. This allows identification of a clearly optimal solution (or solutions) over a wide range of possible future conditions and values for the key governing parameters. As such, monetization can be seen as an intermediate step, one that values all of the elements that make up the option and its impacts in a common unit (in a way that reflects how society values those elements), providing a measure of human welfare (the ultimate definition of economics). The goal of the EESA is to identify the option that maximizes human welfare, optimally, over a range of likely future conditions, within a reasonable range of uncertainty. The EESA is not a CBA. Rather, it uses CBA techniques as one part of a process designed to identify optimal choices for decision makers and explore the conditions under which those decisions may change or be modified.

A more complex sensitivity analysis involves creating a database of possible NPV results for all of the options being considered across the full range of likely parameter values. So, for instance, if the value of carbon (GHG emissions) is taken to vary between zero dollars per tonne of CO_2 equivalent and $100 per tonne (a notional estimate for the SCC), this range could be discretized at $10 increments across that range and NPVs calculated for each option for each value of carbon. Similarly, if water was a key issue, the value of water could be varied from $0/kL to $5/kL at $0.50/kL increments and NPVs calculated for each value of water against the base case values for the other parameters and then again using each of the values of carbon. Adding another axis, the same process could be completed for the value of NOx emissions, and so on, until a database is created that contains NPVs for each option being considered, for each value of carbon, water, and every other parameter across their full ranges in every direction (Figure 3.15). What results is a database of numbers that represents every possible combination of results over a full range of possible future conditions expected for the project. Each of the results is treated as equiprobable; no weighting factors or judgments about where the mean should lie and what the likelihood of various values might be are used. The database simply contains the NPVs that could occur.

Gauging the Implications

A powerful way of examining the resulting data set is to consider the NPV results as a cumulative probability curve. Each option is described by a curve, as shown in

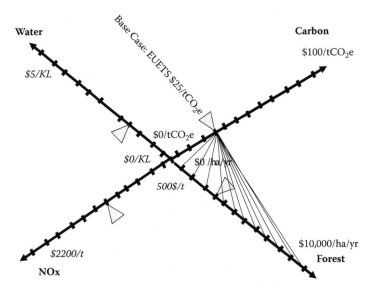

FIGURE 3.15 Multivariate sensitivity analysis. In this case, four main parameters are tested across a wide range of possible values. Carbon is varied from zero (carbon is worth nothing, or climate change is not a problem) to the Stern estimate for the social cost of carbon (SCC; climate change is a serious issue). The base case is chosen as the current EU. ETS phase 2 average. NPVs for each option are calculated first varying each value for forest between, in this case, zero and $10,000/ha/year. Then, carbon value is increased to the next increment, and the procedure is repeated, and so on, until all combinations have been used. This process is then repeated for all values of water and for all values of NOx. The resulting equiprobable NPVs are examined to identify robustly superior options across the full range considered.

Figure 3.16. To the degree that a curve lies further to the right (NPV positive) than the others and is not crossed by the others (as with option 2 in Figure 3.16), it is more environmentally, socially, and economically sustainable under the full-range values for the parameters examined. Where option lines cross, it means that under certain circumstances, one option is better than the other, but at a certain combination of values (which can be identified), the other becomes more sustainable and economic. Where the curves have long tails in the upper range of values (near the top of the graph), the option has a large upside—certain values for key input parameters generate significantly larger NPVs, making those options strongly economic and sustainable. Options that have large negative NPV tails have considerable economic risk—low values of key input parameters will drive down the NPVs, making these options potentially unsustainable.

The data behind this type of analysis can be examined in a variety of ways. However, what is important for decision makers is which options are robustly superior over the range of conditions examined. This can be described by a simple pie chart, as shown in Figure 3.17, which illustrates which option is most economic and sustainable over the range of options considered. By increasing the number of sampling points across the ranges, better definition can be provided.

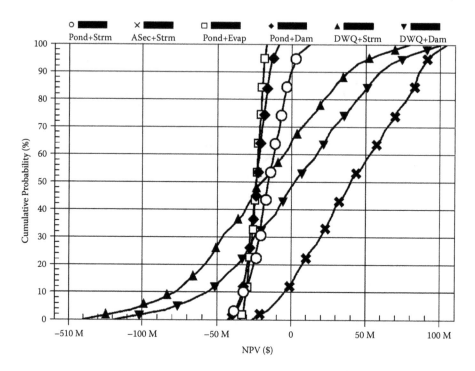

FIGURE 3.16 Chart of the equiprobable NPVs of each option over the full range of variability considered for each parameter. Option 2 (crosses) on average is the superior choice over the full range of conditions considered.

The information provided in this type of analysis helps decision makers not only select options that are economically optimal and sustainable over a wide range of possible future conditions but also cull out options that perform poorly on a consistent basis. Economic and sustainability risks and upside can be identified, and conditions under which certain options begin to perform relatively better than others can

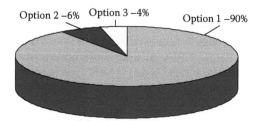

FIGURE 3.17 Most sustainable option over the range of conditions examined. In this example (not related to Figure 3.16) option 1 is the most economic and sustainable under 90% of all of the possible combination of values examined. The specific conditions under which option 2 is best can be identified, and then decision makers can assess whether those conditions are more or less likely to occur.

also be determined, aiding in future planning and contingency development. EESA is not just about optimizing decisions; it can be used to create planning hierarchies that are resilient and balanced. Projects can be future proofed by identifying options that perform best over a wide range of possible future conditions.

Communicating Decisions

Perhaps the most valuable feature of the economic sustainability assessment is its ability to help decision makers communicate their decisions: Why was a certain option chosen? Why were others not? Moving the analysis along the progression from physical quantification to monetization and ultimately back to a comparative ranking (but still based on the physical and the monetary) provides a presentation of results that is simple and visual, in units that all stakeholders understand (money). Often, the results help defuse controversial questions that may otherwise bog down discussions with stakeholders, such as what is the appropriate value for carbon. In many cases, it is the overall combination of values that determines the result, and while variation of a particular parameter may change the *absolute value* of the NPVs, the *choice* of robustly superior option is often insensitive to variations in that parameter (e.g., the analysis could show that it does not matter what value of carbon you use, the best option is still the best option).

People often intuitively know what the most sustainable course of action is, but because they cannot quantify this belief in terms that decision makers and senior executives understand, they cannot demonstrate conclusively why a more sustainable course of action should be taken. The EESA allows optimal economic choices to be identified, quantified, and communicated in a way that can profoundly influence the way decisions are made. Over the last 15 years of developing and applying economic sustainability assessments worldwide, the author has found repeatedly that the language of money, used to express the full implications of various options, allows the rational to be balanced with the emotional and lets people's inherent desire to "do the right thing" find harmony with their need to survive and prosper. Above all, this is a rational way for business and industry to examine in objective terms the full long-term implications of their actions and find clear optima that lets them continue to operate and generate profit for their shareholders while improving the way they interact with society and the environment.

EXAMPLES

The rest of this book is devoted to a series of examples that demonstrate all or parts of the environmental and economic sustainability assessment process, some in more detail, some in less. Some of the examples apply the complete statistical sensitivity analysis, and some examine the implications on a simpler high-medium-low basis. The examples cover a wide range of industry sectors and focus on some of the key environmental and social issues that are of interest today, including water, GHG emissions, air pollution, land contamination and remediation, biodiversity, and cultural and social issues. Each of the examples is constructed from situations and data from a variety of projects and studies, modified and combined to best illustrate key issues and to protect confidentiality.

NOTES

1. Organization for Economic Cooperation and Development/International Energy Agency (OECD/IEA). 2008. *Global Energy Outlook 2008.* OECD/IEA, Paris.
2. Goodstein, E. 2005. *Economics and the Environment.* 4th ed. Wiley, New York.
3. World Bank. 2007. *Environmental Costs of Physical Damage in China.* World Bank and Chinese EPA. World Bank, Washington, DC.
4. Goodstein, E. 2005. *Economics and the Environment* 4th ed. Wiley, New York.
5. Nordhaus, W., and E. Kokkelenburg, eds. 1999. *Nature's Numbers: Expanding the National Economic Accounts to Include the Environment.* National Academy Press, Washington, DC.
6. Worldwatch Institute. 2008. *State of the World, 2008.* Earthscan, London.
7. Talberth, J., C. Cobb, and N. Slattery. 2007. *Sustainable Development and the Genuine Progress Indicator: An Updated Methodology and Application in Policy Setting.* Redefining Progress, Oakland, CA.
8. United Nations Environment Programme (UNEP). 2007. *Global Environmental Outlook 4: Environment for Development (GEO-4).* UNEP, Nairobi, Kenya.
9. Pearce, D., A. Markandya, and E. Barbier. 1989. *Blueprint for a Green Economy.* Earthscan, London.
10. Martin, J. 2006. *The Meaning of the 21st Century: A Vital Blueprint for Ensuring Our Future.* Eden Project Books, London.
11. Costanza, R., R. d'Arge, R. de Groot, S. Farber, M. Grasso, B. Hannon, K. Limburg, S. Naeem, R.V. O'Neill, J. Parvelo, R.G. Raskin, P. Sutton, and M. van der Belt. 1997. The Value of the World's Ecosystem Services. *Nature*, 387: 253–260.
12. Pearce, D., and J. Warford. 1981. *World Without End.* World Bank Press, Washington, DC.
13. Hardisty, P.E. 2008. Analyzing the role of decision-making economics in the climate change era. *Management of Environmental Quality*, 20(2), 205–208.
14. Department of Climate Change. 2008. *Carbon Emissions from Coal Fired Energy.* Government of Australia, Canberra.
15. WorleyParsons. 2008. *EcoNomics™ Risk Assessment Methodology.* WorleyParsons, Sydney, Australia.
16. CNN. 2006. *Eight Expats Abducted from Nigeria Oil Rig.* June 2. http://www.cnn.com.
17. Pearce, D., A. Markandya, and E. Barbier. 1989. *Blueprint for a Green Economy.* Earthscan, London.
18. Hanley, N., J.F. Shogren, and B. White. 2001. *An Introduction to Environmental Economics.* Oxford University Press, Oxford, UK.
19. Spash, C.L. 2009. Social Ecological Economics. *Socio-Economics and the Environment in Discussion. CSIRO Working Papers Series, 2008–2009.* CSIRO, Canberra, Australia.
20. Hardisty, P.E., and E. Ozdemiroglu, 2005. *The Economics of Groundwater Remediation and Protection.* CRC Press, Boca Raton, FL.
21. Pearce, D. and J. Warford, *World Without End.* World Bank Press, Washington, DC.
22. Hardisty, P.E., and E. Ozdemiroglu, 2005. *The Economics of Groundwater Remediation and Protection.* CRC Press, Boca Raton, FL.
23. Ibid.
24. Ibid.
25. Czech, B. 2009. Ecological Economics. In *Animal and Plant Productivity* (ed. R.J. Hudson), in *Encyclopaedia of Life Support Systems (EOLSS)*. Chapter 27. UNESCO, EOLSS Publishers, Oxford, UK. (http://www.eolss.net.login.ezproxy.library.ualberta.ca)
26. Pearce, D., A. Markandya, and E. Barbier. 1989. *Blueprint for a Green Economy.* Earthscan, London.

27. Hanley, N., J.F. Shogren, and B. White. 2001. *An Introduction to Environmental Economics*. Oxford University Press, Oxford, UK.
28. Shell. 2007. *Sustainability Report*. Shell, The Hague.
29. Intergovernmental Panel on Climate Change (IPCC). 2007. *Fourth Assessment Report. The Physical Science Basis*. Cambridge University Press, Cambridge, UK.
30. World Bank. 2007. *The State of the World's Carbon Markets*. World Bank Press, Washington, DC.
31. Pearce, D. 2003. The Social Cost of Carbon and Its Policy Implications. *Oxford Review of Economic Policy*, 19(3), 362–384.
32. Stern, N. 2006. *The Economics of Climate Change—The Stern Review*. Cambridge University Press, Cambridge, UK.
33. Government of Australia. 2008. *Carbon Pollution Reduction Scheme, Green Paper.* July. Government of Australia, Canberra.
34. OECD/IEA, *Global Energy Outlook 2008.*
35. Tol, R. 2007. The Social Cost of Carbon: Trends, Outliers and Catastrophes. *Economics Discussion Papers*, No 2007–44, 1–23.
36. Stern, N. 2006. *The Economics of Climate Change—The Stern Review*. Cambridge University Press, Cambridge, UK.
37. Department for Environment, Food, and Rural Affairs (DEFRA). 2007. *The Social Cost of Carbon and the Shadow Price of Carbon: What They Are, and How to Use Them in Economic Appraisal in the UK*. Economics Group, DEFRA, London, UK.
38. Intergovernmental Panel on Climate Change (IPCC). 2007. *Fourth Assessment Report— Synthesis Report*. Cambridge University Press, Cambridge, UK.
39. Hope, C., and D. Newbery. 2006. *Calculating the Social Cost of Carbon*. Cambridge University Press, Cambridge, UK.
40. Stainforth, D., T. Aina, C. Christensen, M. Collins, N. Faull, D. Frame, J. Kettleborough, S. Knight, A. Martin, J. Murphy, C. Piani, D. Sexton, L. Smith, R. Spicer, A. Thorpe, and M. Allen. 2005. Uncertainty in Predictions of the Climate Response to Rising Levels of Greenhouse Gases. *Nature*, 433(7024): 403–406.
41. Department for Environment, Food, and Rural Affairs (DEFRA). 2007. *The Social Cost of Carbon and the Shadow Price of Carbon: What They Are, and How to Use Them in Economic Appraisal in the UK*. Economics Group, DEFRA, London, UK.
42. Moore, B., and B.H. Braswell. 1994. The Lifetimes of Excess Atmospheric Carbon Dioxide. *Global Biogeocem. Cycles*, 8(1), 23–38.
43. European Commission Directorate General Environment. 2005. *Damages per Tonne of Emissions of PM2.5, NH3, SO2, NOx and VOC from each EU25 Member State and Surrounding Seas*. CAFÉ Programme. AEA Technology, European Commission, Bruxelles.
44. Delucchi, M.A. 2000. Environmental externalities of motor-vehicle use in the U.S. *Journal of Transport Economics and Policy*, 34, 135–168.
45. Evolution Markets. 2009. *NOx Markets Report*. Evo.id, New York, NY. http://www.evo-markets.com/index.php.
46. Evolution Markets. 2009. *SOx Markets Report*. Evo.id, New York, NY. http://www.evo-markets.com/index.php.
47. European Commission. 1999. ExternE: *Externalities of Energy*, Vol. 10, *National Implementation*. Directorate General XII, Science Research and Development, Luxembourg.
48. American Institute of Chemical Engineers (AIChE). 1999. *Total Cost Accounting Manual*. AIChE, Center for Waste Reduction Technologies New York, NY.
49. European Commission Directorate General Environment. 2005. *Damages per Tonne of Emissions of PM2.5, NH3, SO2, NOx and VOC from each EU25 Member State and Surrounding Seas*. CAFÉ Programme. AEA Technology, European Commission, Bruxelles.

50. United Nations. 1992. *Dublin Statement on Water and Sustainable Development.* International Conference on Water and the Environment (ICWE), Dublin.

51. Canadian Council of Ministers for the Environment (CCME). 2009. *Water Valuation Guidance: Development of a National Water Valuation Guidance Document.* Report by WorleyParsons and Eftec. CCME Water Agenda Development Committee, Ottawa.

52. Luisetti, T., R.K. Turner, and I.J. Bateman. 2008. *An Ecosystem Services Approach to Assess Managed Realignment Coastal Policy in England.* CSERGE Working Paper, ECM-2008-04. CSERGE, London, UK.

53. Bateman, I.J., B.H. Day, S. Georgiou, and I. Lake, 2006. The Aggregation of Environmental Benefit Values: Welfare Measures, Distance Decay and Total WTP. *Ecological Economics,* 60(2), 450–460.

54. Greenley, D.A., R.G. Walsh, and R.A. Young. 1981. Option Value: Empirical Evidence from a Case Study of Recreation and Water Quality: Reply. *Quarterly Journal of Economics,* 96(4), 657–673.

55. South Australia Office of Water Security. 2009. *Water Plan 2050.* OWS, Adelaide, Australia.

56. Wade, N.M. 2001. Distillation Plant Development and Cost Update. *Desalination,* vol. 136, issues 1–3, 3–12.

57. Dupont, D.P., and S. Renzetti. 2008. Good to the Last Drop? An Assessment of Canadian Water Value Estimates. *Canadian Water Resources Journal,* 33(4), 363–374.

58. Commonwealth of Australia. 2002. *The Value of Water: Inquiry into Australia's Management of Urban Water.* Report of the Senate Environmental, Communications, Information Technology and the Arts References Committee. Government of Australia, Canberra.

59. Dupont, D.P., and S. Renzetti. 2005. Cost-Benefit Analysis of Water Quality Improvement in Hamilton Harbour, Canada. In Brouwer, R. and Pearce, D.W., *Cost Benefit Analysis and Water Resources Management* (ed. R. Brouwer and D.W. Pearce). Edward Elgar, Cheltenham, UK. pp. 195–222.

60. Adamowicz, W., D. Dupont, A. Krupnick, and J. Zhang. 2007. *Valuation of Cancer and Microbial Disease Risk Reduction in Municipal Drinking Water: An Analysis of Risk Context Using Multiple Valuation Method.* Resources for the Future Discussion Paper, 07-39.

61. United Nations. 2002. *Millennium Ecosystem Assessment (MA).* United Nations, New York, NY. http://www.maweb.org.

62. American Institute of Chemical Engineers (AIChE). 1999. *Total Cost Accounting Manual.* AIChE, center for waste Reduction Technologies New York, NY.

63. United Nations Environment Programme (UNEP). 2007. *Global Environmental Outlook 4: Environment for Development (GEO-4).* UNEP, Nairobi, Kenya.

64. American Institute of Chemical Engineers (AIChE). 1999. *Total Cost Accounting Manual.* AIChE, center for waste Reduction Technologies New York, NY.

65. Forestry Commission. 2003. *The Social and Environmental Benefits of Forestry in Britain.* Forestry Commission, London, UK. http://www.forestry.gov.uk/economics

66. Van Bueren, M., and J. Bennett. 2004. Towards the Development of a Transferable Set of Value Estimates for Environmental Attributes. *The Australian Journal of Agricultural and Resource Economics,* 48(1), 1–32.

67. Costanza, R., R. d'Arge, R. de Groot, S. Farber, M. Grasso, B. Hannon, K. Limburg, S. Naeem, R.V. O'Neill, J. Parvelo, R.G. Raskin, P. Sutton, and M. van der Belt. 1997. The Value of the World's Ecosystem Services. *Nature,* 387: 253–260.

68. American Institute of Chemical Engineers (AIChE). 1999. *Total Cost Accounting Manual.* AIChE, center for waste Reduction Technologies New York, NY.

69. Costanza, R., R. d'Arge, R. de Groot, S. Farber, M. Grasso, B. Hannon, K. Limburg, S. Naeem, R.V. O'Neill, J. Parvelo, R.G. Raskin, P. Sutton, and M. van der Belt. 1997. The Value of the World's Ecosystem Services. *Nature,* 387: 253–260.

70. Turpie, J.K. 2003. The Existence Value of Biodiversity in South Africa: How Interest, Experience, Knowledge, Income and Perceived Level of Threat Influence Local Willingness to Pay. *Ecological Economics*, 46(2), 199–216.
71. McPherson, E.G. 1998. Atmospheric CO2 Reduction by Sacramento's Urban Forest. *Journal of Aboriculture*, 24(4), 215–223.
72. Killicoat, P., E. Purzio, and R. Stringer. 2002. The Economic Value of Trees in Urban Areas: Estimating the Benefits of Adelaide's Street Trees. *Treenet Proceedings of the 3rd National Street Tree Symposium*, pp. 1–10.
73. Costanza, R., R. d'Arge, R. de Groot, S. Farber, M. Grasso, B. Hannon, K. Limburg, S. Naeem, R.V. O'Neill, J. Parvelo, R.G. Raskin, P. Sutton, and M. van der Belt. 1997. The Value of the World's Ecosystem Services. *Nature*, 387: 253–260.
74. McArthur, L., and J.W. Boland. 2006. The Economic Contribution of Seagrass to Secondary Production in South Australia. *Ecological Modelling*, 196(1–2), 163–172.
75. United Nations Environment Programme (UNEP). 2006. *In the Front Line: Shoreline Protection and Other Ecosystem Services from Mangroves and Coral Reefs*. UNEP-WCMC Biodiversity Series No. 24. UNEP, Nairobi, Kenya.
76. Ibid.
77. Venn, T.J. 2005. Financial and Economic Performance of Long-Rotation Hardwood Plantation Investments in Queensland, Australia. *Forest Policy and Economics*, 7, 437–454.
78. Van Bueren, M., and J. Bennett. 2004. Towards the Development of a Transferable Set of Value Estimates for Environmental Attributes. *The Australian Journal of Agricultural and Resource Economics*, 48(1), 1–32.
79. Department for Environment, Food, and Rural Affairs (DEFRA). 2007. *An Introductory Guide to Valuing Ecosystems Services*. U.K. DEFRA, London.
80. Department for Environment, Food, and Rural Affairs (DEFRA). 2004. *Valuation of the External Costs and Benefits to Health and Environment of Waste Management Options*. DEFRA report. U.K. DEFRA, London.
81. Department of Environment, Transport and the Regions (DETR). 1999. *Potential Costs and Benefits of Implementing the Proposed Water Resources Framework Directive*. UK DETR Report 10959, DETR, London, UK.
82. Canadian Council of Ministers for the Environment (CCME). 2009. *Water Valuation Guidance: Development of a National Water Valuation Guidance Document*. Report by WorleyParsons and Eftec. CCME Water Agenda Development Committee, Ottawa.
83. Ibid.
84. Hardisty, P.E., and E. Ozdemiroglu, 2005. *The Economics of Groundwater Remediation and Protection*. CRC Press, Boca Raton, FL.
85. Hardisty, P.E., M. Sivapalan, S. Van Der Linden, and S. Donohoo. 2009. *WorleyParsons DELTA Toolset for Environmental, Social and Economic Life-Cycle Modelling*. WorleyParsons, Sydney, Australia.

4 Water

INTRODUCTION

Freshwater is essential to life on Earth and is a vital input to almost every productive human activity. As described in Chapter 2, water scarcity is increasing globally. Today, hundreds of millions of people do not have access to safe supplies of potable water. Climate change is expected to exacerbate water scarcity over the coming decades, adding to the pressure on resources and increasing potential for conflict between competing users. Past failures to recognize the true value of water have and continue to lead to misallocation and wasteful use of our dwindling supplies.

This chapter provides a series of examples that illustrate how the application of environmental and economic sustainability assessment (EESA) can help decision makers see directly the effect of valuing water appropriately and the important trade-offs that exist among water, energy, carbon (greenhouse gas [GHG] emissions), biodiversity, and other important environmental, social, and economic issues. As demonstrated throughout the following chapters, a key to incorporating water fully and fairly within an EESA is to remember that water is only one of a number of external issues that need to be considered, and that water is linked intricately with many other factors in a directly causative, physical way. If these links are not represented, trade-offs cannot be made, and the power of the assessment is much reduced. The following examples are provided in this chapter:

1. Freshwater resources management in petroleum production
2. Management of produced water in petroleum production
3. Freshwater management in mining
4. Treatment and disposal of treated wastewater
5. Water supply options assessment in an arid region

Each of the examples is treated at a different level of detail, illustrating how the environmental and economic sustainability assessment approach can be applied in more or less-sophisticated ways, considering smaller or larger ranges of variables and stakeholder concerns. The examples in this and the following chapters are all based on elements of real case histories from various parts of the world, but in each case, details of the actual situation have been changed, or information from several actual cases combined, to better illustrate salient points and to protect the confidentiality of the original projects. As such, none of the examples contains the specifics of an actual project but rather has been constructed from elements of real situations.

WATER MANAGEMENT IN INDUSTRY: OVERVIEW

Throughout their life cycle, from exploration to discovery, to production and finally facility abandonment, industry operations have the potential to impact water resources. In the mining and petroleum sectors, for instance, exploration for new reserves involves a number of activities that may affect water resources: well drilling and completion (Figure 4.1), the incursion of roads into remote areas, the construction and operation of exploration camps and associated facilities, and the disposal of a range of wastes from these operations. The impacts may be felt on surface water bodies (lakes, rivers, and wetlands) and on freshwater aquifers.

Even aquifers that are not being used currently, or that might have marginal water quality, may be used in the future. In economic terms, such unexploited water resources have significant option value—society is willing to pay to retain the option of using that resource at some time in the future. This component of total economic value (TEV) was discussed in Chapter 3.

As water resources are impacted, through being used or wasted or from pollution, the costs of the associated damages flow through to society. Damage may manifest itself as a physical decline in the availability of water to support other human activities, such as farming, fishing, aquaculture, or recreation, or as degradation in water quality, which can directly or indirectly affect the health and well-being of ecosystems and people using the water (Figure 4.2).

Table 4.1 summarizes some of the key activities in the industrial life cycle and their potential impacts on water, along with mitigation measures that may prevent or reduce the risk of impact and remediation considerations for each. The literature is consistent in its conclusion that the costs of prevention and mitigation are almost always far less than the costs of remediation (trying to fix it after the damage has been done).[1] In economic terms, the case for prevention is even more compelling: Damage avoided by implementation of prevention measures is measured as an economic benefit.

When considering situations involving damaging impacts to water resources, the examples in this chapter suggest that resource protection is generally more economic

FIGURE 4.1   Oil exploration wastes in an unlined open pit. Oil and produced water may seep into shallow groundwater, causing contamination.

FIGURE 4.2 Testing of groundwater used for irrigation at a small farm in Yemen. Contamination from various industrial sources (including oil and gas operations) can have an impact on groundwater quality over the long term, affecting the livelihood and health of farming communities.

overall than remediation after contamination has occurred, and that as the value of water rises (as it does in conditions of scarcity), both protection and remediation options become more attractive.

WATER USE AND PROTECTION IN OILFIELD DEVELOPMENT IN NORTH AFRICA

BACKGROUND

At present, it is common practice in large parts of North Africa to use freshwater from the regionally extensive Cretaceous aquifer as injection water for reservoir pressure maintenance. This groundwater, which is also increasingly being exploited in the region for domestic, industrial, and agricultural use, is of high quality and is readily available in many of the oil-producing regions, making it a convenient source of water for pressure maintenance. In many of the oil-producing regions there are no other water-producing horizons that are as convenient (shallow), cheap (easy to drill to and chemically compatible with oil-producing formations), and prolific. Because of this, there has been almost no documented effort to find deeper, higher-salinity, sources of water for reservoir injection. The environmental, social, and economic sustainability of this practice, which takes a valuable and scarce resource (fresh potable water) in one of the most arid parts of the world and permanently eliminates it from the hydrosphere by sequestering it within the oil reservoir, is examined.

TABLE 4.1
Typical Industrial Impacts on Water Resources

Life-Cycle Stage	Activity	Potential Impacts on Surface and Groundwater	Mitigation	Remediation
Resource exploration surveys	Road construction, sampling, mapping	Alter drainage and recharge patterns, reduce quantity of water available for other users, impact on water quality	Select routes, construction techniques, and survey methods to protect water resources	Road reclamation, revegetation, access prevention
	Seismic surveys, charge detonation	Alter groundwater/spring flow regime	Use alternative methods (vibration)	Typically impractical
Resource explorations	Well drilling	Cross connection of aquifers, contamination of aquifers, use of large volumes of freshwater; contamination of freshwater	Well construction and completion methods designed to protect and isolate freshwater; appropriate waste management plans	Aquifer and surface water remediation methods available but expensive, time consuming, and rarely completely effective
	Well testing	Contamination of groundwater and surface water with drilling fluids	Drilling fluid management; lined sumps; use of biodegradable and nontoxic drilling fluids and additives	Soil cleanup by excavation and treatment (can be costly); groundwater remediation rarely completely effective, typically expensive
Pipelines	Construction	Alter drainage and recharge patterns—surface and groundwater	Follow best practice EIA, adjust routes and construction techniques to minimize impacts to freshwater bodies	Revegetation of construction disturbance; reduce erosion and sedimentation
	Spills	Contamination of groundwater and surface water with hydrocarbon liquids, slurries, or other chemicals	Use appropriate pipeline design and materials, regular monitoring, emergency response systems	Surface spill response and cleanup; subsurface remediation of spills by soil excavation and treatment; remediation typically expensive and rarely completely effective

Facility construction	Access construction	Clearing and construction of access and transport systems to the facility may impact surface water bodies and dependent ecosystems and societies	Select transport corridor (roads, rail lines) routes to avoid important surface water bodies; detailed examination of ecosystem vulnerability to change in water flow, drainage, recharge, and circulation patterns from roads, bridges, culverts, and tunnels	Once impacted, water resources and dependent ecosystem difficult to restore without reestablishing original conditions
	Construction	Land clearing and fundamental changes in water drainage, flow, and recharge patterns can affect surface and groundwater, particularly if the scale of the changes is large	Select facility location to reduce impacts on water resources; design facility to reduce footprint and effects on local hydrological conditions; select site conditions protective of underlying aquifers; avoid construction above vulnerable and important freshwater aquifers	Once impacted, water resources and dependent ecosystem difficult to restore without reestablishing original conditions
Facility operations	Accidental spills or discharges	Contamination of groundwater and surface water with hydrocarbon liquids, process chemicals, tailings and other wastes and by-products	Fully implement best practice facility environmental management systems; regular facility and systems maintenance and inspection; minimize use of buried lines and tanks; accurate and regular volume reconciliation	Surface spill response and cleanup; subsurface hydrocarbon remediation by soil excavation and treatment; aquifer remediation typically expensive and rarely completely effective; risk-based approaches usually more economic

(continued)

TABLE 4.1 (CONTINUED)
Typical Industrial Impacts on Water Resources

Life-Cycle Stage	Activity	Potential Impacts on Surface and Groundwater	Mitigation	Remediation
	Water supply management	Use of large volumes of freshwater for production process conflicts with competing users of other stakeholders, causing declines in supply availability and rising unit cost of water; may negatively impact ecosystems, particularly in arid areas	Critically examine water use within all stages of the process, consider water efficiency, reuse, recycling and reduction measures; consult with local stakeholders to avoid conflicts; examine use of lower-quality water resources instead of freshwater; examine long-term implications and possible effects of climate change on water use and availability	Replenishment of water resources, once depleted, takes time, and dependent on the hydrological system; long-distance transfer of water expensive and often energy intensive; water replenishment through desalination also expensive and energy intensive
	Wastewater management	Disposal of contaminated water may impact groundwater and surface water quality, damaging ecosystems and increasing water stress and health impacts on other users Groundwater contamination may cause long-term damage to aquifers	Treat wastewaters to appropriate standards before disposal; follow best practice for disposal; minimize or eliminate surface disposal; protect freshwater aquifers; secure deep-well disposal an attractive option for liquid wastes that are difficult to treat	Once contamination has occurred, remediation perhaps time consuming and technically challenging, particularly for groundwater; costs for remediation typically much higher than for proper treatment and disposal
Facility abandonments	Residual contamination improperly handled	Legacy issues from improper facility abandonment can cause long-term contamination of surface water and groundwater	Detailed preabandonment site investigation; remedial planning based on risk assessment; protection of water resources from long-term impact	Removal of sources of future risk to water resources; site remediation done after a facility already abandoned typically more expensive and harder to fund (reopening closed projects, new funding sources)

Water Resources in North Africa

North Africa is one of the driest places on the planet, with average annual rainfall varying between 10 and 500 mm. However, only 5% of the region by area receives more than 100 mm of rain a year. Accordingly, there is very little in the way of fresh surface water resources in the region, other than the Nile Basin in Egypt. The region, however, is blessed with abundant sources of fresh, high-quality groundwater, particularly in the arid southern areas. The presence of a significant and extensive aquifer system beneath the desert regions of North Africa was first surmised scientifically in the 1930s based on earlier work done by explorers who noticed patterns linking the various desert oases of the region.[2] The hydrocarbon exploration of the 1960s and 1970s confirmed the presence of significant groundwater resources, predominantly in the Cretaceous sandstones (including the Nubian sandstone),[3] and allowed the identification of the main sedimentary basins governing groundwater flow and distribution on a regional scale.[4] Specific hydrogeological evaluations soon followed, including hydrogeochemical analysis and groundwater modeling of the Kufra and Sarir regions of Libya.[5]

These studies, and others, formed the scientific basis for the Great Man Made River Project (GMMR) in Libya, which systematically evaluated and quantified the extent of the resource[6] and developed a system of pumping wells and pipelines to extract groundwater and deliver it to the population centers on the Mediterranean coast.[7] The groundwater occurs in a series of overlapping sedimentary basins. Four main aquifer systems exist: (1) Paleozoic sedimentary basins, (2) Mesozoic, (3) Tertiary, and (4) Quaternary sediments forming shallow systems.

Modern and current recharge occurs only in the north of the region, along the coast of the Mediterranean. The rest of the basins receive no effective modern recharge. Two major recharge episodes are responsible for the large amounts of groundwater currently residing within the Nubian and other major regional aquifers. The first occurred over a prolonged period prior to about 20,000 years ago. An arid period followed between about 20,000 and 12,000 years ago. This was followed by an intense wet period, especially in the south. The current arid climatic conditions established themselves about 5,400 years ago. Essentially, this means that the water within the regional aquifers, and the water being considered for use in reservoir pressure maintenance in this example, is between 5,500 and 12,000 years old (sometimes called "fossil water"). This water is an inheritance from the past.[8]

An Ancient Groundwater Resource

Currently, groundwater flows from south to north within and across the five major basins, reflecting the influence of the paleo-recharge sources in the south. Groundwater is accordingly generally fresher in the south and increasingly more mineralized to the north (saline and brackish), eventually commingling with seawater as it discharges to the Mediterranean Sea. The aquifer currently being used most for pressure maintenance by the petroleum industry is the Cretaceous sandstone, which reaches a thickness of more than 3,000 m at its center and extends across Libya and into neighboring Egypt, Chad, and Sudan, covering over 350,000 km² in Libyan territory alone. Groundwater quality before the onset of widespread development was excellent, with total mineralization (TDS, total dissolved solids) concentrations of 100 to 400 mg/L—pure enough to drink without treatment.[9] The Kufra basin in Libya alone is estimated to contain reserves of fresh groundwater of over 20,000 km³ (20 million m³).[10]

One of the key qualities of the Nubian sandstone aquifer is its tremendous production potential. The permeability and flow characteristics of the aquifer rock, combined with the large saturated thickness and confined nature of the system, provide wells with very high transmissivity. This is one of the key reasons that oil producers have focused on this aquifer as a source of water for reservoir pressure maintenance; the water is readily accessible at a relatively shallow depth, and wells can produce at very high rates. Unfortunately, widespread use over the last 30 years has inevitably led to impacts. Water levels in the Cretaceous aquifer have been declining steadily, and in 2000 were between 15 and 25 m below predevelopment static levels. At the same time, overall water quality has also begun to suffer.[11] This reflects the fact that in the major southern groundwater basins, where water quality is best, there is little or no current recharge. In effect, groundwater extracted from these aquifers is being "mined." As discussed, groundwater recharging in the distant southern outcrop areas is flowing only very slowly toward the points of use and indeed will take many centuries before its impact is felt. From an economic point of view, and certainly considering the planning horizon of this oil development project, it is assumed that groundwater is a nonrenewable resource in this part of the world.

Water Use and Availability in Libya

Libya is among the most arid countries in the world. With little in the way of fresh surface water resources, 97% of freshwater used in Libya comes from groundwater.[12] The remainder is provided by surface water and from desalination plants situated along the Mediterranean coast. Despite having access to significant groundwater resources, Libya finds itself in a water deficit position. The latest data available, compiled for 1998, showed a total national water supply of 2,655 Mm³/year (millions of cubic meters per year), predominantly from the GMMR project, which exploits the Cretaceous Nubian sandstone aquifer (2,553 Mm³/year). Set against this supply, national demand for water in 1998 was 3,923 Mm³/year, of which the vast majority (3,335 Mm³/year) was for agriculture (with 425 Mm³/year for urban domestic supplies). This resulted in an annual water deficit of 1,267 Mm³ in 1998.[13] Water demand has increased steadily since these data were compiled, and the annual water deficit has grown considerably since then. Table 4.2 shows data on water use in Libya by sector and basin.

The data in Table 4.2 are predicated on the concept of safe aquifer yield. If the withdrawal rate exceeds this safe yield rate, the aquifer is deemed to be susceptible to long-term damage, typically in the form of steadily declining piezometric surfaces. *Safe yield* is defined as the amount of water that can be withdrawn from an aquifer annually without producing an undesired result.[15] In normal circumstances, it is the recharge rate that determines what a safe yield is. As long as withdrawal is matched by available recharge, groundwater levels do not decline. Even when pumping exceeds recharge, the aquifer can establish a new equilibrium at a lower water level, often by eliminating previously active natural discharges.[16] This explains in part why oases

TABLE 4.2

Water Balance and Consumption Data for Libya, 1998 (Mm³/yr)[14]

	Gabal El Akhdar	Al Kufra	Al Gfara	El Hamda	Marzak	Total
Available groundwater	250 (S)[a]	741	250 (S)	400 (S)	912	2,533
Available surface water	16	2	26	17	–	61
GMMR net[b]	+113	–113	+110	–	–110	0
Desalination	5.1	0.5	–	12	–	17.6
Treated wastewater	1.8	1.2	7.5	4.9	8.8	24.2
Total supplies	385	632	394	433	810	2,656
Agricultural use	80	492	1,476	540	746	3,335
Urban use	119	30	188	57	58	453
Industrial use	5	109	10	5	7	136
Total use	204	632	1,675	602	810	3,924
Water balance[c]	+181	–	–1282	–168	–	–1,268

[a] (S) represents withdrawal rates that are considered "safe."

[b] Positive values represent net inflow into that basin from GMMR water; negative values mean export from that basin.

[c] Positive values mean the basin is in a surplus position—it produces more water than it uses. Negative values mean the basin is in a deficit position—it uses more water than the aquifers are safely capable of producing.

and springs can dry up and stop flowing in areas where large-scale groundwater pumping is occurring. Many workers in the field have noted that undesirable results are subjective and must be measured against the benefits produced.[17] This concept is explicitly tested here by examining the proposed oilfield development in the context of the environmental and social benefits and costs involved. In a zero-recharge environment typical of the conditions considered in this example, continued pumping at a rate that exceeds the lateral inflow rate from southern recharge areas will cause a change in aquifer storage, manifested by declining water levels. This is already being seen throughout the region in areas being exploited for groundwater supply.

Water Use in the Petroleum Sector and Regulatory Context

Throughout the region, the overall level of regulatory oversight of the use of water in petroleum production is low. In Libya, for instance, an interdepartmental government committee was formed in 2006 specifically to examine the issue of water resources and the petroleum industry in Libya. This committee consists of representatives of the General Water Authority (GWA), the National Oil Corporation (NOC), and the Environment General Authority (EGA). It is common practice in Libya and throughout the region, at present, to use Nubian sandstone and analogous aquifers as a source of makeup water for reservoir pressure maintenance. To date, it appears that no records of such water use have been kept. Oil companies have simply drilled their own water wells on location and pumped whatever they needed for injection purposes. The Libyan and other governments to date have not charged the oil companies for the use of this water. It is estimated that the amount of fresh groundwater being used by the oil industry in Libya exceeds 75 Mm^3/year in Libya alone.[18]

In the medium to longer term, there is every likelihood that freshwater use for oilfield injection will be regulated, either through a direct charge on water use or as a complete ban. The eventual regulatory position will almost certainly depend on the degree of continued aquifer depletion and water quality deterioration, the price of oil, and the future value of water. Operators demonstrating a socially and environmentally responsible stance on this issue are more likely to be viewed positively by the regulators and may enjoy the benefits of improved regulatory goodwill and cooperation.

Environmental and Economic Sustainability
Assessment Objective and Options

The objective of the EESA in this example is to determine the most environmentally, socially, and economically efficient way to produce oil in the region in the context of reservoir pressure maintenance alternatives. The following alternatives were examined:

- *Option 1:* Default position—business as usual (BAU) in the region for many years, using as much freshwater as required for optimal reservoir performance, over the 25-year expected life of the field.
- *Option 2:* Use less freshwater than the BAU case.
- *Option 3:* No injection of water (freshwater or produced water). Produced water is handled as wastewater and managed using evaporation ponds.
- *Option 4:* No freshwater use (all produced formation water [PFW] reinjected into formation) over 25 years.
- *Option 5:* Find and use an alternative supply of low-quality water obtained from a horizon deeper than the Cretaceous (allows the BAU oil production profile to be achieved).
- *Option 6:* Add a community-based project involving use of Cretaceous aquifer water in a bottled water operation, as a social offset, and inject freshwater under BAU conditions.
- *Option 7:* Replacement of freshwater used for pressure maintenance by operation of a solar desalination plant located at the coast and inject freshwater under BAU conditions

Option Development and Costing

Each of the options is described in detail in this section and summarized in Table 4.3. In developing these options, local information was used when available,

TABLE 4.3
Options Description Compared to BAU

Case	Description	Production Period (yr)	Total Oil Production (bbl)	Aquifer Water Use (Mm³)
Option 1	Default position (BAU), unrestricted use of groundwater as required for optimal reservoir performance	30	Baseline	Baseline
Option 2	Reduced freshwater injection profile	30	−110	−22
Option 3	No injection of water (freshwater or produced water); produced water is handled as wastewater (evaporation ponds)	30	−220	−146
Option 4	No freshwater use (all produced formation water reinjected into formation)	30	−185	−146
Option 5	Find and use an alternative supply of low-quality water obtained from a horizon deeper than the Cretaceous	30	0	−146
Option 6	Community-based project involving use of aquifer water in a bottled water operation	30	0	0
Option 7	Replacement of freshwater used by operation of a solar desalination plant located at the coast	30	0	0

and best-available information on costs for similar systems in analogous environments was used when specific North African data were not available. This analysis has been completed on a comparative basis using the BAU option as the baseline in terms of oil production and water use. The data reveal clearly the importance of pressure maintenance: The greater the volume of water injected overall, the higher the oil production is.

Option 1: Base Case—Unrestricted Freshwater Use

The option of unrestricted freshwater use provides for optimal production over a 30-year project life by using as much water as required. All PFW is reinjected into the producing formation and is supplemented as required by fresh groundwater to meet optimal reinjection conditions. Fresh groundwater is used early in the life of the reservoir to maintain pressure and improve performance when PFW flows from the reservoir are low. As the water cut increases over time, PFW is substituted for freshwater, and groundwater use drops. Again, this is the BAU option and is used as the baseline for comparing options. It is worth noting that in this example we assume that, as with many countries in the region, a substantial part of the revenue generated by the sale of oil is collected by the government as royalty payments. While this has a significant impact on the operator's cash flow, it is important to note that in an economic analysis, benefits and costs to all of society are included. Royalties, in this case, are a transfer payment and therefore are not considered as a cost in a social economic analysis (as they would be in a financial analysis).

Option 2: Lower Groundwater Use Option

Option 2 is similar to option 1 except that the rates of oil production, PFW production and reinjection, and groundwater use have been modified. The overall water injection profile starts at a lower level than option 1 but reaches a higher maximum injection rate at its peak and then falls away quickly. This results in an overall predicted groundwater use of about 22 Mm^3 less than option 1 overall. Overall oil production is also lower than option 1. By changing the water injection profile to save water, oil production is reduced by about 110 million barrels over the life of the project. The undiscounted direct cost, however, is almost identical to the option 1 case.

Option 3: No Groundwater Use Option

Option 3 examines the effects on production, costs, and benefits if no external water is used. PFW is not reinjected but is treated at the surface and disposed of in a way that is assumed not to have an impact on the environment (evaporation). Oil production falls off quickly as reservoir pressures fall. This scenario produces 220 million fewer barrels of oil than option 1 (BAU) and over 682 million barrels of PFW requiring treatment and management at the surface.

The undiscounted direct cost of production for option 3 is US$620 million less than BAU, however. In addition to the direct costs of extracting the oil from the ground, the costs of treating the PFW at the surface must be added. It is assumed that PFW will be treated in a conventional oil–water separator system and pumped to a series of lined holding ponds where remaining oil is removed, and evaporation is allowed to take place. Given the high TDS content of the PFW expected, complete evaporation of the water must be achieved so that the salt can finally be sequestered in a lined on-site landfill. Throughout the Middle East and other parts of the world, the surface disposal of treated PFW (oil removed to about 50 mg/L TPH [total petroleum hydrocarbons]) has resulted in significant impacts on local ecosystems and contamination of shallow groundwater resources by salt.[19] In this example, the costs of managing PFW at the surface to provide for complete protection of local water resources and ecosystems assumes a maximum PFW production rate is approximately 5 Mm^3/year. Using an evaporation rate of 5,000 mm/year,[20] an estimated evaporation area of 1 km^2 is required. Assuming an initial TDS of 15 g/L, evaporation will produce 1.6 million tonnes of salt, which will require periodic removal from the ponds and sequestration in a purpose-built landfill. Approximately 500,000 m^3 of landfill space would be required over the life of the project. Costs for pond and landfill construction have been estimated using U.S. base costs[21] adjusted for price differentials in North Africa.

The capital cost of excavating the required evaporation ponds and lining them with high-density polyethylene (HDPE) is estimated to be on the order of US$16 million. For the required area, a total cost of US$19.2 million is estimated. The capital cost of installing a basic oil–water separation system is estimated at US$0.5 million. Landfill excavation and lining would cost on the order of US$3.2 million. Total

CAPEX is estimated at US$22.9 million. Incurred in year 9, at a 3.5% discount rate, this represents a present value (PV) cost of US$16.8 million. The annual operation and maintenance (O&M) costs for the treatment and disposal system are estimated to reach a yearly maximum of about US$0.35 million (including monitoring, engineering, and labor costs) and decline with time as the rate of PFW production declines. Nondiscounted total O&M costs are estimated at approximately US$8.9 million. The total estimated nondiscounted cost for PFW management at the surface is about US$31.8 million.

Option 4: No Freshwater Use with PFW Reinjected into the Producing Formation

Option 4 is the same as option 3 except PFW is reinjected into the formation rather than treated at the surface. The cost of production is estimated to be about US$415 million less than BAU. There are no additional costs for surface management of PFW, however.

Option 5: Alternative Source of Injection Water

Instead of injecting high-quality freshwater into the oil reservoir, another source of poorer-quality groundwater can be developed and used in its place. As discussed, Nubian sandstone groundwater has typically been used because it is plentiful, chemically benign, and cheap to access (shallow). In geologically analogous areas, such as in the Arabian Peninsula, where vast areally extensive Cretaceous sandstone aquifers also exist (such as the Mukalla aquifer in Southern Yemen), deeper horizons are present that can yield significant amounts of brackish water. Available oilfield exploration data from the subject area suggest that similar horizons exist below the Cretaceous aquifer, which could be exploited. However, identification and proof of a deeper lower-quality water resource would require additional investment in data review, exploratory drilling, and aquifer testing (which could be readily coupled with existing oil exploration programs). It is assumed that this new water resource would be able to provide all of the necessary pressure maintenance fluids to allow option 1 production levels to be achieved with only a basic level of pretreatment.

The undiscounted direct cost of production is the same as for option 1, but in addition it is assumed that a nominal budget of US$100 million is invested in a systematic exploration program to identify and develop the alternative source of poorer-quality groundwater, effectively conserving the freshwater for more beneficial uses.

Option 6: Community-Based Bottled Water Operation

The water from the aquifer is of a quality that is entirely suitable for high-quality bottled water. The bottled water market worldwide, including in North Africa, is a major business. In 2004, the market consumption in Libya alone was 11 million liters, with a retail value of approximately US$8 million (retail price $0.75/L).[22] In Africa and the Middle East, the bottled water market in 2004 was 12 billion liters a year, valued at US$2.1 billion. Per capita consumption in the region was 11 L/year and has been growing at more than 5% per year. Western European consumption, in comparison, was 112 L/year per person in 2004.

Option 6 involves establishing a community-based bottled water operation in the nearby community. The intent would be to provide economic and social development for the local community as compensation for the use of the nonrenewable water resource. This project could be coupled with any of the main oil development options discussed. In this case, it is assumed that option 1 would be used. The objective of the project would be to realize some of the much higher value of the Nubian sandstone aquifer, in part to defray its use in option 1 for oil reservoir pressurization.

A bottled water plant capable of producing 10 million liters annually would be established in the neighboring community at an estimated cost of about US$5 million. It is assumed that the capital cost would be provided by the company. One of the major challenges of such an operation would be transporting the product to the main markets on the coast. In this case, establishment of a marketing network, distribution chain, transport system, and a customer base would require considerable effort. Annual O&M costs would be on the order of US$0.15/L, transport costs US$0.10/L, and marketing and management costs approximately US$0.15/L. Total ongoing O&M costs are thus estimated at about US$0.40/L. Estimated wholesale price is US$0.50/L.

Option 7: Solar Desalination Mitigation Replacement

This option provides for the replacement of the freshwater used in the oilfield development through solar desalination of seawater at the coast in parallel with ongoing oilfield operations. The option is essentially an environmental offset, replacing the nonrenewable resource used for pressure maintenance. Solar desalination was assumed to provide the most environmentally benign way of replacing the freshwater lost. Conventional desalination plants are highly energy intensive and are typically powered by oil or natural gas. The GHG emission footprint of such operations is considerable, and in this case it is assumed that it is more appropriate to invest in the development of emerging solar technologies, whose GHG emissions are very low, to power desalination. As discussed in Chapter 5, the contribution of the GHG damage value to the replacement cost equation can drive up the real cost of water considerably, depending on what assumptions are made about the eventual economic impacts of global climate change over time.

This option would not start to produce freshwater until year 27 of the project, and for the first 12 years would produce at a fairly even rate of about 7 Mm³/year. After year 39, production would fall off quickly. On this basis, the solar desalination plant would need to be sized for this initial production rate.

A solar desalination plant operating on the Mediterranean coast would require a capacity of about 15,000 m³/day to fully replace the Nubian sandstone aquifer water used over the 30 years of oilfield operation. Total water production over the 30 years would be about 150 Mm³. Typical solar desalination installation capital costs range from about US$750 to $1,250 /m³/day.[23] A median capital cost of US$15 million is assumed. O&M costs for such a facility are estimated conservatively at US$0.18/m³ or about $0.85 million per year.[24]

Approach

The approach for this analysis is to attempt to capture the maximum likely benefits that would accrue to both the proponent (private benefits) and society (external benefits) should various project alternatives be enacted. To do this, a conservative approach (from the economic point of view) has been adopted with each external (societal) monetizable benefit valued using a method that will tend to overstate (rather than understate) the benefits. Thus, likely costs are compared with conservatively high benefits. In adopting this approach, the assessment is biased toward the regulatory position. This approach can be used for a relatively rapid, high-level, and relatively low-effort assessment and does not include the more detailed sensitivity analysis included in some of the further examples in this chapter.

Benefits Valuation

For this simple assessment, only two main benefit categories are considered: the value of oil produced (oil is valued by society as a source of energy; the proponent generates revenue from its sale) and the TEV of the fresh groundwater water used or saved during the production (recognizing that it is a nonrenewable resource in the context of this analysis).

The benefit of oil production to society as a whole is based on a long-term assumption about the market price of oil. In the base case, the oil price is estimated as US\$28.70/barrel, which based on recent market trends is an underestimate but fits with the conservative nature of the assessment. Everything is compared to BAU as a datum. Options that produce more oil revenue than does BAU are credited with a positive benefit. Those that suffer a decline in production are assigned the appropriate level of disbenefit. The differential cost of production for each benefit is also factored in to produce a net change in revenue compared to BAU.

In some of the development scenarios, significant amounts of nonrenewable freshwater are used to maintain reservoir pressure. This water, once injected into the producing formation, is essentially destroyed in terms of its ability to be used for any other economic activity that might produce benefits for society. Used for reservoir pressure maintenance, the water creates direct economic benefit simply as a replacement fluid for the extracted oil, allowing more oil to be produced. However, an important distinction is that while the inherent economic value of freshwater is tied to the fact that there are *no substitutes* for its life-giving services, reservoir performance enhancement can actually be accomplished by injecting other fluids, including lower-quality water (brines and brackish water), gas, and even interfacial-tension-reducing compounds such as CO_2. In fact, reservoirs can be produced without any form of pressure maintenance, albeit with lower ultimate recovery. Freshwater used for oil recovery is no longer available for any other beneficial use and is lost to society forever. Thus, freshwater use in this application, in one of the most arid places on the planet, must be seen in terms of the replacement value of the freshwater: What would it cost to replace the freshwater used if no other reasonable alternative for reservoir management is available or is used?

As discussed, most governments in the region do not charge oil companies for freshwater used in oil production. Operators simply drill the wells they need and pump as much water as they require, using it as they see fit. Therefore, in a typical financial (internal economic) analysis, the value of the water used is not considered. As an external good, it is considered to be valueless. However, it is clear that this freshwater is a very valuable resource in its own right, and that any analysis that does not consider this value is inherently lacking if society and the environment are considered.

From this perspective, the replacement value of the freshwater is considered to be equivalent to the current cost of desalination by conventional means, with a premium added for the external costs associated with the GHG emissions resulting from the desalination process. The cost of desalination varies between about US$0.70/m^3 and US$5.30/m^3, depending on the scale of the facility (larger-capacity facilities produce water at lower unit costs).[25] On this basis, for the capacity of facility being considered in this analysis, a value of US$1.25/m^3 has been chosen. To this base cost, a premium representing the external environmental damage value of the CO_2 emitted during the desalination process is also added. In the base case, the cost of CO_2 is put at US$19/tonne, which at the time of analysis was the European carbon market price for carbon. Based on published data, oil-fired power production typically generates about 0.74 kg of CO_2 per kilowatt hour, and desalination, which is very energy intensive, requires about 4.5 KWh/m^3 of water produced. On that basis, a carbon premium of about US$0.06/m^3 of water produced is added to the unit replacement value of water. Thus, the base case water replacement value is estimated at US$1.31/m^3.

If option 5 (alternative source of water) is chosen and a comprehensive effort to locate saltwater or brackish aquifers that can be used for pressure maintenance is successful (it is assumed it will be), it is possible that the standard practice of using freshwater for oil reservoir pressure maintenance could be discontinued throughout the country, if not the region. Given the growing worldwide concerns over water and current regulatory trends in the region, the discovery of a viable alternative source of water for the oil industry could push all operating companies to change their practices. The clear benefit to society would be that the valuable freshwater currently being used for oil development could be used for a variety of other uses for which there is no substitute (potable supply, agriculture, etc.). To value this additional benefit, it is assumed that implementation of option 5 would cause a rapid shift to the new source of water across the oil industry in the country. Based on available data, it is estimated that about 58 Mm^3 of freshwater would be saved each year. This represents a direct economic benefit of US$75.98 million/year at the TEV (replacement value) estimate discussed.

If option 6 (bottled water project) is implemented, benefits will accrue from the production of a valued commodity, which brings a high level of utility to its customers. For an operation at 10 million liters per year, using the wholesale price of US$0.50/L and the US$0.40 O&M cost discussed, the operation would produce net benefits of about US$1 million/year. However, the fundamentals of BAU remain otherwise unchanged.

If option 7 is chosen (solar desalination), the investment in developing and main-streaming solar technology for this application would likely bring wider benefits to society. Alternative energy is already fast becoming an area of great interest and investment, primarily because of the elimination of greenhouse emissions during the operational phase. A 15,000-m³/day solar desalination plant would be one of the largest in the world, and the investment of the proponent in such a facility would cer-tainly create a drive within the regulatory agencies and among other oil companies to develop similar offset schemes or to recognize that freshwater used in oil produc-tion has value that must be recognized. Thus, it can be assumed that replacement options will generate external benefits similar to option 5: the realization of the value of the freshwater used by the oil industry, which has a replacement value of about US$76 million/year (at the base case water value). However, unlike option 5, the benefit in spurring the industry and regulators to realize and account for the value of water that can be directly attributed to option 7 is only equivalent to the value of water replaced over the number of years before regulators would have required such costing anyway. In other words, if regulators require the industry to pay for its fresh-water use by 2020, then implementation of option 7 in 2012 will provide 8 years of the annual US$76 million benefit.

For options 5, 6, and 7, there are clearly significant potential benefits to the pro-ponent's reputation in terms of improved community and regulatory relations. Given the high-level nature of this example, these benefits are not explicitly valued. It should be noted, however, that such benefits accrue to an individual firm or group of firms undertaking the effort. As such, these are considered as private benefits only and do not show up in an overall social economic assessment (society as a whole does not benefit inherently if the reputation of a single firm declines or improves compared to its competitors).

Costs and Benefits: Base Case

Table 4.4 shows the PV of benefits and costs, and the total NPV (net present value) of each of the options considered, using BAU as the basis for comparison. A social discount rate of 3.5% was used in the base case analysis. Even at the relatively low long-term base case oil price, all of the proposed options are strongly economic, with substantial net overall benefits to society. Even though the BAU pressure maintenance water use option (option 1) uses over 145 million m³ of water (enough to supply a city of a million people for 4 years), and even considering a relatively high water value (replacement cost), the revenues from oil production are so large that there is little impact on the overall economics of any of the options. However, regarded in terms of relative performance compared to the current standard practice (BAU), it is clear that certain options are more economic overall (and thus more sustainable) than others.

Under base case condition, option 5 (alternative water use) is over US$1.4 billion better than the BAU water use scenario (option 1). By replacing high-value freshwater with a low-quality alternative source for pressure injection, society as a whole would be considerably better off, despite the operator paying more initially. Of course, real-izing these benefits would require an investment in finding and exploiting such an alternative water resource. For much the same reasons, the next most economic and

TABLE 4.4

Costs and Benefits for the Base Case Compared to BAU

Option	Development Option	Planning Horizon (Years)	PV Net Benefit Oil Production (Millions of $)	PV Freshwater Cost (Millions of $)	PV Additional Cost of Option Implementation (Millions of $)	PV Additional External Benefits (Millions of $)	PV Net Benefit (Millions of $)
1	BAU water use	30	Baseline	Baseline	0	0	Baseline
2	Reduced water use	30	-1,833[a]	-18[b]	0	0	-1,815[c]
3	No water use: surface PFW disposal	30	-3,600	-111	21	0	-3,510
4	No water use: PFW reinjection	30	-4,893	-111	0	0	-4,728
5	Alternative water	30	0	-111	100	1,397	1,408
6	Bottled water CSR	30	0	0	5	18	13
7	Solar desalination replacement	30	0	-111	27	425	509

[a] A negative benefit compared to the BAU datum is indicated. This means that the option produces lower benefits than BAU.

[b] A negative freshwater cost compared to BAU is indicated. This option saves freshwater, compared to BAU and so makes the overall NPV of this option better compared to BAU.

[c] This option is US$1.815 billion worse than BAU over the 30-year planning horizon when the value of freshwater is taken into account.

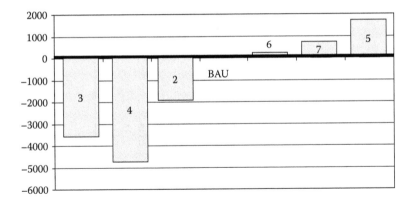

FIGURE 4.3   Marginal NPVs (compared to BAU) of each option, with options ranked in order of lowest to highest marginal capital cost (compared to BAU) for base case values (30-year NPVs at 3.5% in millions of U.S. dollars).

sustainable option is 7, in which freshwater use is replaced through innovative and low-carbon-footprint (low external environmental damage cost) solar desalination. The analysis also reveals the economic case for pressure maintenance. Not using any water for pressure maintenance results in a significant drop in overall benefit (as much as US$4.7 billion) due to a significant decline in oil production.

Figure 4.3 shows the base case marginal NPVs (compared to BAU) for each option, with the options presented from left to right in order of increasing additional capital cost. In this case, the most expensive option (option 5, the identification and use of an alternative low-quality source of water) also produces the maximum overall societal net benefit compared to BAU, an improvement of more than $1.4 billion over BAU.

Sensitivity Analysis

Case of High Water Value

Base case values used were chosen to be conservative. A simple sensitivity analysis explores the implications if higher values for key parameters are used. A case of high water value is developed by using a higher unit replacement cost for water based on a value of US$2.08/m^3, in part reflecting a notionally higher value of the atmospheric carbon emissions that would result if that water had to be replaced through conventional desalination (oil or gas fired). The results are shown in Table 4.5.

Increasing the value of the freshwater alters slightly the overall net benefit of each option that conserves aquifer water but does not change the ranking of options determined in the base case analysis. The overall ranking of the options remains the same. The overall conclusions of the analysis are in fact quite insensitive to freshwater value. In fact, if freshwater were valued at US$10/m^3, the same overall conclusions and ranking of options would still apply. This result reflects a fundamental of the EESA process that is seen repeatedly in real situations: Simply by placing a reasonable value, or range of values, on a critical external resource, BAU approaches are revealed as inferior choices for society and in many cases for industry as well.

TABLE 4.5
Costs and Benefits Compared to BAU: Case of High Water Value

	Development Option	Planning Horizon (Years)	PV Net Benefit Oil Production (Millions of $)	PV Freshwater Cost (Millions of $)	PV Additional Cost of Option Implementation (Millions of $)	PV Additional External Benefits (Millions of $)	PV Net Benefit (Millions of $)
1	Optimal water use	30	Baseline	Baseline	0	0	Baseline
2	Reduced water use	30	−1,833	−31	0	0	−1,802
3	No water use: surface PFW disposal	30	−3,600	−177	21	0	−3,444
4	No water use: PFW re-injection	30	−4,893	−177	0	0	−4,662
5	Alternative water	30	0	−177	100	2,218	2,295
6	Bottled water CSR	30	0	0	5	18	13
7	Solar desalination replacement	30	0	−177	31	674	820

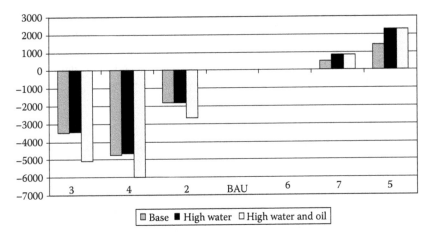

FIGURE 4.4 Marginal NPVs (compared to BAU) of each option, with options ranked in order of lowest to highest marginal capital cost, for base case, high-water case, and the high water and oil value case (30-year NPVs at 3.5% in millions of U.S. dollars). Option 5 remains the best choice for society as a whole under all cases, with option 7 next best. Option 6 is almost equivalent to BAU in terms of overall NPV. Reducing oil production by not using any freshwater (options 2, 3, and 4) means society is worse off overall compared to BAU.

Case of High Oil Value and High Water Value

The same analysis was run with a new higher oil price of US$70/barrel, reflecting a higher long-term expected value of oil. Figure 4.4 shows the comparison of the options under the base, high-water, and the high water and oil value cases. As expected, the boost in oil price results in all options being commensurately more economic. However, the overall findings of the analysis again do not change. The new water source option 5 remains the most economic overall by a significant margin. The no freshwater use option 3 is the least economic (but still, considered alone, is still strongly economic; the value of oil revenues to the project is still sufficient for a profitable operation). The water replacement option 7 is next most attractive. Even with a high oil price, the water savings produced by options 5 and 7 in particular are valuable enough to society to make BAU a suboptimal choice. In fact, the additional oil revenues from a rising oil price effectively make the investment required to protect the freshwater resource increasingly small in relative terms.

Variation of Discount Rate

In the analyses presented, a social discount rate of 3.5% was used. This rate is considerably lower than typically used in the private sector and reflects the inherent longer-term view of society, especially with respect to the conservation and management of natural resources for future generations. Typically, the private sector requires much higher returns on capital to satisfy markets and shareholders and tends to use discount rates of 8% and higher. All other things being equal, a higher discount rate tends to put more emphasis on the nearer-term costs and benefits and puts less value on the future. The sensitivity of the results to discount rate was tested by evaluating the case of high water value using a 10% discount rate. As expected, the higher discount rate considerably reduces the NPVs of all project options considered. However, the overall ranking of the options does not change. Option 5 remains the most economic and sustainable option overall, option 3 is the least economic and sustainable, and the replacement option is also worthy of consideration.

IMPLICATIONS

In this example, the environmental and economic sustainability assessment has shown that across a wide range of economic conditions, considering the key external environmental and social aspects of the project, all of the project options evaluated are economic: They produce net positive benefits for society as a whole. By definition, all project options considered result in an overall improvement of human welfare, as defined in economic terms, considering the major external and private costs and benefits. This result is driven by the high value of the oil extracted. In fact, the value of oil is so high in proportion to the production costs that the value of the freshwater used in the various options, while considerable, is dwarfed in comparison. This result explains why the status quo in the region today remains the widespread use of nonrenewable fresh groundwater for reservoir pressure maintenance. From a purely internal perspective, and given the present regulatory stance, private financial analysis favors unlimited use of freshwater for pressure maintenance. It is low risk, maximizes internal return, and meets all local regulatory requirements. However, as discussed, such a course of action uses significant quantities of a nonrenewable, highly valuable resource for which there is no substitute. The assessment clearly shows that there are alternatives that can considerably improve the overall economics and sustainability of the project by conserving or replacing freshwater.

Water management options can be considered in terms of their marginal economic sustainability. Options 5, 6, and 7 all provide improvement on BAU. Companies wishing to pursue a more socially and environmentally responsible development policy have the opportunity to actually significantly increase overall economic benefit (and human welfare) and the long-term sustainability of their operations by finding a workable alternative to injection of fresh potable water. Injecting salty, low-value water, for which there is no other beneficial use, would conserve the freshwater without reducing overall oil production. The very small effect on profitability would likely be more than offset by expected increases is oil revenues as the price of oil rises over time and the improved relationship with the regulatory authorities and the community engendered by this progressive and responsible new approach. This action would likely set a precedent in the region and could possibly spur regulators to impose similar requirements on other operating companies, resulting in significant benefits across the region in terms of conserving valuable freshwater.

Specific limitations of the analysis are that only two main benefit categories are monetized, water and oil. There are additional benefits that have not been monetized (such as wider community impacts and benefits from the project and the wider external costs of implementing the development project during the construction phase). In a more detailed assessment, these could be included. Nevertheless, this example shows that a robust assessment can be carried out looking at a few key parameters, supported by some basic sustainability assessment, to reveal some important implications.

PRODUCED WATER MANAGEMENT IN OILFIELD OPERATIONS

BACKGROUND: PRODUCED WATER

Disposal of the huge volumes of PFW that accompany oil production is a major issue for the petroleum industry worldwide. For every barrel of oil that goes to market, as many as ten barrels of PFW are produced. Worldwide, billions of barrels of produced water, typically containing elevated concentrations of salts, dissolved and free hydrocarbons left over from the separation process, and trace levels of potentially toxic metals, are disposed. Produced waters may also contain production chemicals such as biocides, corrosion inhibitors and emulsifiers, and sometimes naturally occurring radioactive elements. All of these compounds, from salts to carcinogenic aromatic hydrocarbons like benzene, have potential to harm water resources and the environment if not properly managed. As fields mature and water cuts rise, produced water issues take on increasing economic and environmental significance.

At present, there exists no uniform standard guidance for PFW management. If surface disposal, storage for evaporation, or surface treatment are being considered, the potential impacts of produced water on the surface environment and shallow groundwater should be assessed and quantified as part of a risk assessment. Treating relatively fresh produced waters to a standard suitable for irrigation of salt-tolerant crops is an alternative currently being researched by several operators. But, as discussed in the first example that follows, surface disposal over long periods can result in significant accumulations of salt in shallow soils and rock, which may have long-term environmental and economic implications.

EXAMPLE: PRODUCED WATER DISPOSAL AND GROUNDWATER PROTECTION

In parts of the Middle East, and historically in other parts of the world, some PFWs are still disposed to the surface, into rivers, into near-surface geological horizons that are in hydraulic connection with freshwater aquifers, or into other parts of the hydrosphere. A simple rough-cut economic sustainability analysis of produced water disposal options for an oilfield in the Middle East illustrates the advantages of prevention over remediation.

Until recently, produced water at this facility was treated in evaporation ponds for liquid hydrocarbon separation before being discharged to the ground. The intention was to evaporate the water by spreading it over a large area of ground, given the very arid hot conditions at this location. Several million cubic meters of produced water have been disposed to the surface, much of which has infiltrated into the subsurface and migrated laterally away from the disposal site. Produced water has eventually found its way into the neighboring wadis, where good-quality groundwater is used by local farmers (Figure 4.5). Modeling of predicted movement of produced water indicates that if the situation remains unchanged and surface disposal continues apace, over 500,000 m^3 of potential groundwater abstraction from the wadi and underlying aquifer could be eliminated on an annual basis. Considered over a 20-year (one-generation) planning horizon at a discount rate of 5% (appropriate for assessing social concerns) and using a replacement value of water as the current

FIGURE 4.5 Farmers in the nearby wadis depend on groundwater to irrigate their crops and for domestic supply.

treatment cost for desalinated water of US$1.10/m^3, the PV benefits of action of preventing produced water impact on the wadi system are estimated to be on the order of US$7 million. This estimate may be justified by the fact that, in this area, no other ready source of freshwater exists, and replacement options for freshwater are extremely limited.

Halting additional surface disposal of produced water could be achieved by reinjecting PFW into the producing horizon. This would require the drilling of reinjection wells but has the added advantage of helping to maintain declining reservoir pressures, which has become more of an issue as the field has matured. The PV cost of providing sufficient injection capacity for 20 years of operation is estimated at approximately US$7.5 million. However, this cost would be more than recouped, at current prices, by the anticipated additional oil production provided by improved reservoir pressure management. The cost of implementing a ground-water containment system that would prevent produced water from continuing to enter the wadi system would be on the order of $3 million to install and operate over the next 20 years. Thus, the economic implications of produced water disposal at this particular oilfield are put into context. Cost savings achieved in the early years of production by the use of surface disposal (the least-cost solution) was in fact not an economic decision when the best interests of the whole of society are considered. The combination of benefits from reservoir maintenance (to the producer) and prevention of additional damage to water resources (to the local people) projected over the next 20 years (and using a replacement value of water) more than offsets the cost of injection system installation and operation or the implementation of an environmental remediation program to mitigate the existing damage from surface disposal. As the value of water increases, the reinjection approach becomes steadily more economic.

Background

At a major oil production operation in the Arabian Peninsula, PFW is managed by injection into deep disposal wells, where it is sequestered from the biosphere. However, the water produced from this reservoir is actually of reasonable quality, with relatively low mineralization (TDS levels of between 3,000 and 5,000 mg/L), and once hydrocarbons have been removed, it could be suitable for supporting basic agriculture. This example explores the possibility of reusing some of the PFW to support irrigated agriculture in the vicinity of the production facility to help in the overall economic development of the area and its people. The area in question experiences very low rainfall and arid conditions, and water is very scarce. The millions of barrels of PFW that are produced each day by the oilfields of the area may present a significant opportunity to provide additional water to the local farmers and communities if that water can be safely treated and reused. Equally, PFW management may have a direct impact on existing water resources in the area. The region is blessed with some very-high-quality aquifers, which have supported life in the area for millennia. Because PFW contains hydrocarbons, salts, and occasionally trace metals, it has the potential to severely degrade groundwater quality if not properly managed.

As part of this effort, a pilot program was implemented to test the applicability of low-technology, low-cost solutions for removing hydrocarbons from the PFW and to demonstrate the use of the treated water for irrigation. Trials centered on the use of constructed wetlands to remove hydrocarbons and dissolved metals and the use of the effluent to grow a selection of salt-tolerant crops. Results indicated that, if properly managed, an irrigated agriculture project could cultivate a combination of date palms (highly salt tolerant) and alfalfa (a moderately salt-tolerant fodder crop). Careful control and management of cropping and soil salinity would be required. The trials indicated that such an approach was technically feasible and could be implemented at full scale for an average cost of US$0.50/m^3 of PFW.

Similar large-scale trials conducted in Oman suggested costs of US$0.11/m^3.[26] In contrast, current deep-well disposal costs are on the order of US$0.30/m^3.

Options Development and Costing

A series of water reuse options was developed, representing a broad range of possible approaches for project development. These options include a base case involving the status quo (current standard practice of deep-well injection of PFW), a series of cases examining large-scale irrigated agriculture projects (both on the plateau closer to the production facility and further away in the nearby wadi), and a much smaller community-based demonstration project. The specific options considered were

Option 1: Status quo with all PFW disposed into secure geological horizons at depth.

Option 2: Small (4-ha) community demonstration project immediately adjacent to the production facility

Option 3: Large project (900-ha) in area close to the production facility

Option 3w: Same as option 3 but with fresh groundwater added to reduce salinity
Option 4: Large project (500 ha) in the wadi
Option 5: Same as option 3 but with high-tech water treatment
Option 6: Same as option 4 but with high-tech water treatment

The basis of design, costs, and qualitative advantages and disadvantages of each option are summarized in Table 4.6. In all cases involving irrigation, environmental liability would be associated with PFW salt being introduced into the biosphere. For the larger-scale projects, at water application rates of 5,000 m³/ha/year, over 450,000 tonnes of salt would accumulate in soils over 20 years. This could pose a serious potential environmental risk associated with wind-blown salt deposition, possible impacts on groundwater, and increased salinization of runoff and recharge waters. On this basis, at the end of the project, salt-impacted soils would have to be excavated and permanently sequestered in a contained lined and capped cell. This decommissioning cost is required to eliminate the environmental liability and renders the option equivalent to option 1 (status quo) in this respect, allowing direct comparison. Table 4.6 includes decommissioning costs based on standard U.S. costs for excavation, soil handling, and HDPE liner installation,[27] adjusted for local conditions.

Benefits Identification and Valuation
The following benefit categories were considered based on a risk analysis of the options:

- The benefit of agricultural production from the irrigated agriculture project based on market rates
- The economic value of groundwater that could be damaged if the PFW were released to the environment instead of being treated and or disposed in secure geological formations (external cost of damage)
- The economic value of desert ecosystems that would be damaged if the PFW were released to the environment instead of being treated and or disposed in secure geological formations (external cost of damage)
- Benefits associated with development of best practice in produced water management in the region

Estimates of the benefits of agricultural production are based on a long-term assumption about the market prices of various crops and the inputs into their production. For ease of understanding, in this analysis the benefits of agricultural production are calculated net of all costs of input except water. Data on crop yields, labor and other input costs, and water requirements are provided in Table 4.7. As with all inputs to this analysis, the benefits streams are estimated on a year-by-year basis and discounted as annual flows at the same rate as the costs. Agricultural benefits (net of input costs other than water) are calculated for various crop selections for the size of areas in each option, using current market values for each crop.

The base case (option 1, BAU) provides currently for full protection of groundwater resources from the potential impacts of contamination from PFW if it were to be disposed of directly at the surface. As discussed in the introduction, surface disposal is still widely practiced in other parts of the world, and in some cases

TABLE 4.6
20-year PV Costs, Advantages, and Disadvantages of Each Option

	Option	Description and Key Assumptions	Advantages and Disadvantages	Water Treatment Costs (Millions of $)	Decommissioning Costs (Millions of $)
1	Deep-well disposal (BAU)	Currently 16 operating disposal wells, disposing of all of the PFW from the facility.	Average disposal cost is low (US$0.30/m³); method represents best practice and is fully protective of the environment and water resources.	0	0
2	Small community project	Local community- and government-based project; total 4 ha of irrigated plots; government and community would take over after production agreement expires. Uses 20,000 m³/yr of PFW; wetland-based treatment.	Provides impetus for the future use of PFW in the country; small scale lowers potential liability of residual accumulation of salt in the soils of irrigated plots; government and community capacity would be sufficient to take over and develop the concept.	0.10	0.05
3	Large irrigation project	Project involving 900 ha of irrigated area within 2 km of production facility; wetland treatment system producing 4.5 Mm³/yr of water; dates and fodder crops produced with local labor; government takes over after production agreement expires.	Significant contribution to economic development of the area, making use of a resource that would otherwise be lost; scale of the project may be beyond capability of government and community to manage successfully alone; significant remnant environmental liability posed by 450,000 tonnes of salt accumulated in soils in the irrigated area.	5.97	9.01
3w	Large irrigation project + wells	Project is identical to option 3 except that groundwater wells are installed into the regional aquifer and 12,000 m³/day of freshwater blended with treated PFW to reduce salinity.	Improved agricultural yields and crop diversity from better-quality irrigation water; soil salinization effects reduced; decommissioning still required to sequester salt-contaminated soils.	3.89	9.01

(continued)

TABLE 4.6 (CONTINUED)
20-year PV Costs, Advantages, and Disadvantages of Each Option

	Option	Description and Key Assumptions	Advantages and Disadvantages	Water Treatment Costs (Millions of $)	Decommissioning Costs (Millions of $)
4	Large irrigation project in wadi	Project involving 500 ha of irrigated area within 5 km of production facility; wetland treatment system producing 2.5 Mm³/yr of water; dates and fodder crops produced with local labor; government takes over after production agreement expires	Significant contribution to economic development of the area; scale of the project may be beyond capability of government and community to manage successfully alone; may displace other farming activity in wadi; wadi more sensitive to salt impact (soil and groundwater); significant remnant environmental liability from 250,000 tonnes of salt in soils.	3.40	7.50
5	Large irrigation project with advanced treatment	Project is identical to option 3 except that dissolved air flotation (DAF) treatment is used to treat PFW before irrigation; improves overall water quality (DAF replaces wetland).	DAF treatment reduces evaporative losses during treatment considerably, resulting in better-quality water for irrigation, improved crop yields, and reduced soil salinization; higher capital costs and specialized equipment would require significant technical assistance from operator to community; significantly increased power consumption; less likely to be successfully run and maintained over the long term by the community and government.	28.85	9.01
6	Large irrigation project in wadi with advanced treatment	Project is identical to option 4 except that DAF treatment is used to treat PFW before irrigation.	Project has the advantages and disadvantages of options 4 and 5; DAF treatment provides better-quality water, improving crop yields, but the increased environmental risks associated with wadi operations and the residual requirement for decommissioning remain.	16.25	7.50

TABLE 4.7
Irrigated Agricultural Crop Data and Revenue

Crop Type	Wheat	Sorghum	Alfalfa	Date Palm
Water demand (m³/ha)	5,500–6,000	3,650–5,000	15,000	230
Yield (t/ha)	2–3.5	1–1.5	132	30–80 kg/palm
Max salinity tolerance (mg/L)	4,900	3,900	High	High
Crop value ($/ha)	1,142	1,114	500	$1/kg
Seed cost ($/ha)	16.50	6.60	1.35	5,000
Crop use	Human food	Animal fodder	Animal fodder	Animal fodder
Plowing cost ($/ha)	25	25	25	25
Harvest cost ($/ha)	50	50	50	100
Growing season	Oct–Feb	Mar–Apr/Jul–Aug	Sept–Feb	Sept–Feb
Revenue (net of water cost) ($/ha/yr)	1,050	1,032	673	6,625

demonstrable and measurable impacts on groundwater have occurred, with resulting loss of economic value.[28] The economic benefit to society of the injection program, therefore, is the value of the damage to the aquifer that will be averted over time because of that action. Assuming that over time PFW, if released into the near surface environment, would have an impact on shallow wadi and deeper bedrock groundwater by 1:1 displacement, ignoring the effects of dilution or dispersion and allowing for a 5% recharge rate,[29] PFW released to surface could contaminate as much as 2 Mm³ of groundwater a year. In reality, detailed hydrogeological studies would have to be carried out to determine the expected impacts of such uncontrolled PFW disposal.

Given the arid environment and scarcity of water in the region, the unit value of water can be taken to be the cost of replacing a similar amount of freshwater. The replacement value of freshwater is considered to be equivalent to the current cost of desalination by conventional means, with a premium added for the external costs associated with the GHG emissions resulting from the desalination process. On the same basis as the first example of this chapter, a base case value of US$1.31/m³ for freshwater has been chosen. Using this unit value of water, the economic benefit of BAU reinjection is, by any measure, considerable. Using the very rough assumption of 2 Mm³/year, an annual benefit of US$2.6 million is estimated. Over 20 years, the PV of this benefit would be on the order of US$33 million at a 3.5% discount rate. BAU is strongly economic in terms of water resource protection alone. However, for this comparative assessment, it is more useful to consider the BAU case as the economic datum. Since all of the other options involve an off-take of a relatively small proportion of the overall PFW being managed by the facility, all of the options considered still capture essentially the same economic benefits of the damage prevented by PFW deep-well injection. The question then becomes whether the use of PFW for irrigated agriculture generates sufficient other economic benefits to warrant the costs involved.

Groundwater may also be affected by the introduction of PFW through irrigation as excess applied water infiltrates into either shallow wadi aquifers or deeper hydraulically connected bedrock aquifers. Using the same very basic assumptions as for estimating impact to underlying aquifers, an estimate of the social disbenefit (equivalent to an external cost) arising from aquifer damage can be estimated for each of the irrigation options.

In this part of the Arabian Peninsula, ecological diversity and abundance are mainly concentrated in the wadis, where a protected, cooler microclimate sustained by periodic rainfall accumulations maintains a unique arid ecosystem. A study of the value of desert ecosystems in California in the United States provides a base case value of US$28/year/household per 2 million hectares.[30] Based on available population information, it is assumed that 100,000 households might suffer a loss if PFW use has a negative impact on the wadi ecosystem. This is clearly an overestimate based on the considerable socioeconomic differences between this region and the study area and will significantly overstate the value of these wadis. Nevertheless, as discussed in Chapter 3, this provides a conservative overestimate of the value of the wadi ecosystems as a starting point for the analysis, skewing the results in favor of society. Only the two wadi-based options (4 and 6) would impact this ecosystem. Assuming that the area of wadi affected would have been equivalent to the maximum irrigated area considered for this study, for the amount of water used (4.5 Mm^3/year over 900 ha for option 3) and using the previous valuation, the ecosystem damage disbenefit for option 3 is estimated as US$1,270/year. Despite the high unit valuation, the annual disbenefit is small.

Several of the project options being explored in this analysis could, if enacted, result in potential improvements to current best practice in the region. With such improvements can come tangible economic benefits. If any of the PFW reuse options are chosen and the reuse of PFW is demonstrated to be feasible and economic, there is a real possibility that PFW reuse could become standard practice throughout the region (where PFW chemistry is similar). The clear benefit to society would be that PFW could be judiciously used in other areas and on other projects to produce economic benefits. This benefit would only be realized if the proposition was demonstrated in practice to be both technically feasible and economic.

Base Case Economic Sustainability Analysis

Preliminary high-level cost and benefit functions were developed for each PFW reuse option considered. Table 4.8 provides the net benefit for each of the options for the base case at a 3.5% discount rate over a 20-year planning horizon. To provide a more instructive comparison of relative costs and benefits, the aquifer protection benefit is based on the maximum amount of water used in the reuse options. Positive PV net benefit values indicate that the benefits of the option exceed the costs, and thus it is economically worthwhile to implement.

All of the proposed PFW reuse options are uneconomic. The benefits of agricultural production (and all of the associated local jobs and economic activity) are not sufficient to overcome the direct costs of operation, decommissioning, and the external environmental costs associated with impacts on water and biodiversity. All of the water reuse options except the small corporate social responsibility demonstration project (option 2) have substantial negative net benefits.

TABLE 4.8

Costs and Benefits for the Base Case

Option	PV Benefit Crop Production (Millions of $)	PV Water Damage or Use Disbenefit (Cost) (Millions of $)	PV Ecosystem Damage Disbenefit (Cost) (Millions of $)	PV Additional Water Treatment Costs (Millions of $)	PV Decommissioning Costs (Millions of $)	PV Net Benefit (Relative to BAU) (Millions of $)
1 Deep-well disposal (BAU)	0	0	0.	0	0	0
2 Small community project	0.01	0	0	0.10	0.05	−0.14
3 Large irrigation project	7.35	4.27	0.01	5.97	9.01	−11.91
3w Large irrigation project + wells	8.82	85.87	0.01	3.89	9.01	−89.96
4 Large irrigation project in wadi	4.08	2.33	0.01	3.40	7.50	−9.16
5 Large irrigation project with advanced treatment	7.35	4.27	0.01	28.85	9.01	−34.79
6 Large irrigation project in wadi with advanced treatment	4.06	2.33	0.01	16.25	7.50	−22.03

Sensitivity Analysis

The base case results imply that none of the PFW reuse options examined are economic or sustainable. The economic benefits produced from using the relatively good-quality (but still salt-laden) formation water are not sufficient to warrant the capital and operation costs of the schemes and the additional social and environmental impacts associated with risks to groundwater. The cost of residual damage to soil from salinization only reinforces this conclusion. The current PFW deep-well injection system (BAU), which represents best practice in the petroleum industry, is highly protective of the environment and water resources and is in itself highly economic and sustainable. The overall environmental and social benefits of deep-well disposal (in terms of avoided damage to water resources and the environment) are significantly greater than the costs of operating the system. Sensitivity analysis is used to explore whether these conclusions hold over a wider range of possible conditions.

If internal and external costs are held fixed at base case values, net revenue generated by agricultural production would have to be at least 2.5 times as high as the current market values to provide an economically neutral result for the least uneconomic option (large wadi irrigation project). It is unlikely, however, that the value of lower-grade salt-tolerant agricultural products would rise independently to this extent compared to other goods and services in the economy. If the value of water, for instance, is doubled from the base case value to $2.62/m^3$, all of the PFW reuse options become even more uneconomic and unsustainable.

Implications

The analysis provides optimistic estimates of the benefits of implementing various PFW reuse schemes. Even with high agricultural revenues, none of the major water reuse schemes proves to be economic when all of society is considered. These are therefore unsustainable options environmentally, socially, and economically. This implies that, under current economic conditions, continuing with proper and environmentally responsible disposal of PFW is not only protective of the environment but also produces net economic benefits for society as a whole. The costs of downhole disposal incurred by the operating company are more than outweighed by the benefits of aquifer damage avoided. In contrast, the reuse of PFW must involve some significant costs for water treatment and end-of-life decommissioning to be technically and environmentally feasible, but the agricultural production resulting from this irrigation is not sufficient to cover these additional costs. Large-scale PFW reuse is not economic in this case; significant cost and management effort would be required by the company, and long-term environmental and social liabilities would accompany the project.

Limitations

This example illustrates a high-level conceptual assessment, based on limited available data, intended to be completed quickly and at relatively low cost. Throughout, simplifying assumptions are used but are applied consistently across all options examined. By maintaining consistency, a like-for-like comparison of options is provided that delivers a robust conclusion over a wide range of values for the key parameters. Only three main external asset categories are monetized for this simple assessment: water, crop production, and desert ecology. A more detailed study could have been performed based on additional groundwater investigations, hydrogeologic modeling, and more extensive research on agriculture in the area. These and other factors could be examined in greater detail if a more comprehensive assessment were required, including GHGs associated with energy use, examination of the effect of changing energy costs, and possible displacement of local farmers by the large-scale irrigation operation.

The Value of More Data

Based on the results discussed, it is unlikely that additional data collection efforts would have been worthwhile. The scale of the economic negative in the base case assessment is such that any uplift in benefits would have to be considerable to have any effect on the overall decision. This illustrates another distinct advantage that EESA can bring to the decision-making process. By examining the *scale* of costs and benefits required to change a fundamental conclusion, managers can determine if it is worthwhile to fund expenditures on data collection and research required to refine understanding of a particular issue. This could help to avoid the all-too-common prospect of spending large sums of money to study a particular issue in great detail when in fact the resulting data would have no effect on the decision.

In this example, for instance, trebling the value of water would have no effect on the ranking of options; it would only reinforce the decision. Putting all of the issues into perspective, using a common unit of measure, allows this kind of data-worth analysis to be undertaken and allows decision makers to focus limited resources on studying the dominant, important issues. A decision can be made without additional detailed studies of the hydrogeology of the area and the real TEV of water in this particular location. While this information would be useful for more detailed planning and design, it is not required to identify the best course of action. The EESA reveals this explicitly and in many situations can direct and optimize spending on technical studies and data collection.

WATER MANAGEMENT IN MINE DEVELOPMENT

The following is an example of a more detailed assessment than provided previously in this chapter. The example considers a wider and more strategic range of options and a greater number of external issues, representing a larger cross section of stakeholders, and includes a more comprehensive exploration of the sensitivity of the results.

BACKGROUND

A new open-pit mine is proposed in an arid part of the world. The prospect is one of several in the region, where ore extraction requires significant dewatering both prior to and during ongoing mine operations. Disposal of this water is an important component of mine planning. At other mines in the region, water from dewatering has been disposed into the immediate surface environment, with regulatory approval.

However, the large volumes of water involved have produced impacts to the local environment, which is normally characterized by dry conditions. This has in turn produced increasingly negative reactions from local communities and has driven regulators to request that alternative disposal methodologies be considered for new developments. The issue is made doubly important because the company is planning more expansion in the area, of which the current project is only the first. An EESA was used to determine an effective and sustainable water management strategy for the mine.

The depth of the ore body is significant, and mining operations are expected to extend below the regional water table. Preliminary estimates from hydrogeological assessments of the area suggest that the annual volume of dewatering will initially be between 5 and 25 billions of liters GL/year, with volumes decreasing rapidly to between 1 and 5 GL/year by the end of the 20 years of mining operation. The groundwater is known to be fresh and is potable or near potable in quality. Hence, the surplus represents a substantial resource that must be effectively and efficiently managed during future mine development and operation. Preferred options for management of this groundwater will be those that are both profitable and provide the highest level of use or reuse of the water.

The key objective of the EESA was defined as identification of the most economical and sustainable way to manage water at the mine. The stakeholder group identified a number of options for water disposal, determined key constraints, listed key externalities, and set the financial parameters to be considered in the assessment.

OPTIONS DESCRIPTION

Costs for each option were developed based on three expected annual dewatering profiles of between 5 and 25 GL/year initially, decreasing exponentially to between 2 and 5 GL/year by the end of the 20-year mine life, as shown in Figure 4.6. The high dewatering profile was used for the base case assessment

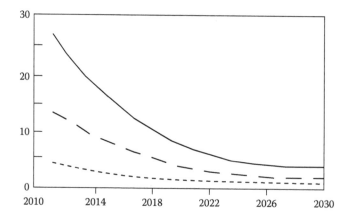

FIGURE 4.6   Expected dewatering rates (GL/yr) for high, medium, and low scenarios.

and equates to approximately 10 GL/year on average. The following options were considered:

- *Option 1: Continuous disposal to creek (BAU)*: The creek receives constant discharge of all excess water over the life of the mine, reflective of past practice at other sites in the area. The advantages are that the method is familiar, can be implemented immediately, and is low cost. However, as discussed, the method creates environmental impacts on the creek ecosystem (changing it from a normally dry ephemeral watercourse to a permanently inundated channel) and creates risks to heritage and indigenous sites of spiritual value within the creek valley system. This method is coming under increased scrutiny and opposition from regulatory bodies and the local community.
- *Option 2: Periodic disposal to creek (shoulder period discharge)*: This involves storing excess water in order to release it down the creek in biannual flow events during "shoulder periods." This would involve the construction of a storage dam to hold approximately half the annual excess dewatering volume, which would be released in a semicontrolled manner just prior to and just after the main wet season. This would produce biannual 10-year average recurrence interval (ARI) flows. This flow regime more closely approximates the natural ephemeral nature of the creek and so would reduce ecological damage resulting from continuous water flow and inundation. This "biomimicry" approach would reduce environmental impacts associated with the creation of permanent watercourses at relatively low cost.
- *Option 3: Continuous disposal to multiple creeks*: This involves constant discharge of all excess water over the life of the mine into a number of creek beds in the area of the planned mine rather than a single creek. It is assumed that the discharge is divided equally between the creeks. The advantages of this option are that it is low cost and easy to implement. By reducing the flow to any given creek system, the overall damage to each may be reduced. However, inundation would still occur in each, and in effect, damage would become more widespread. It is unlikely that community or regulatory pressure for change would be alleviated by this method.
- *Option 4: Pit lake disposal with shoulder period discharge*: This involves disposal of excess water into pit voids, creating pit lakes. A succession of pit lakes would be created as each mining void is exhausted of ore. During mining of the first pit (approximately 5 years), excess water would be disposed of using the periodic shoulder discharge disposal option. The storage dam for periodic discharge would be built on an existing mine area and would not be required after 5 years. Available data indicate that acid mine discharge (AMD) will not be an issue. This option would result in localized storage of water, ensuring retention in the environment from which it originated, and would eliminate BAU surface discharge after the initial 5-year period.
- *Option 5: Bulk water supply to other users in the area*: This would involve transporting of excess water across 20 km to an existing nearby mine that

is short of water for general processing use. Water would be sold to the user, generating revenue for the mine operator. The option would eliminate the environmental impact associated with surface discharge at the site and capture market value for the water.

- *Option 6: Aquifer recharge*: Aquifer recharge involves reinjecting excess water into the source aquifer. A series of injection wells would be installed and excess water injected into the aquifer at an appropriate distance from the mine to prevent flow back to the pits. This option would eliminate water discharge to the surface environment and would replace the extracted water back into the same aquifer, albeit at some distance from the mine. The regulatory authorities approve of this method, but there is significant technological risk (the capacity of the aquifer to receive injected water is uncertain), and the process is highly energy intensive.
- *Option 7: Disposal to evaporation pond*: This option involves creation of a large, shallow basin where water will be allowed to evaporate. The pond would need to be approximately 5 km² in area and be at most 1–2 m deep to ensure rapid evaporation. This option would eliminate environmental damage to the creek and thus prevent damage to indigenous heritage sites within the creek valley. However, the basin would produce a significant footprint.
- *Option 8: Disposal to existing dam reservoir*: Option 8 involves the construction of a pipeline to take water to a dammed reservoir 50 km away, which is currently used to recharge regional groundwater aquifers. This would prevent surface discharge to the creek and damage to cultural assets. However, a costly pipeline construction effort would be required.
- *Option 9: Solar thermal power station cooling*: This option would use a small part of the anticipated annual dewatering volume for cooling water in a new purpose-built 250-MW solar thermal power station in the vicinity of the mine. Solar thermal power stations turn sunlight into renewable electrical energy through the generation of steam. This technology is examined in more detail in Chapter 6. The steam circuit requires large amounts of water for cooling. For a notional 250-MW power station, an estimated 1 GL/year of cooling water would be required. For a conservative analysis, it is assumed that the remainder of the water would be disposed using the BAU option. It is assumed that the electrical energy could be sold on the local network. The main advantage of this option is that it turns some of the water that is otherwise discarded to a high-value purpose (low-GHG renewable energy production). However, the capital costs of the power plant are significant, and the measure does not prevent the discharge to the creek and the resulting environmental and social damage.

OPTION COSTS

The engineering quantification and cost basis for implementation of each option are summarized in Table 4.9. A summary of the overall CAPEX and operating expenditure (OPEX) for each option is shown in Table 4.10.

TABLE 4.9
Financial and Engineering Components of Each Option

Option	Description	Capital Costs	Operating Costs
1	Creek disposal	10-km pipeline 200-kW supply pump 10-km service road	Electricity Pump repair and maintenance
3	Multiple-creek disposal	25-km pipelines to 2 creek heads 200-kW supply pump 260-kW supply pump 25-km service roads	Electricity (only one pump operating at any time) Pump repair and maintenance
4	Pit lake disposal	7-km pipeline 120-kW supply pumps	Electricity Pump repair and maintenance Pipelines and infiltration trenches
6	Aquifer reinjection	20-km pipeline 200-kW supply pumps (2) Buffer dam Injection bores (20) Backflush pumps (20) Wellheads (20) 8-km internal pipes 20-km service road 8-km HV transmission line 8-km internal bore field power distribution Step-down transformers (20)	Electricity Pump repair and maintenance Bore field repair and maintenance
7	Disposal to evaporation pond	508 ha of land clearing 240 ha of pond lining 88,000 m^3 of earthworks 11.6 km of anchor trenches Overflow structure and piping 7-km pipeline 120-kW supply pumps (2)	Electricity Pump repair and maintenance Weed killing
8	Disposal to dam	6-km pipeline 49-km pipeline to dam Buffer dam 180-W supply pumps (2)	Electricity Pump repair and maintenance

TABLE 4.10
CAPEX and OPEX Summary (in 2010 U.S. Dollars)

| | | | | Energy OPEX (Millions of $) | |
| | | CAPEX | OPEX (Millions of $), | Year | Year |
Option	Description	(Millions of $)	Nonenergy	2011	2030
1	Creek disposal (BAU)	28.9	0.01	1.0	0.15
2	Shoulder period discharge	32.4	0.004	0.74	0.10
3	Multiple-creek discharge	38.2	0.01	1.20	0.20
4	Pit lake disposal	53.7	0.80	0.60	0.09
5	Bulk water supply	62.9	0.01	1.5	0.20
6	Aquifer reinjection	78	0.10	3.9	0.60
7	Evaporation ponds	105.4	0.01	0.59	0.09
8	Dam disposal	187.4	0.01	0.89	0.10
9	Solar thermal + BAU	1,223	1.10	0	0

BENEFITS ASSESSMENT AND VALUATION

The following benefit categories were considered in the analysis. Each externality is discussed in more detail:

- The *disbenefits associated with GHG emissions* (disbenefits may also be considered as external costs).
- The *TEV of water*, broken down into three components: the direct use value (used or potentially useable by humans), the ecological support value, and the option value (value to society from having the resource available at some time in the future to be used).
- *Value of damage associated with proliferation of weeds.* Creating permanently flowing watercourses in channels that would otherwise be only ephemerally flooded would cause a significant and rapid shift in the flora of the watercourse and promote growth of weeds, which would replace the indigenous fauna.
- *Indigenous native title and heritage value*, expressed as a disbenefit, where a physical disruption, either by land clearing or flooding with water, would eliminate, damage, or destroy that value.
- *Loss of ecological resources*, specifically the loss of native bushland from clearing associated with activities directly related to water management options considered.
- *Loss of biodiversity*, specifically associated with the loss of endemic species from the permanent flooding of watercourses, expressed as a disbenefit.
- *Loss of ecosystem support value of streams.* To the degree that any of the natural ephemeral watercourses in the area are affected by the water disposal options, there will result a loss of ecosystem support provided by those watercourses.

- *Loss of amenity value of watercourses.* The extent to which watercourses represent a unique social amenity for the inhabitants of the region and may be altered or damaged by large-scale water discharge.

Greenhouse Gas Emissions

There is a limit to the capacity of Earth to absorb GHGs into the atmosphere without harmful effects to human populations. With this in mind, caps have been set on the total amount of GHG emissions in given areas, such as the European Union. Permits, which give the holder the right to emit a portion of the total allowable emissions, are traded like other commodities in open markets. The market price represents the value of the emissions based on supply (the cap initially set based on current scientific knowledge) and demand (the desired amount of emissions), a balance between the interests of the people as a whole and the individuals or groups who wish to emit GHG. A long-term average market value from the European market was used as a base case value in this analysis at $25/tCO_2e$ (tonnes CO_2 equivalent). The upper end of the spectrum for carbon value was chosen as the social cost of carbon (SCC), estimated at $85/tCO_2e$.[31] Chapter 5 provides a more detailed discussion of GHG and carbon management. The low value for GHG was designated as zero. GHG emissions associated with each option were estimated based on energy use.

Total Economic Value of Water

The TEV of water was defined in Chapter 3, and for this assessment was broken down into three components: the direct use value (used or potentially usable by humans), the ecological support value, and the option value (value to society from having the resource available at some time in the future to be used). For this particular analysis, this designation is useful because, to a greater or lesser degree, different options being considered result in the realization or loss of each of these benefits. It is assumed that option value is retained for alternatives that replace, all or partially, the water into an analogous suitable aquifer, from which, theoretically, the water could be accessed and produced at a later point in time. Ecosystem support value is associated with the assumed contributions of the groundwater, in its naturally occurring state or in some similar state, to maintaining local ecosystems. To the degree that the water is removed from the area, as with the dam option, for instance, this value is lost. Use value is triggered explicitly by the removal of the water from the aquifer and its delivery to a specified user or its storage in a location where it is available for use, in some way, by humans.

Each of these three components of the TEV of water are either realized/retained or eliminated, as the case may be. Due to the lack of detailed data on the hydrology of the area at this stage of development, no partial realization or elimination has been assumed. Within the sensitivity analysis, therefore, the TEV of water is varied around a base case estimate of the value of water of $0.35/m^3$, currently used by the regional authorities. As the TEV of water is varied, the relative contributions of the three categories of TEV are assumed to remain constant: use value 60%, option value 15%, and ecological value 25%.[32,33] In addition, certain options involve a degree of evaporation. An evaporation coefficient has been applied to the TEV of water as required. Table 4.11 shows the apportionment of the TEV of water categories for each of the options considered.

TABLE 4.11

Apportionment of Water TEV Water Components and Evaporative Losses by Option

TEV Attribution	Direct Use	Ecological Value	Option Value	Losses (% of Baseline)	Proportion of TEV Achieved[a]
Baseline	No	Yes	Yes	0	0.4
Business as usual	No	Yes	No	10	0.21
Shoulder period discharge	Yes	Yes	No	15	0.19
Multiple-creek disposal	No	Yes	No	10	0.21
Pit lake disposal (+ SPD)	Yes	Yes	Yes	15	0.94
Bulk water supply	Yes	No	No	5	0.58
Aquifer recharge	No	Yes	Yes	1	0.396
Evaporation pond	No	No	No	100	0
Disposal to dam	Yes	No	No	20	0.52
Solar thermal + BAU (combination of the above)	Yes	—	—	15	n/a[b]

[a] Proportion of TEV out of 1.0.

[b] Not applicable. The solar thermal option uses only 1 GL/yr of the total water withdrawal, with the remaining water managed using the BAU option. Alone, therefore, this option does not manage the water issue.

Given the arid environment and scarcity of readily accessible freshwater in the region, the high unit value of water can be taken to be the cost of replacing a similar amount of freshwater. The replacement value of freshwater is considered to be equivalent to the current cost of desalination by conventional means, with a premium added for the external costs associated with the GHG emissions resulting from the desalination process. For this analysis, a unit value of $1.65/m^3$ was chosen.

Proliferation of Weeds

The surface discharge options will create a continuously wet or inundated environment that is expected to cause substantial ecological changes. Chief among these changes is assumed to be the relative and disproportionate colonization of weeds—plants that normally could not thrive in the naturally arid environment of the project area. Recent work on the economic benefits/disbenefits of the presence of weeds and their eradication in Australia[34] (which, although in a different part of the world, has similar conditions to the site) yielded an annual benefit from weed eradication, prorated to site conditions, of $3,386/year, based on a per site annual range of $927 to $5,846. This value is expressed as a disbenefit that occurs when conditions will lead to weed formation. In this case, this disbenefit will apply only to the BAU option, the multiple-creek discharge option, and the solar power plant option (which assumes that the water not used for cooling is discharged to the creek as per BAU).

Indigenous and Heritage Value

The value of indigenous and heritage assets protected or damaged was a key consideration for this assessment. This involves valuation of the benefits to the indigenous population and society in general of retaining cultural and indigenous heritage, which we have termed *cultural value* (CV).[35] Benefits of CV include existence, option, bequest, education, or prestige value. For this example, CV was deemed to consist of two main subvalues, cultural space value (CSV) and cultural heritage value (CHV). The former identifies that impact on native title and heritage sites involves the loss of a physical place used for cultural activities that could be substituted with place with similar physical characteristics. The latter identifies the loss of the connection with *a particular* cultural space that cannot be substituted. Thus,[36]

$$CV = CSV + CHVI + CHVNI$$

where CHVI is indigenous cultural heritage value and CHVNI is nonindigenous cultural heritage value.

Cultural Space Value

Valuing disbenefits due to loss of cultural space involves identifying the amount of cultural land impacted by each option and pricing a land area with similar physical characteristics. In this case, an area of about 500 ha was impacted. It is assumed that a similar area, deemed an acceptable replacement by indigenous landholders, was identified, purchased, and provided as a gift to the affected parties. Estimations of the price of land in the region were obtained in a review of currently available land parcels and verbal conversations with real estate sales offices.

Cultural Heritage Value

There is very limited information on the monetary valuation of indigenous cultural heritage anywhere in the world. However, research has shown considerable discrepancy in the qualitative value placed on indigenous cultural heritage by indigenous people and nonindigenous people.[37] As a result, cultural heritage value has been split into CHVI and CHVNI.[38] For CHVNI, there is a reasonable body of available research on the value that nonindigenous people place on cultural heritage assets, from castles to theaters to ancient rock art. For CHVI, a risk-based valuation of possible sanctions due to breaches of cultural heritage norms and regulations was used as a proxy for value placed on indigenous cultural heritage by indigenous people themselves. There are several possible sanctions that could be placed on the development by regulators, such as delays to schedule, fines, and loss of license to operate. It is assumed that these actions, enshrined in regulation and law, reflect wider society's perception of lost cultural heritage value. Whether the sanctions are activated depends, to a significant degree, on the actions of the indigenous people. Thus, by choosing to activate a sanction, indigenous people are effectively revealing their preference for cultural heritage value. It is important to note that valuing cultural heritage is still in its infancy worldwide.

Ecological Footprint

Despite its arid character, there is significant ecological diversity and abundance in the project area. Any options that involve significant land clearing, such as the evaporation ponds, will cause direct ecological damage. For this analysis, it is assumed that these ecosystems would not otherwise have been destroyed or damaged. Valuation estimates for the surface ecology in the project area were provided by several sources, which provided estimates of the willingness to pay (WTP) for preservation of similar native vegetation in Australia, ranging from a low value of US$0.30 per household per square meter per year (hh/m²/year)[39] to a high value of $0.45/hh/m²/year.[40]

These estimates were based on the number of households willing to pay for the preservation and use of these ecosystems and so are highly dependent on where the population catchment boundary is drawn. There are approximately 5,000 households in the immediate vicinity of the site and about 800,000 in the region. As with all of the other high values used in the sensitivity analysis, the intention was to explore the effect of values that are considerably higher than would be reasonably assumed today but could conceivably be realized at some point in the future if sufficient public scrutiny or changing market conditions came to pass. For each option that involves land clearing or the elimination of natural ecosystems, estimated impacted areas were calculated (see options descriptions), and the range of number of households was applied across the range of annual WTP values.

Loss of Creek Valley Biodiversity

The permanent flooding of watercourses will displace the natural biodiversity of these ecosystems. This value is separate from the introduction of weeds and from the ecological footprint, which accounts for the value of noncreek bushland overall, including biodiversity within the cleared lands. This value is expressed as a disbenefit for options where permanent flooding of ephemeral watercourses will occur.

The disbenefit is expressed as a WTP per household per year, with a base-case value of US$11.79/hh/year, based on a study that examined the value of rare and endangered species and people's WTP for their protection.[41] Clearly, transferring this benefit to the creek in question is highly conservative because (1) there is little information on the presence of endemic or endangered species in the potentially impacted waterway and (2) this WTP estimate theoretically applies to people's overall desire for protection and so would have to be considerably downscaled to be effectively applied to this very small area. Such a scaling would, unless evidence of unique endemic species in the creek were available, render this value so small as to be almost negligible.

Again, as discussed in Chapter 3, the intent is to provide a conservative analysis of the social benefits from the point of view of society. This intentionally skews the analysis in favor of society. The sensitivity analysis simply varies this unit value by doubling and halving for high and low estimates, respectively. Thus, as with other types of risk assessment, if the overall effect on the analysis from this gross overstatement of biodiversity value is insignificant, then it is clear that other factors drive the results, and such a conservative assumption only serves to demonstrate that external factors have been fully and completely considered.

Ecosystem Support Value of Streams

To the degree that any of the natural ephemeral watercourses in the area are affected by the water disposal options, there will be a resultant loss of ecosystem support provided by those watercourses. Even ephemeral streams maintain base flow of groundwater, which in turn helps to maintain the unique and delicate ecosystems found in these features. Economists have valued this support service in a number of locations, and the benefits transfer method[42] has been used to provide an estimate of the disbenefit that would occur should this function be damaged. A base case value of US$30/hh/year was used,[43] applied to the individual asset, with the same range of affected households used as for the ecological footprint benefit. For sensitivity analysis, a low value of US$25/hh/year[44] and a high value of twice the base case value were used. As with the biodiversity benefit, it is important to note that the benefit transfer assumption made was highly conservative and represents a worst case for the operator by selecting high values for the external ecological assets at risk.

Community Amenity Value of Streams

This analysis was completed on the margin and did not include the economic costs and benefits of the mine development as a whole, only of the water management component. So, for example, no revenue from mine operations was included. Equally, the costs associated with wider environmental and social impacts of the wider mine development were not included in the analysis.

The community amenity value (CAV) estimated as part of the analysis applies specifically to watercourses that may be altered or damaged by large-scale water discharge and represents a unique social amenity for the inhabitants of the region. It was assumed that this value will be applied some distance from the actual mining operations, in association with the waterfalls along the creek in particular, which are popular camping and recreation spots. It was estimated only insofar as water discharges may affect the community's ability to access and enjoy these areas in their current state and did not include other possible impacts of mine operations, such as air quality changes, on these features. If the character of these places is fundamentally altered, it was assumed that a loss of public amenity would occur. This is consistent with a wide number of social amenity-based economic studies worldwide.[45] In this case, a disbenefit would be registered in the analysis. A base case value of US$171/hh/year was used, based on a WTP study of a stream in the United States that included not only recreation value (the focus here) but also bequest, option, and existence value.[46]

Benefits Summary

Based on the information provided, the range of expected values for each of the main benefit categories is provided in Table 4.12. As can be seen, the unit values for benefits vary over a considerable range. Low estimates have been deliberately chosen to reflect the likely minimum of the range and the high to bracket the likely uppermost value and to provide an indication of the likely future value trend. It is highly probable that all environmental assets will steadily increase in value over time given the increasing scarcity of, and growing demand for, these natural resources as world population continues to rise. The medium value is thought to represent the current best estimate of the value of the externality.

TABLE 4.12

Monetized Unit Benefit Values for Externalities (2010 U.S. Dollars)

Benefit Category	Units	Low	Base Case	High
GHG	$/tco$_2$-e	0	25	85
TEV	$/m³	0.35	1.65	2.65
Weeds	$/yr/this site	927	3,400	5,846
Heritage value: land replacement	$/ha[a]	3.5	7	14
Heritage value: nonindigenous value of cultural heritage	$/hh/yr[b]	1	9.3	21.5
Ecological footprint	$/m²/hh/yr	0.30	0.36	0.45
Loss of biodiversity	$/hh/yr	5.90	11.79	23.58
Ecosystem support value of streams	$/hh/yr	25	30	60
Community amenity value of streams	$/hh/yr	120	171	560

[a] Medium land replacement values are based on data from real estate agents dealing in land in the affected area. The low and high values are estimated as half and double the medium value, respectively.

[b] hh denotes household. This benefit is directly related to the number of households affected directly or indirectly by a change in the condition of the asset or resource.

PROPORTION OF BENEFITS REALIZED BY EACH OPTION

As discussed, not all benefits or disbenefits are realized by each option, and in some cases benefits are only partially realized by a given option. For example, actions taken to direct water back into the ground using shallow aquifer recharge (discharge to creek) result in evaporative losses that reduce the option value component of the TEV of water. So, while a net disbenefit does occur, it is relatively small compared to, say, the evaporation pond disposal option (in which the entire volume of water [and associated TEV] is lost).

Another key element in the analysis is that the flows of benefits and disbenefits are time dependent. Ecosystem values, for instance, are typically expressed in dollars/unit/year (where units could be hectares or households affected, for instance), representing the ongoing flow of benefits that living systems provide to the planet and humankind. Thus, the analysis must accommodate time-varying benefit flows. Decommissioning, for instance, occurs at the end of project life (assumed to be 20 years), so any environmental damage or benefit associated with the decommissioning activities will occur in year 20 and after. Benefits (or disbenefits) are thus apportioned over time for each issue based on the anticipated level of damage that will occur for the given option based on the unit of valuation. So, for instance, air emissions disbenefits are calculated by multiplying the anticipated mass of discharge in each year over the planning horizon by the unit value and applying a discount rate.

Benefits are then summed over the planning horizon. In the case of estimating disbenefits associated with impacts on the ecology, environmental impact assessment and modeling results are typically used. In the absence of such analysis, basic assumptions associated with available predictive modeling prepared for the project, or simple geometric relationships based on observed impacts at other analogous sites elsewhere, can be used to develop benefit apportionment profiles. These profiles can readily be adjusted to reflect new information as it becomes available.

ECONOMIC SUSTAINABILITY ASSESSMENT RESULTS

Scope and Basis of the Assessment

This is a marginal economic analysis. It considers only the costs and benefit associated with the various options designed to manage water within the development and does not look at the wider mine development itself. The baseline for comparison and monetization is the current undisturbed condition (predevelopment). If an external asset is damaged by implementation of a particular option, this damage appears as a disbenefit (negative benefit). If the value of the asset is maintained as it is (undamaged), then there is no effect, and no benefit or disbenefit is created. So, for example, if a water resource is left intact, in place, the current ecological support and option values of the water remain, and there is no benefit or disbenefit imputed into the analysis. If bushland is cleared, a negative benefit (disbenefit) is imputed.

Base Case Analysis

The base case analysis uses base case assumptions for all parameter values as indicated in Table 4.12 and assumes a social discount rate of 3.5%. The results for the base case are presented in Table 4.13 and graphically in Figures 4.7 and 4.8. Table 4.13 provides the NPV results in 2010 U.S. dollars for each of the options considered (except the solar thermal option, which is discussed separately), broken down by each component of internal and external value.

As shown, pit lake disposal and bulk water supply, both options for which there is no water disposal to the existing environment and the water can potentially be utilized, have socially positive NPVs—they are economic and therefore sustainable. From a purely financial perspective, as expected, all of the options that consider water only as a waste by-product of mining (something to be disposed as cheaply as possible) represent a net cost to the operator. However, given the anticipated future increasing value of water, and assuming that there are notional opportunities to use this water elsewhere, water disposal can be turned into a long-term value-creation prospect for the operator and the rest of society. As shown in Figure 4.8, the social external economics only reinforces this view. BAU water disposal options not only cost the operator significantly in terms of CAPEX, OPEX, but also result in terms of direct costs to society (from damage caused to the environment and cultural assets). BAU is also likely to represent future schedule and project risk associated with regulatory delays and community pressure.

Perhaps the most important conclusion revealed by the analysis is that all of the options, except pit lake disposal and bulk water supply, are moderately to strongly NPV negative. BAU is significantly negative for all concerned. The clear implication is that this issue requires fresh thinking and the exploration of different options. In this case, finding a way to make use of the water promises a significant improvement in overall human welfare.

Base Case without Heritage

The results of the base case assessment without the inclusion of heritage externality values are presented in Table 4.14 and in Figure 4.9. As discussed, the difficulties and

TABLE 4.13

Base Case Assessment Results (Millions of 2010 U.S. Dollars, 20-year PVs at 3.5% Discount Rate)

	Creek Disposal	Shoulder Period Discharge	Multiple-Creek Disposal	Pit Lake Disposal (+ SPD)	Bulk Water Supply	Aquifer Recharge	Evaporation Pond	Disposal to Ophthalmia Dam
Financial								
CAPEX	-29	-32	-38	-54	-63	-78	-105	-187
OPEX (nonenergy)	0	0	0	-3	0	-2	0	0
OPEX (energy)	-6	-5	-8	-4	-9	-24	-4	-5
Sale of water	0	0	0	0	92	0	0	0
Backfill costs avoided	0	0	0	20	0	0	0	0
Total financial	-35	-37	-46	-41	20	-104	-109	-193
Externalities								
Value of greenhouse gas contributions	0	0	0	0	0	-1	0	0
TEV of water	-52	-57	-52	146	49	-1	-108	33
TEI of weeds	0	0	0	0	0	0	0	0
Ecological footprint	0	0	0	0	0	0	0	0
Loss of biodiversity	-1	0	-1	0	0	0	0	0
Streams (ESV)	-2	0	-2	0	0	0	0	0
Streams (CAV)	-12	0	-12	0	0	0	0	0
Heritage value	-105	-17	-58	-17	-55	-49	-62	-55
Total externalities	-172	-74	-125	130	-7	-50	-170	-23
NPV	-207	-111	-171	89	14	-155	-279	-216

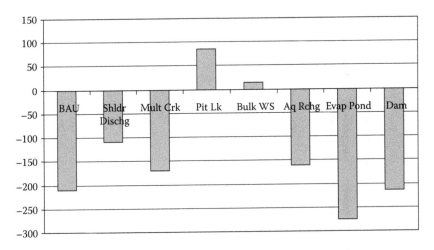

FIGURE 4.7 Net present value over 30 years of water management options (in millions of 2010 U.S. dollars at 3.5% discount rate) for the base case analysis. The significant external costs associated with BAU can be clearly seen. Despite being the lowest-cost option, its overall social NPV is strongly negative. Two options are NPV positive (pit lake and bulk water supply options). These options lie in the "sweet spot" between the extremes of cost, not the cheapest and not the most expensive.

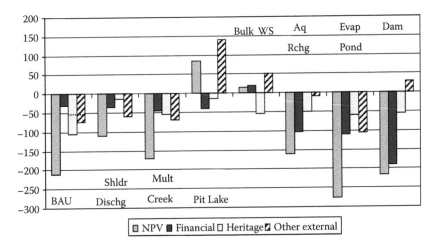

FIGURE 4.8 Thirty-year NPVs for each water management option broken down into financial, social (external), and heritage components of value for the base case analysis (millions of 2010 U.S. dollars, at 3.5% discount rate). The bulk water supply option is the second-best alternative from a complete social perspective and is also the only option that adds private benefit for the operator.

TABLE 4.14

Base Case Assessment Results: No Heritage (Millions of 2010 U.S. Dollars, 20-year PVs at 3.5% Discount Rate)

	Creek Disposal	Shoulder Period Discharge	Multiple-Creek Disposal	Pit Lake Disposal (+ SPD)	Bulk Water Supply	Aquifer Recharge	Evaporation Pond	Disposal to Dam
				Financial				
CAPEX	−29	−32	−38	−54	−63	−78	−105	−187
OPEX (nonenergy)	0	0	0	−3	0	−2	0	0
OPEX (energy)	−6	−5	−8	−4	−9	−24	−4	−5
Sale of water	0	0	0	0	92	0	0	0
Backfill costs avoided	0	0	0	20	0	0	0	0
Total financial	−35	−37	−46	−41	20	−104	−109	−193
				Externalities				
Value of greenhouse gas contributions	0	0	0	0	0	−1	0	0
TEV of water	−52	−57	−52	146	49	−1	−108	33
TEI of weeds	0	0	0	0	0	0	0	0
Ecological footprint	0	0	0	0	0	0	0	0
Loss of biodiversity	−1	0	−1	0	0	0	0	0
Streams (ESV)	−2	0	−2	0	0	0	0	0
Streams (CAV)	−12	0	−12	0	0	0	0	0
Total externalities	−67	−57	−67	147	49	−2	−108	33
NPV	−102	−94	−113	108	69	−106	−217	−160

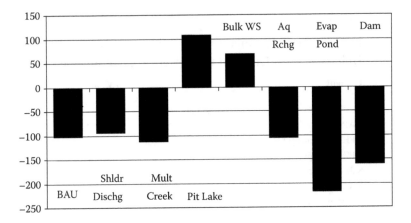

FIGURE 4.9 NPVs of water management options excluding heritage component, base case 30-year analysis (in millions of U.S. dollars at 3.5% discount rate).

sensitivities associated with heritage issues in this context are such that it is useful to view the economic analysis results without the heritage component. It is clear that while heritage values can be considerable in dollar terms, under base case assumptions, their inclusion makes very little difference to the relative attractiveness of the various options under consideration. Removal of the heritage component causes the bulk water supply option to be more economic relative to the pit lake disposal option. However, the economic ranking of the other options does not change from the base case.

Solar Thermal Power Plant Option

The results of the base case assessment of the solar thermal power plant cooling option are presented in Table 4.15 and in Figure 4.10. The solar thermal option is provided as a theoretical investigation of the wider potential sustainability issues on site and in the region. It is not, on its own, a viable water management option for the mine. The volumes of water required by the solar thermal plant are not sufficient to deal with the quantity of disposal required at the site, so this option must be coupled with one of the other water disposal options described. In this case, it was assumed that BAU disposal was used for the remaining dewatering flows.

Despite the considerable capital cost (in excess of $1 billion), the longer-term value of power production means overall strong financial and economic performance of this option. In fact, the externalities associated with the assumed worst-case disposal option for water not used in cooling (option 1, BAU creek disposal) are almost completely offset by the reduction in power costs and GHG emissions achieved by bringing clean and secure energy online (at base case values). Choosing a better disposal option for the remaining water based on the analysis shown in Figure 4.9 would yield an even more NPV-positive result. The solar option provides the strongest overall economic performance of all options considered. Figure 4.10 reveals the important contribution of the sale of produced energy on the overall economics of the solar option.

TABLE 4.15
Base Case Results: Solar Option (Millions of 2010 U.S. Dollars at 3.5% Discount Rate), Heritage Not Included

Element	PV, 30 yr
Financial	
CAPEX	−1224
OPEX (nonenergy)	−16
OPEX (energy)	0
Sale of water	0
Backfill costs avoided	0
Site energy savings	392
On sell of energy	1,100
Renewable energy certificates	284
Total financial	537
Externalities	
Value of greenhouse gas contributions	0
GHG savings	37
TEV of water	−43
TE impact of weeds	0
Ecological footprint	0
Loss of biodiversity	0
Streams (ESV)	0
Streams (CAV)	−2
Total externalities	−8
NPV	529

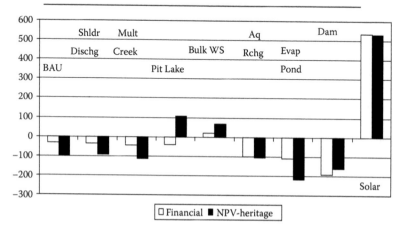

FIGURE 4.10 NPVs of water management options, 30-year analysis, (in millions of U.S. dollars at 3.5% discount rate), with solar thermal power plant option included. This is not strictly a water management option alone and must be coupled with one of the other eight options (BAU in this case). NPV-heritage refers to base case conditions without heritage benefits. From this point in the example and in the remaining figures of this example, heritage issues are not included.

SENSITIVITY ANALYSIS

Any analysis of this type is inherently subject to uncertainty. CAPEX estimates are ±30% accurate, and the valuation and estimation of benefits is often subject to even larger changes, as discussed in Chapter 3. However, the key to the analysis is to reveal not absolutes in terms of dollars, but better and worse decisions overall compared to the range of possible decisions that could be made. From this perspective, sensitivity analysis is important because it allows the overall conclusions of the analysis to be tested across a wide range of parameter inputs. If a decision is favorable or economic over a wide range of parameter inputs, compared to other possible decisions, then despite the overall uncertainty in the actual dollar figures, the decision can be identified as superior to its competitors. This is particularly useful when considering the sustainability of options. By definition, sustainability is concerned with the future, which is inherently uncertain. By varying key input parameters over a wide but reasonable range, the implications of a range of possible futures can be examined.

To examine the effect of key parameters across their full range for each option, the NPV results were calculated for each option across the full range of assumed values for each parameter against every other possible combination of the other values, with each resulting NPV treated as equally likely to occur (equiprobable).

Sensitivity to the TEV of Water

Figure 4.11 shows the sensitivity of the NPV of each option to variation in the TEV of water (solar option not included), with all other factors fixed at base case values. Under the case of a high TEV of water, options that provide direct or potential direct use of water generated an increase in NPV, all other factors remaining equal. The pit lake option preserves ecological and option value and

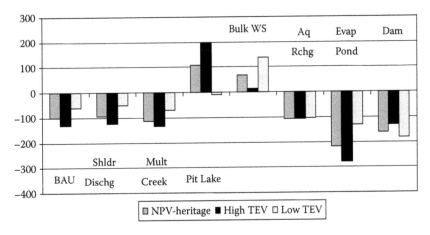

FIGURE 4.11 Sensitivity of NPV of each option to TEV of water (in millions of U.S. dollars at 3.5% discount rate over 20 years), with all other factors remaining fixed at base case values. NPV-heritage refers to base case conditions without heritage benefits.

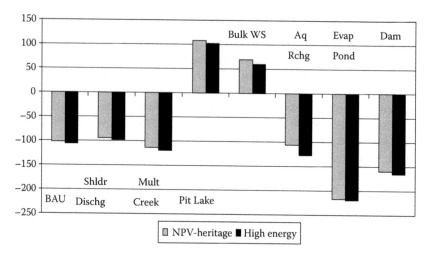

FIGURE 4.12 Sensitivity of NPV of each option to high energy price escalation (in millions of U.S. dollars over 20 years at 3.5% discount rate), with all other factors remaining fixed at base case values. NPV-heritage refers to base case conditions without heritage benefits. Aquifer recharge option is most affected due to its relatively high energy intensity.

creates a potential direct use value, making it the most economical option under that condition. The bulk water supply option remains the second most preferable option. For these same reasons, under the case of a low TEV of water, bulk water supply becomes the best option, and the pit lake option becomes the second most preferable option (although uneconomic overall, with negative NPV). Bulk water supply improves because it assumes a constant sale price for water. Under the low case, this means that the sale price exceeds the direct use component of the TEV. If water sale price varies with the overall change in the TEV of water, however, the pit lake option remains the most economic and sustainable, with bulk sale the next-best choice.

Sensitivity to Energy Price Escalation

Figure 4.12 shows the impact of future real relative changes in energy prices on the relative performance of the options. Variation in the rate of escalation of energy price makes a small difference to each of the options but does not change the ranking of options from the base case results. The base case assumes no real changes in current relative energy costs. The high case assumes a 10% annual escalation. With escalation, energy-intensive options such as the aquifer recharge option and multiple-creek disposal fare worse, all other factors remaining equal. However, as can be seen, energy price changes will not affect the fundamental ranking of the options: The current base case disposal option remains resolutely uneconomic for the operator and for society. The alternative creek and aquifer recharge options also remain strongly uneconomic, and the pit lake and bulk water use options remain the most favorable.

Conclusions and Implications

This example presents the conventional financial implications of various water management options for a new mine but also includes explicitly monetized values for the external social and environmental issues associated with those options. The analysis yields some important conclusions and implications for decision makers:

1. From a traditional financial (internal CAPEX and OPEX only) perspective, the current BAU option (surface discharge to an ephemeral watercourse) is the least-cost solution. In contrast, some of the other options considered are much more expensive (aquifer recharge, favored by regulatory authorities, and disposal to an existing dam reservoir).
2. Monetization of the externalities associated with water disposal at the site reveals that these issues can have a substantial impact on the overall economics and sustainability of various possible water management options. Key externalities that materially affect the overall social economics are water and heritage issues. The amenity value of the watercourse and associated cultural and recreational places within the valley system have a sizable impact on the NPVs (in the millions of dollars NPV over the planning horizon). GHG and the other biodiversity values do not have a significant impact on the analysis under the full ranges of valuation assumptions.
3. Perhaps the most important conclusion revealed by the analysis is that current BAU practice is strongly NPV negative. In fact, from an overall societal point of view, implementation of this option would result in a significant reduction in overall human welfare. This option is uneconomic and unsustainable.
4. Two options appear to be consistently more beneficial overall, and thus more economic, than the rest. Turning the pits into amenity features (pit lakes) after mining has finished, finding a way to use the water off site, or both, hold the promise of significant improvements in overall human welfare over any of the other options. Sensitivity analysis shows that these two options outperform all of the others, no matter which combination of values and assumptions is used. This result points to the need to develop these options further, beyond the conceptual level of this analysis, to determine if the promise indicated here can be realized.
5. The solar thermal power option is interesting. Despite the significant capital cost of this option, the longer-term value of power production means overall strong financial and economic performance. Externalities associated with the assumed worst-case disposal option for water not used in cooling (option 1, BAU creek disposal) are offset against considerable GHG elimination benefits. Choosing a better disposal option for the remaining water would yield an even more economic and sustainable result. The solar option provides the strongest overall economic performance of all options considered.

While it is extremely unlikely that a mine operator would unilaterally build such a facility in response to this water management issue, the analysis illustrates that if this high-quality water can be put to beneficial use, significant benefits for society can result. In this case, perhaps a consortium of industrial and government organizations could together develop such a renewable power project, which could help meet long-term renewable energy targets, reduce GHG emissions, provide long-term energy security for the region at a guaranteed price (as conventional fossil fuel prices rise over time), and derive value from a valuable resource (water).

6. With escalation of conventional energy prices, energy-intensive options such as the aquifer recharge option fare worse, all other factors remaining equal. However, energy price changes do not affect the fundamental ranking of the options; the current base case disposal option remains resolutely uneconomic, and the pit lake and bulk water use options remain the most favorable.

7. The analysis shows that the greater the level of public knowledge and concern for the area and for the way in which the mine is developed, the less attractive the BAU options become. Looking into the future, it must be considered more likely than unlikely that more people will be willing to pay greater amounts to protect unique natural and cultural resources. Nevertheless, even if the number of households interested in this issue increases, the overall conclusions do not change.

8. The findings of the assessment are robust over a wide range of valuation assumptions, mine water production rates, and assumed conditions. While there is significant uncertainty regarding the absolute values of NPVs for each option, the *relative* economic performance of the options appears to be quite consistent. This illustrates one of the key points of Chapter 3: The EESA method begins with physically based parameterization of the options, moves into a monetization step, and then emerges finally to present a business decision-making support result. The absolute values of the NPVs is not of primary interest here; it is the relative ranking of options and how they fare over a wide range of likely future conditions that is of real use to decision makers.

LIMITATIONS

This analysis is of course subject to many assumptions, economic and technical. In this example, the mine is in an early stage of development, and there are clearly uncertainties in several key data input areas, notably including the exact volume of water that will be produced from dewatering operations and the level of actual physical and ecological damage that results from unrestricted surface discharge. Once more data become available in these and other areas, as the project matures, the EESA can be updated to reflect new information. At this early stage, the assessment is designed to evaluate a range of options for water management on a like-for-like basis, monetizing all key internal and external factors, using a common set of assumptions, based on best available information at the time.

EXAMPLE: DETERMINING A SUSTAINABLE WASTEWATER TREATMENT AND DISCHARGE STRATEGY

OVERVIEW: TREATMENT AND DISCHARGE OF WASTEWATER

A major water utility manages the treatment and disposal of wastewater throughout a large jurisdiction. Some regional wastewater treatment plants (WWTPs) discharge wastewater, treated to various standards, to local watercourses either year round or for part of the year. Increasingly, stakeholders are expecting the water utility to treat all wastewater to drinking-water quality standard before discharge to the environment, based on the commonly held assumption that high levels of treatment must inevitably bring increased benefits to the environment (Figure 4.13). However, this approach is expected to involve significant cost as it requires either extremely high levels of treatment or construction of dams for wastewater storage and reuse.

An effective and sustainable treatment and disposal strategy for the WWTPs must address the energy requirements of the treatment and disposal process and the value of the water being disposed and find economic and realistic long-term solutions that do not cause adverse impacts to the environment. Under conventional financial analysis, decisions on whether to move to higher levels of treatment would be based purely on the costs of construction and operation set against tariff-based revenues. No value would typically be placed on the wider costs or benefits to society from the provision of the service. This would result in an incomplete picture of the real costs and benefits of the various possible courses of action.

FIGURE 4.13 In some parts of the world, raw sewage continues to be discharged to natural watercourses. Wastewater treatment plants worldwide provide an essential service, without which the natural environment and human health would suffer significantly. However, as we seek to improve levels of treatment by upgrading existing facilities or setting even more stringent water quality standards to achieve even greater gains in environmental protection, costs rise. (Photo courtesy of WorleyParsons.)

To provide a wider decision-making perspective and to examine the overall social economics and sustainability of the issue, an EESA was used to compare current treatment and disposal methods to a range of other viable and practical alternatives. The analysis not only included relevant environmental and social externalities but also took into consideration the likely real changes in those values across the planning horizon. Given the energy-intensive nature of many wastewater treatment methods, the economic sensitivity of each approach to real relative changes in energy costs was also examined. Similarly, given the large quantities of water disposed by the WWTPs of the utility and the increasing value being placed on water in what is an increasingly arid part of the world, the contribution of the economic value of water to each approach was examined.

BASIS OF ANALYSIS

The analysis presented is a marginal economic analysis. It considers only the costs and benefits associated with the various options designed to manage the treatment and disposal of wastewater from the plant and does not consider the wider WWTP operations or revenues. The baseline for comparison and monetization is the current minimum operating condition for which facultative pond treatment is required (thus capital costs of such a treatment methodology are sunk costs). If an external asset is damaged by implementation of a particular option, this damage appears as a disbenefit (negative benefit). If the value of the asset is maintained as it is (undamaged), then there is no effect, and no benefit or disbenefit is created. So, for example, if a water resource is left unaffected, the current ecological support and option values of the water remain, and there is no benefit or disbenefit imputed into the analysis.

The objective of the EESA was to identify the most economic and sustainable wastewater treatment and disposal method for the test site.

TREATMENT AND DISCHARGE OPTIONS

A range of options to be tested was developed by a multistakeholder team comprising representatives from engineering, environment, sustainability, finance, and project management disciplines. The alternatives were tested at a major WWTP site for which appropriate background data were available. A total of six treatment–discharge couplet options were generated for comparative analysis. Each is described in order of increasing capital cost:

- *Option 1: Facultative pond treatment with stream disposal:* This option represents the minimum treatment and disposal methodology that could be deployed while maintaining the required protection of human health. It is the least technically complex and lowest-cost option and currently represents the minimum standard permitted by the regulator.
- *Option 2: Advanced secondary treatment with stream disposal:* This involves treatment of wastewater to a higher standard and disposing to the environment. It involves higher CAPEX than facultative pond treatment used in option 1 and is also more energy intensive, but it delivers a higher-quality

effluent, which may be more acceptable to external stakeholders, and reduces the overall impact of nutrient loading on the environment.

- *Option 3: Facultative pond treatment with disposal to an evaporation pond:* Evaporation ponds have been built in several locations in the state to meet expectations of zero discharge to the environment. All treated wastewater is contained within the pond, and under normal operations there is no continuous discharge occurring to local waterways. This option has high CAPEX and has a large physical footprint.
- *Option 4: Facultative pond treatment with disposal to storage dam and sale of water:* Storage dams have also been used as a means of preventing continuous discharge to the local environment. They have the benefit of enabling the water to be retained for sale to nearby users, depending on the availability of a suitable customer. Dams require less area than evaporation ponds but are significantly more expensive.
- *Option 5: Tertiary treatment and reverse osmosis with stream disposal:* This involves treating the wastewater to drinking water standard with tertiary treatment and reverse osmosis. It is not an option that has been extensively used by the utility in the past, primarily due to its high cost and significant energy requirements. The main benefit of this option arises from the reduced nutrient input into the environment.
- *Option 6: Tertiary treatment and reverse osmosis with disposal to storage dam and on-sell of water:* This is similar to option 5 except that disposal is to a storage dam for holding and subsequent sale. This option is the costliest of those considered due to the high treatment costs and the relatively high costs of storage. It provides the highest level of treatment and the highest level of protection to the environment.

BENEFITS

The approach for this analysis was to attempt to capture the maximum likely benefits that would accrue both to the utility (private benefits) and to society (external benefits) should various alternatives be enacted. To do this, a conservative approach (from the economic point of view) was adopted. External (societal) benefits were monetized using a method that would tend to overstate (rather than understate) the value. Thus, in the analysis likely costs were compared with conservatively high benefits, or disbenefits, as the case may be. The benefit categories considered in the analysis included

- The *TEV of water*, broken down into three components: the direct use value (used or potentially usable by humans), the ecological support value, and the option value (value to society from having the resource available at some time in the future to be used). It was assumed that the extent to which each of these three components of TEV is realized by each option depends on the relative quality of the water resulting from the treatment level for each option. Within the sensitivity analysis, therefore, the TEV of water was varied around a base case estimate of the value of water of US\$0.35/kL.[47] As the TEV of water was varied, the relative contributions of the three categories of TEV remained constant, using the same formula used in the previous example: use value 60%, option value 15%, and ecological value 25%.[48] Given the arid environment and scarcity of readily accessible water in the region, the high unit value of water can be taken to be the cost of replacing a similar amount of freshwater. The replacement value of freshwater was considered to be equivalent to the current cost of desalination by conventional means, with a premium added for the external costs associated with the GHG emissions resulting from the desalination process. The water replacement value used for this example was estimated at $1.65/m³.

- *Loss of biodiversity*, specifically associated with the loss of endemic species from the introduction of poor-quality water into watercourses, expressed as a disbenefit. The introduction of poor-quality water into watercourses in particular will displace the natural biodiversity of these ecosystems, so this will be expressed as a disbenefit for options for which this occurs. Disbenefit is expressed as a WTP per household per year, and the base value is US$11.79/hh/year, on the same basis as the previous example in this chapter.
- *Loss of ecosystem support value of streams.* To the degree that any of the natural ephemeral watercourses in the vicinity of each WWTP are affected by the disposal options, there will result a loss of ecosystem support provided by those watercourses. A base case value of $30/hh/year was used, on the same basis as the previous example in this chapter.
- *Loss of amenity value of watercourses* indicates the extent to which watercourses that represent a unique social amenity for the inhabitants of a region may be altered or damaged by water discharge. The base case value of $171/hh/year was used, on the same basis as the previous example in this chapter.
- *GHG emissions*, associated with treatment and pumping operations. A long-term average market value from the European market was used in this analysis as a base case value ($25/tCO$_2$e). The high value for GHG was set at $85/tCO$_2$e,[18] based on recent estimates of the SCC, as discussed in the previous example.
- *Service provision benefit.* The extent to which the utility provides an acceptable level of wastewater treatment service to the community is an important external factor. A base case value of $17/hh/year has been applied, based on a study which examined social preferences for improving water quality from wastewater treatment.[49]

Based on the information provided, the range of expected values for each of the major benefit categories is provided in Table 4.16. The unit values for benefits vary over a considerable range. Low estimates have been deliberately chosen to reflect the likely absolute minimum of the range, with the high to bracket the likely uppermost value and to provide an indication of the *likely future value trend*.

Other environmental assets could also have been valued. These include the value of tourism and the bequest value represented by social amenity features maintained by waterways (WTP to pass on something to future generations). However, for this level of analysis, these external (social) benefit categories were deemed to be unsuitable for monetization.

TABLE 4.16
Monetized Unit Benefit Values (2008 U.S. dollars)

Benefit Category	Units	Low	Medium	High
GHG	$/t CO$_2$-e	0	25	85
TEV of water	$/m^3	0.35	1.65	2.65
Species protection	$/hh/yr	5.90	11.79	23.58
Ecosystem support value of streams	$/hh/yr	25	30	60
Community amenity value of streams	$/hh/yr	120	171	560
Service provision	$/hh/yr	−100%	17	+100%

PROPORTION OF BENEFITS REALIZED BY EACH OPTION

Not all benefits were realized by each option, and in some cases benefits were only partially realized by a given option. In this analysis, the basis for this variable apportionment arose from two sources: the variable treatment quality of water disposed in each option and the variable water volumes disposed to the environment at each location. For example, disposal of facultative pond quality water is a worse outcome for waterways, in terms of loss of biodiversity and ecosystem services, than the disposal of advanced secondary treated water. So, while a net disbenefit occurred in both cases, it was relatively smaller for the latter case.

Apportionment Due to Treatment Quality

The main parameters of regulatory concern when considering treated wastewater quality are nitrogen and phosphorus. Intrinsic in this concern is that the treatment regime chosen maintains minimum health standards with respect to discharge of fecal matter and other hazard-related biological waste matter. This assumption was carried through in this assessment, such that the risk of minimum health standards being breached was assumed to be zero. This obviated the need to include any explicit monetization of human health value within this analysis. As such, the analysis was a marginal one—the overall human health benefits of treating wastewater, which are clearly significant, were not included. This analysis examined the social economics of additional expenditure specifically incurred to prevent additional damage to the natural environment from the discharge of treated wastewater.

The assessment used the concentrations of nutrients (nitrogen and phosphorus) in discharged wastewater as a key characteristic of the treatment effectiveness of each option. A hypothetical baseline was set using nitrogen and phosphorus concentrations typical of untreated wastewater. This baseline was assigned a relative quality of 0%, while the nitrogen and phosphorus concentrations corresponding to drinking-quality water were assigned a relative quality of 100%. Treated water from each option was then assigned a relative quality within this range, based on nitrogen and phosphorus concentrations, as shown in Table 4.17.

This relative quality breakdown was used to assign the contribution of water-quality-related benefit (or disbenefit) to each option. For TEV of water, each option was attributed the unit value of TEV scaled by its relative treatment quality. For the species

TABLE 4.17

Relative Treatment Quality Scaling

Treatment Quality	Typical Nitrogen Concentration (approximate, mg/L)	Typical Phosphorus Concentration (approximate, mg/L)	Relative Quality (%)
Raw sewerage	50	12.5	0
Facultative pond	45	10	15
Advanced secondary	15	2	90
Tertiary + reverse osmosis (RO) (drinking water quality)	1	0.1	100

protection and stream benefits, the baseline standard was chosen to be discharge of advanced secondary treated water, that is, that the quality of water produced by this treatment methodology resulted in no net change to environmental conditions. Thus, each treatment method was assigned a scale factor based on its relative treatment quality achieved. Options for which no disposal occurred were similarly treated as analogous to the baseline condition (no change to preexisting environmental conditions in the waterway). For the service provision benefit, the baseline was assumed to be where no disposal was occurring to stream. Thus, options with stream disposal were scaled relative to this baseline by the relative treatment quality. Table 4.18 shows the relative scaled breakdown of each benefit and its application to each option.

Apportionment Due to Location

Disposal of treated wastewater is an emotive community issue, and WWTPs are often blamed for poor environmental conditions in waterways even if only contributing a small percentage of the total nutrient load. Hence, it is important to include a parameter that scales the flows of benefits (where stream disposal occurs) based on the overall contribution of the WWTP to nutrient loading in the receiving waterway. The site being assessed has a current treatment volume of 6,000 kL/day (based on servicing of 8,600 households), with an anticipated annual growth rate in flow of 2.0%. The average annual contribution to receiving waterway nutrient loading at this facility was estimated as 10%, based on the methodology given and the available current data.

OPTIONS COSTS

The total estimated capital and operating costs for the six treatment–disposal options assessed are provided in Table 4.19. The options are ranked from one to six in order of increasing capital cost, with option 1 the cheapest and option 6 the most expensive.

ASSESSMENT RESULTS: BASE CASE

The results of the base case analysis (using the central base case assumptions for values of all parameters over 30 years at 3.5% discount rate) are presented in Table 4.20 and graphically in Figure 4.14.

TABLE 4.18
Benefit Apportionment to Each Option Arising from Variable Treatment Quality

Option	Relative Quality (%)	GHG	TEV of Water (%)	Species Protection (%)	Streams (ESV)[a]	Streams (CAV)[b] (%)	Service Provision (%)
1. Facultative pond + stream discharge	15	—	+15	−75	−75	−75	+15
2. Advanced secondary + stream Q	90	—	+90	0	0	0	+90
3. Facultative pond + evaporative pond	15	—	0	0	0	0	+100
4. Facultative pond + dam + on sell	15	—	+15	0	0	0	+100
5. Tertiary + RO + stream Q	100	—	+100	+10	+10	+10	+100
6. Tertiary + RO + dam + on sell	100	—	+100	+10	+10	+10	+100

[a] ESV = ecological support value
[b] CAV = community amenity value

TABLE 4.19
CAPEX and OPEX Summary (Millions of U.S. Dollars, 2008)

Option	Description	CAPEX	OPEX Nonenergy	OPEX Energy Year 2009	Year 2038
1	Facultative pond treatment + stream discharge	0	0.28	0.13	0.97
2	Advanced secondary treatment + stream discharge	7.7	1.93	0.17	1.26
3	Facultative pond treatment + discharge to evaporation pond	12.5	3.44	0.13	0.97
4	Facultative pond treatment + discharge to dam + on sell	19.7	5.51	0.13	0.97
5	Tertiary treatment + RO + stream discharge	24	3.72	0.82	5.97
6	Tertiary treatment + RO + discharge to dam + on sell	43.7	8.95	0.82	5.97

TABLE 4.20

Base Case Results (Millions of U.S. Dollars, 2008)

	FP Treatment + Stream Discharge	Secondary Treatment + Stream Discharge	FP Treatment + Evaporation Pond	FP Treatment + Dam + On Sell	Tertiary Treatment + Stream Discharge	Tertiary Treatment + Dam + On Sell
Financial						
CAPEX	0	−7.7	−12.5	−19.7	−24	−43.7
OPEX (nonenergy)	−0.4	−2.6	−4.6	−7.4	−5.0	−12.0
OPEX (energy)	−6.6	−8.6	−6.6	−6.6	−40.6	−40.6
FP quality water on-sell	–	–	–	3.5	–	–
DWQ water on-sell	–	–	–	–	–	77.8
Total financial	−7.0	−18.8	−23.7	−30.2	−69.5	−18.5
Externalities						
Value of GHG contributions	−1.6	−2.1	−1.6	−1.6	−9.9	−9.9
TEV of water	3.9	23.5	0	3.7	27.7	0
Loss of biodiversity	0	0	0	0	0	0
Streams (ESV)	−0.5	0	0	0	0.06	0
Streams (CAV)	−2.7	0	0	0	0.4	0
Service provision	0.5	3.2	3.5	3.5	3.5	3.5
Total externalities	−0.3	24.6	1.9	5.6	21.7	−6.4
NPV	−7.3	5.8	−21.8	−24.6	−47.8	−24.9
BCR	0.38	1.27	0.14	0.30	0.40	0.77

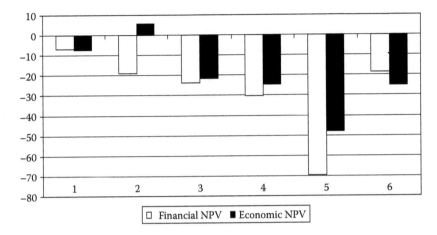

FIGURE 4.14 Thirty-year financial and economic NPVs of the six wastewater treatment and disposal options, under base conditions, ranked right to left from lowest to highest CAPEX (in millions of 2008 U.S. dollars). Note the significant impact of externalities. Only one option emerges as economic and sustainable overall (option 2, advanced secondary treatment with stream discharge).

From a complete social, environmental, and economic perspective, only the secondary treatment with disposal to stream option (option 2) has a socially positive NPV and is economic. From a purely private financial perspective, as expected, all of the options represent a net cost to the utility. Furthermore, in a conventional financial analysis, BAU (Option 1) is optimal because it is the least-cost option under base case conditions and assumptions about future energy prices and TEV of water (low and nil, respectively). This is instructive. From a conventional financial perspective, the values of the external environmental and social costs and benefits are not included, and so, essentially, the analysis becomes a least-cost exercise. The cheapest option will appear to be the best. But clearly, here, and in many other similar types of issues, there are other things that matter. Water has significant real value to society, particularly in a relatively arid and increasingly drying climate. If significant amounts of energy are needed to achieve a high level of wastewater treatment (as in options 5 and 6, notably), then as energy prices rise in real terms, these options will become less attractive to society as a whole, particularly if that energy is largely fossil fuel based, as it continues to be in most parts of the world. It is only by examining all of the relevant costs and benefits over the long term, across the full asset life cycle, that the true full picture of the overall value of an option can be assessed.

Given the anticipated future internalized cost of carbon and rising energy prices, it is clear that the higher treatment (energy intensive) and zero discharge (high construction cost) methods cannot be justified except if a significant premium is achieved through notional opportunities to use or sell this water. In other words, if water has a high real value to society, higher levels of treatment may be justified. Indeed, this reveals that treated wastewater can potentially be turned into a long-term value-creating prospect for society, something that many are beginning to realize worldwide.

Perhaps the most important conclusion revealed by the analysis is that both the current minimum-cost option (pond treatment and stream discharge) and the regulator- and community-favored options (no stream discharge or drinking water quality discharge) are NPV negative; society, on the margin, loses more than it gains with either of these two approaches, for different reasons. The lowest-cost option does not achieve a level of environmental protection that optimizes environmental or water value, and the higher-cost options (options 5 and 6) require more in the way of energy, steel, and concrete and produce too much in the way of secondary damage (GHG emissions) to justify the improvements in environmental and water quality. In fact, the more expensive options, from an overall societal point of view, result in a significant reduction in overall human welfare (strongly NPV negative). However, there is an optimum that lies somewhere between these two extremes where society as a whole benefits.

SENSITIVITY ANALYSIS

The discussion so far has been based on median base case values considered to be reasonable and likely over the planning horizon. However, it is evident that in a rapidly changing world, the costs of commodities in the market economy (energy, water, concrete, steel, and so on), as well as the estimated values of externalities (carbon, other components of the TEV of water, biodiversity, and the like), are almost certainly subject to volatility over the 30-year planning horizon. Thus, it is vital to be able to examine the sensitivity of the full social economics of these options, on a comparative basis, to changes in the values of the key input parameters.

If a decision is favorable (economic) over a wide range of parameter inputs, compared to other possible decisions, then despite the overall uncertainty in the actual monetized NPV results, the decision can safely be identified as superior to its competitors. This is particularly useful when considering sustainability of options. By definition, sustainability is concerned with the future, which is inherently uncertain. By varying key input parameters over a wide but reasonable range, the implications of a range of possible futures can be examined.

Using the method described in Chapter 3, the NPV results are calculated for each value across its full range (as provided in Table 4.16), against every other possible combination of the other values, for each option. What results is a multidirectional statistical sensitivity analysis based on a database of what are treated as equiprobable NPV outcomes for each option, across a range of values of key parameters listed. These are displayed as a cumulative probability curve in Figure 4.15. This approach to sensitivity analysis was detailed in Chapter 3.

Figure 4.15 reveals that option 2 (identified as optimal under base case conditions) is NPV positive under about 85% of the entire range of possible future values assumed. Option 5, the next most attractive option, is NPV positive under less than 40% of conditions. In fact, option 2 is the most NPV positive of all of the options, on average, under all conditions examined. As shown in Figure 4.16, option 2 provides the most environmentally, socially, and economically sustainable result in 92.8% of the cases examined (928 of 1,000 points randomly sampled across the distribution). This is a powerful result. Essentially, this shows that option 2 is the most economic (in the widest, social, environmental, and financial sense) and thus inherently the

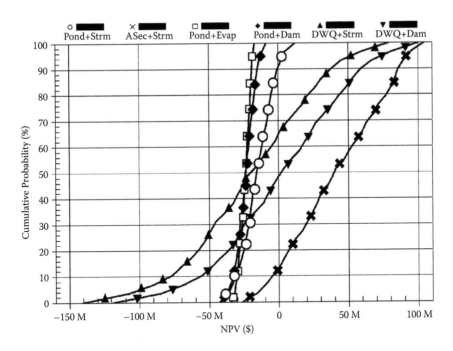

FIGURE 4.15 Plot showing the cumulative probability of NPVs of each option. Each option is represented by one of the curves, as per the legend. Option 2 (ASec+Strm) is the most NPV positive, on average, under the full range of values examined.

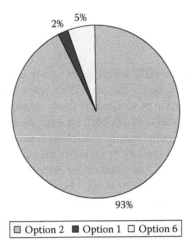

FIGURE 4.16 Detailed breakdown of one thousand samplings of NPVs for each option across the full range of parameters tested to determine which option is the most environmentally, socially, and economically sustainable (highest NPV). Option 2 (ASec+Strm) has the highest NPV in 928 of 1,000 cases. While monetization is the basis for comparison in like-for-like terms, what is presented here moves away from the monetary to stress a comparative ranking to answer the question: Which options are most economic and sustainable under the widest range of possible future conditions?

most *sustainable* of all of the options considered under the widest range of possible future conditions examined.

What is revealed, for instance, is that the value of carbon used in the assessment is not important; whatever value of carbon materializes in the future (with the range tested), option 2 is the best choice among those examined. Similarly, under a wide range of energy cost escalation scenarios (from zero to 10% per annum real change in the value of energy compared to other goods and services in the economy), option 2 is the best choice. This is a risk management result; it allows decision makers to choose solutions that are robust across a wide range of future conditions.

CONCLUSIONS

Application of the environmental and economic sustainability assessment in this example leads to the selection of a more sustainable and inherently more profitable and robust project alternative than BAU. An NPV-positive result is both economic and sustainable. An NPV-negative result is neither—society gives up more than it gains and thus will not want to continue supporting the proposition over the long term.

In this example, the cheapest, BAU, approach is not economic or sustainable. But, investment in the more energy-intensive options, which provide the higher level of treatment being advocated by regulators, is also uneconomic and unsustainable on the margin. In between these extremes there exists an economic optimum at which all parties benefit. Treatment level is increased, and residual impacts on water quality and the environment are significantly reduced, but the costs for doing so are commensurate with the benefits achieved. The assessment allows the optimum to be identified, justified, and communicated as the clearly superior option over a wide range of future possible conditions.

SUMMARY

Sustainability as a notion is widely accepted but poorly practiced. In engineering and project delivery, this is particularly the case. Even when decision makers intuitively know the right and best course of action, it is rarely taken, usually because the "economics" do not work. Typically, we fall back on what we know and what has worked in the past. But, every metric of planetary well-being is telling us that we can no longer continue on this course. The impacts of global change are being felt everywhere, by everyone, and are accelerating. The cause of our present predicament is a treacherous combination of outdated economic thinking, obsolete decision-making systems, and previous-century engineering practices and design codes. Unless we explicitly include all of the issues that matter in decision making, environmental, social, and financial, we will continue to make decisions that produce profits for few but not benefits for all. BAU thinking and analysis will continue to support BAU decisions.

NOTES

1. Hardisty, P.E., and E. Ozdemiroglu. 2005. *The Economics of Groundwater Remediation and Protection*. CRC Press, Boca Raton, FL.
2. Hellstrom, B. 1940. The Subterranean Water in the Libyan Desert. *Geografiska Annaler*, 22, 206–239.

3. Edmunds, W.M. 2006. Libya's Saharan Groundwater: Occurrence, Origins and Outlook. In D. Maltingly, S. McLaren, E. Savage, Y. Al-Fasatwi, and K. Gadgood. (eds.). *The Libyan Desert: Natural Resources and Cultural Heritage.* Society for Libyan studies, London, UK. pp. 45–64.

4. Conant, L.C., and G.H. Goudarzi. 1967. Stratigraphic and techtonic framework of Libya, *AAPG Bulletin*, 51(5), 719–730.

5. Wright, E.P., A.C. Benfields, W.M. Edmunds, and R. Kitching. 1982. Hydrogeology of the Kufra and Sirte Basins, Eastern Libya. *Quarterly Journal of Engineering Geology, London,* 15(2), 85–103.

6. Pim, R.H., and A. Binsarti. 1994. The Libyan Great Man Made River Project. Paper 2. The Water Resource. *Proceedings of the Institute of Civil Engineers: Water, Maritime and Energy*, 106(2), 123–145.

7. Pallas, P., 1980. Water resources of the Socialist People's Libyan Arab Jamahiriya. In A. Salem, and P. Buswreil. (eds.). *The Geology of Libya*, vol. II. Academic Press, London. pp. 539–594.

8. Edmunds, W.M. 2006. Libya's Saharan Groundwater: Occurrence, Origins and Outlook. In D. Maltingly, S. McLaren, E. Savage, Y. Al-Fasatwi, and K. Gadgood. (eds.). *The Libyan Desert: Natural Resources and Cultural Heritage.* Society for Libyan studies, London, UK. pp. 45–64.

9. Ibid.

10. Great Man Made River Authority (GMMRA). 2006. Great Man Made River Authority Web site http://www.gmmra.org/en/.

11. Edmunds, W.M. 2006. Libya's Saharan Groundwater: Occurrence, Origins and Outlook. In D. Maltingly, S. McLaren, E. Savage, Y. Al-Fasatwi, and K. Gadgood. (eds.). *The Libyan Desert: Natural Resources and Cultural Heritage.* Society for Libyan studies, London, UK. pp. 45–64.

12. General Water Authority. 2000. Personal communications. Government of Libya, Tripoli.

13. Ibid.

14. General Water Authority, 2006. *Water Use Data.* Government of Libya, Tripoli.

15. Todd, D.K. 1959. *Ground Water Hydrology.* Wiley, New York.

16. Freeze, A., and J. Cherry. 1979. *Groundwater.* Prentice-Hall, Upper Saddle River, NJ.

17. Domenico, P.A. 1972. *Concepts and Models in Groundwater Hydrology.* McGraw Hill, New York.

18. Mr. A. Karoun, Environmental Protection Department, National Oil Company of Libya. 2006. Personal communication. Tripoli, Libya.

19. Hardisty, P.E. 2005. The Economics of Environmental Protection: A Middle East Oil and Gas Industry Perspective. *Middle East Economic Survey*, 48(28), 28–29.

20. General Water Authority (GWA). 2000. *Water Resources Strategy in Libya, 2000–2025.* Report AL-WR-296 of the General Water Authority, Tripoli. GWA, Tripuli, Libya.

21. Rast, R.R. and K. O'Gallagher. 2007. RACER—the solution to environmental remediation cost estimating. *Federal Facilities Environmental Journal*, 9(3), 47–67.

22. Zenith International Ltd. 2006. Personal communication.

23. United Nations Environment Programme (UNEP). 2007. *Global Environmental Outlook 4: Environment for Development (GEO-4).* United Nations Environment Programme, Nairobi, Kenya.

24. Ibid.

25. California Coastal Commission. 1993. *Seawater Desalination in California.* CCC, Saeramento, CA.

26. Sluijterman, A.C., Y. Al-Lawatiya, S. Al-Asmi, P.H.J. Verbeek, M.A.S. Schaapveld, and J. Cramwinckel. 2004. *Opportunities for Re-use of Produced Water Around Desert Oilfields.* Society of Petroleum Engineers, SPE report 89667. SPE, Richardson, TX.

27. Rast, R.R. and K. O'Gallagher, 2007. RACER—the solution to environmental remediation cost estimating. *Federal Facilities Environmental Journal*, 9(3), 47–67.
28. Hardisty, Economics of Environmental Protection.
29. United Nations Environment Programme (UNDP). 2004. *Water Resources Management Studies in Hadramawt Region, Yemen*. Report RFPS-63, UNDP/NWRA YEM 97/200. UNEP, New York, NY.
30. Richardson, R.B. 2005 *The Economic Benefits of California Desert Wildlands*. Monograph. Wilderness Society, San Francisco.
31. Stern, N. 2006. *The Economics of Climate Change—The Stern Review*. Cambridge University Press, Cambridge, UK.
32. Greenley, D.A., R.G. Walsh, and R.A. Young. 1982. Option Value: Empirical Evidence from a Case Study of Recreation and Water Quality: Reply. *Quarterly Journal of Economics*, 96(4), 657–673.
33. Ibid.
34. Sinden, J.A., and G. Griffith. 2007. Combining Economic and Ecological Arguments to Value the Environmental Gains from Control of 35 Weeds in Australia. *Ecological Economics,* 61(2–3), 396–408.
35. Bowen, J., M. Sivapalan, and P.E. Hardisty, 2008. *Cultural Value Applied to EcoNomics Assessments*. WorleyParsons Internal Report. WorleyParsons, Sydney, Australia.
36. Ibid.
37. Rolfe, J., and J. Windle. 2003. Valuing the Protection of Aboriginal Cultural Heritage Sites. *The Economic Record*, 79, Special Issue, June, s85–s95.
38. Bowen, J., M. Sivapalan, and P.E. Hardisty, 2008. *Cultural Value Applied to EcoNomics Assessments*. WorleyParsons Internal Report. WorleyParsons, Sydney, Australia.
39. Lockwood, M., and D. Carberry. 1998. *Stated Preference Surveys of Remnant Native Vegetation*. Johnstone Centre Report No. 104. Charles Stuart University, Albury, Australia.
40. van Bueren, M., and J. Bennett. 2004. Towards the Development of a Transferable Set of Value Estimates for Environmental Attributes. *The Australian Journal of Agricultural and Resource Economics*, 48(1), 1–32.
41. Loomis, J.B., and D.S. White. 1996. Economic Benefits of Rare and Endangered Species: Summary and Meta-analysis. *Ecological Economics*, 18(3), 197–206.
42. Department for Environment, Food, and Rural Affairs (DEFRA). 2007. *An Introductory Guide to Valuing Ecosystems Services*. U.K. Department for Environment, Food, and Rural Affairs, London.
43. Le Goffe, P. 1995. The Benefits of Improvements in Coastal Water Quality: A Contingent Approach. *Journal of Environmental Management,* 45(4), 305–317.
44. Sutherland, R.J. and R.G. Walsh. 1985. Effect of distance on the preservation value of water quality, *Land Economics*, 61(3). The Board of Regents of the University of Wisconsin System.
45. DEFRA, *Introductory Guide*.
46. Sutherland, R.J. and R.G. Walsh. 1985. Effect of distance on the preservation value of water quality, *Land Economics*, 61(3). The Board of Regents of the University of Wisconsin System.
47. Western Australia Department for Planning and Infrastructure. 2007. *Guidance on Water Valuation in the State of Western Australia.*
48. Greenley, D.A., R.G. Walsh, and R.A. Young. 1981. Option value: Empirical evidence from a case study of recreation and water quality *Quarterly Journal of Economics*, 96(4), 657–673.
49. Kontogianni, A., I.H. Langford, A. Papandreou, and M.S. Kourtos. 2001. *Social Preferences for Improving Water Quality: An Economic Analysis of Benefits from Waste-Water Treatment*. Working Paper GEC 01–04. Centre for Social and Economic Research on the Global Environment, School of Environmental Sciences, University of East Anglia, Norwich, UK.

5 Greenhouse Gases and Climate Change

INTRODUCTION

Climate change is among the most important sustainability issues facing humankind in the twenty-first century. As discussed in Chapters 2 and 3, there is a growing consensus that the risks posed by climate change will exacerbate many of the other problems that the world currently faces, including water scarcity, poverty, political instability, dwindling natural resources, and loss of biodiversity.[1]

Around the world, industries of all kinds face immense challenges if they are to significantly reduce their GHG emissions.[2] A big part of the solution is simply to examine energy and emissions issues in a wider context to appreciate the potentially massive savings available from collaboration and efficiency within and between industries.[3]

In Oman, for instance, a thriving petroleum sector produces liquefied natural gas (LNG) for export and crude oil for domestic use and sale on the world market. At the same time, a thriving economy has seen demand for electricity in the country jump considerably over the last several years, to the point at which there are local power shortages. In response, the government has promised to increase power supplies in the next few years to meet demand. But, because available gas reserves are locked into LNG contracts, current plans call for construction of a 1,000-MW coal-fired power station, which would burn imported coal. Compared to gas-fired power generation using local gas, this is a highly emissions-intensive and climate-damaging option. Power demand management options are not being considered presently; meanwhile, oil producers continue with the decades-old practice of flaring the gas associated with oil production (representing as much as 1,000 MW of equivalent power generation potential). It is clear that a strategic coordinated examination of energy and petroleum development strategy could offer massive energy, cost, and carbon emissions savings to the country and industry, but unlocking this potential requires a new, more comprehensive view that takes into account the likely possible future changes in energy prices and the cost of carbon and examines the wider environmental and social costs and benefits of various strategies to move beyond conventional thinking.

The environmental and economic sustainability assessment (EESA) is ideally suited to provide a more detailed understanding of the implications of climate change, particularly as they pertain to decision making in business and government. The risk assessment process attempts to identify possible risks associated with a project or activity and assess them based on the probability of occurrence and the magnitude of the expected effect. Risks with very high probability and low impact are deemed unacceptable and are mitigated against. Risks of catastrophic effect (which could jeopardize the project, put the company out of business, or result in significant fatalities), and even very low likelihood, are also typically deemed unacceptable and are mitigated. The risks posed by climate change on business, industry, and indeed society as a whole (and the wider imperative for environmental sustainability) can be examined from two perspectives: first, the risks of operating in a carbon-constrained world that is responding to the need to mitigate climate change (mitigation risk); and second, the risks associated with adapting to a world increasingly affected by a changing climate (climate risk). In this chapter, each of these types of risk is discussed in turn, with a few short examples provided. The rest of the chapter provides examples of environmental and economic sustainability assessments focused on carbon management issues.

CARBON MITIGATION RISK

Carbon mitigation risks for industry flow from the increasing pressure being applied by governments, regulators, investors,[4] nongovernmental organizations (NGOs), community groups, and private citizens. Industry is increasingly expected to be part of the effort to prevent the worst of the damage from climate change by significantly reducing emissions of greenhouse gases (GHGs) such as carbon dioxide and methane.

The scale of the mitigation challenge is monumental, largely because we have wasted the last 30 years: We have known about the problem but have not taken meaningful action.[5] To stabilize the concentration of GHGs in the atmosphere at a level that gives us a reasonable probability of avoiding the worst effects of climate change, we need to decarbonize the economy of the world by as much as 60% to 80% by 2050.[6] Achieving this emission reduction target gives us a reasonable chance of keeping global CO_2 concentrations below 550 ppm (preindustrial levels were 280 ppm).[7] This is widely acknowledged as the point at which the risks of dangerous runaway climate change become unacceptable.[8]

The scale of this change means that appropriate price signals will need to be put in place to progressively drive up the cost of carbon. Managing the introduction of widespread carbon pricing, in one form or another, is a key challenge for business, industry, and governments. Carbon-intensive businesses will need to make profound changes to avoid large cost increases and subsequent effects on profitability and competitiveness. Introduction of cap and trade schemes will also mean that emissions will be restricted overall, preventing expansion and growth in emissions. Companies will have to develop strategies for expansion and growth that work within these new limits. They will have to adapt to changing economic conditions, the emergence of new patterns of investment, and the changing relative importance and success of various sectors of the economy (depending on their ability to adapt and prosper as carbon prices rise).

How business and industry manage the risks, uncertainties, and opportunities associated with the rising cost of carbon will have a major impact on their success in the coming years and decades. The economic costs and benefits of actions taken by businesses to reduce emissions need to be carefully considered as the marginal abatement cost of carbon (MACC; now on the order of US$5 to $25/tCO_2$e [tonnes of CO_2 equivalent] or less) climbs inexorably toward the social cost.[9]

One area in which many manufacturing and resource businesses can achieve significant revenue-positive reductions in GHG emissions, and thus future internalized carbon costs, is in energy efficiency. Many firms require that process and equipment modification to achieve reductions in energy consumption meet financial hurdle rates that are actually higher than for new capital projects. In many instances, energy efficiency projects examined without carbon costs cannot provide internal rates of return that meet these hurdle rates and are therefore rejected. Revenue-positive sustainability can and should be considered in the light of the progressive predicted internalization of the marginal cost of carbon and its trajectory toward the social cost of carbon (SCC). In this way, profits from energy savings in the near term can help to defray the costs of further, more difficult emission reductions in the medium term.[10] Table 5.1 summarizes some of the key risks, consequences, and mitigation measures that may apply to industry.[11]

PRICING CARBON IN BUSINESS DECISIONS

Including carbon management in effective decision making requires that carbon emissions be given a price. That price can be embedded in financial and economic analysis of projects and used to understand present and future implications of various capital investment decisions. However, there are several different ways to examine the value of carbon. The SCC, which represents the true social or damage value of each additional tonne of GHG emitted, was discussed in detail in Chapter 3. A range of market-based prices set within various trading schemes was also summarized in Chapter 3.

Another way to express the cost of carbon is the MACC, which is the cost of reducing emissions rather than the value of the damage caused by the emissions. The MACC also differs from the market price of carbon, which is determined directly or indirectly by policy objectives. The MACC is based on the cost of technological and process measures to eliminate or reduce emissions. Studies have developed MACC curves for the global economy[12] and for various countries, including Australia, the United States, the United Kingdom, and Germany. National MACC curves of this type are necessarily high level and look at all sectors of the economy, from residential energy savings to commercial building upgrades, to power generation alternatives. In each case, these curves reveal a common pattern of significant available negative cost abatement opportunities, primarily from energy conservation and efficiency measures. While these overall trends are generally instructive, national MACC curves are not particularly useful for decision making within particular industries or for particular projects or investment decisions. Here, industry must begin developing its own specific MACC curves to better understand the scale of the abatement opportunities that exist within specific businesses, industries, or projects.

One area in which many industries can achieve significant revenue-positive reductions in GHG emissions, and thus future internalized carbon costs, is in energy and heat efficiency. Many energy and heat efficiency opportunities available right now are cost negative.[13] In a study of a mine expansion in Australia, a project-specific MACC curve was developed.[14] A range of engineering design, equipment selection, transport, and operational alternatives was examined for the potential to reduce GHG emissions at

TABLE 5.1
Climate Change Mitigation Risks for Industry

Risk	Likelihood	Consequence	Mitigation Measures
Carbon taxation	High and increasing over next 10–20 years Already in place in some jurisdictions	Increased production costs, reduced profitability and competitiveness	*Reduce carbon intensity per unit of production* Energy and process efficiency measures Fuel switching, supply chain management, alterative energy sources and suppliers, carbon capture and storage, carbon offsetting, move production to jurisdiction with lower carbon tax
Carbon emission limits or caps	High over next 10–20 years Already in place in some jurisdictions	Limitation on expansion of existing facilities, limitations on new facilities, reduced growth	*Reduce overall carbon emissions* Retrofit existing facilities to lower carbon footing (as above) Design and build new facilities for optimal energy efficiency Purchase permits or allowances from other firms Move production to higher emissions jurisdiction
Shareholder and investor scrutiny	High and rising over next few years	Greater difficulty in securing financing, higher borrowing costs, reduced profits	*Develop strategic plan to reduce shareholder risk from exposure of operations to carbon constraints* Implement carbon intensity reduction and overall emission reduction plans as above Participate in carbon disclosure programs and other business sustainability indices
Public relations and corporate reputation damage	High and rising over next few years	Declining reputation in the market and with customer, declining sales, reduced profits	*Develop and enact corporate sustainability policies to manage risk of negative public sentiment* Take actions that will be seen as part of the solution not as part of the problem Communicate with customers and community stakeholders on climate change, elicit feedback, incorporate into overall mitigation strategy

each stage in the project life cycle. Compared to business as usual (BAU; how the mine would have been designed and operated if recent standard practices were employed), total GHG emissions can be reduced by over 2.5 $MtCO_2e$/year, and average product GHG intensity reduced by 35%, by selecting only *negative cost* measures (Figure 5.1).

But, this type of carbon abatement cost study, similar in nature to the national carbon abatement cost curves discussed, paints only a partial picture. While these

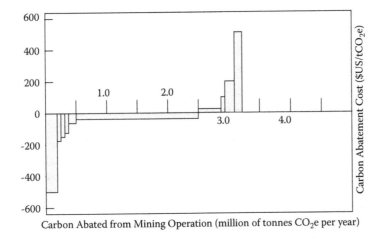

FIGURE 5.1   Marginal abatement cost for carbon. (After Hardisty, P.E. 2009. Analysing the role of decision-making economics for industry in the climate change era. *Management of Environmental Quality.* 20(2), 205–218.

measures produce net cost savings over the longer term, they all require some level of capital investment to be realized. Industry and business will therefore assess the payback of these investments, much as they would any other capital investment, before deciding on implementation. Many firms require process and equipment modification projects to meet or exceed financial hurdle rates used for new capital projects. In many instances, energy efficiency projects examined without carbon costs cannot provide internal rates of return that meet these hurdle rates and are therefore rejected. These calculations almost always exclude any accounting for environmental or social externalities, which might have made the overall economics look significantly different.[15] The result is that many environmentally worthwhile projects are rejected by industry because they fall into the net present value–internal rate of return (NPV-IRR) trap discussed in Chapter 3—they are profitable (or cost negative, as discussed) but not profitable enough to meet existing IRR targets.

Thus, the MACC, despite becoming more widely used in industry, does not provide a full (or even necessarily accurate) picture, on two counts. First, the cost of any decision must be rationally compared to the benefits that are produced from that expenditure; it is not only a question of cost. Many of the benefits of carbon reduction, although certainly not all, are tied to the need of society to reduce and stabilize overall GHG emissions. These considerations could significantly alter the perceived economics of many environmental improvement projects.

Second, the MACC typically does not include the intricate physical links between efforts to limit carbon and other key sustainability objectives, such as water management and energy conservation. As discussed in Chapter 3, these and many other key considerations are linked physically, and efforts to improve performance in one area may trigger changes in consumption or efficiency in another. The examples provided in this chapter illustrate this link and the trade-offs that must be managed in finding the environmental, social, and economic optimum.

EXAMPLE: GHG MANAGEMENT IN THE GAS INDUSTRY

Business decision making can be profoundly affected by a comprehensive under-standing of the implications of the convergence over time between the marginal and social costs of carbon. A simple example from the natural gas industry illustrates how businesses can improve sustainability performance profitability (and reduce the financial risks of impending taxation) by finding cost-negative carbon reduction opportunities and examining the effect of the SCC on energy efficiency opportu-nities. This then allows organizations to consider long-term investments in carbon abatement from a whole-project life-cycle perspective and adapt project design to better manage climate change mitigation risks.

In many parts of the world, natural gas contains a substantial amount of CO_2, which has to be removed before the gas can be sent to market. Australia, Canada, Norway, and Algeria, for instance, have natural gas with high CO_2 content in some areas. Under current standard industry practice, CO_2 is removed from the produced gas stream and vented to atmosphere after separation. A gas development exploit-ing reserves containing approximately 10% CO_2, using a conventional design at full operational capacity and venting CO_2 to the atmosphere, produces approximately 7 million tonnes of CO_2 each year.[16]

Table 5.2 shows the impact on project NPV of the current BAU emission rates for various possible carbon pricings and the impact of implementing a CCS (carbon capture and sequestration) program at the facility, which would virtually eliminate the emissions associated with CO_2 removal from the gas stream. Note in Table 5.2 that the NPV impacts with a nonzero carbon price are compared to those for which carbon is assumed to have no value. Thus, the venting option results in a net negative NPV to society due to the emissions over 30 years compared to what would have been cal-culated as the NPV with no value placed on carbon. Considered separately, the CCS option has a strongly negative NPV with carbon priced at zero but generates significant net benefit at the SCC due to the beneficial reduction in CO_2 emissions. Contrastingly, BAU is attractive when carbon is unpriced. But, if the SCC, or part of it, is recognized explicitly, BAU is revealed as uneconomic and unsustainable for society.

TABLE 5.2

Impacts of Carbon Value on Full-Project NPV (30 years at 3.5%) (Billion U.S. Dollars)

Option	NPV Impact, Unpriced Carbon $0/tCO$_2$e	NPV Impact, Alberta Tax Rate $85/tCO$_2$e $10/tCO$_2$e	NPV Impact, SCC $85/tCO$_2$e
BAU: vent to atmosphere	0	−1.3	−11.0
CCS of CO_2 in gas stream	−2.1	−0.8	+8.9

ADAPTATION RISKS

OVERVIEW

The second category of climate change risks to business and society as a whole is adaptation. The predicted impacts of climate change include widespread changes in rainfall and runoff patterns, melting glaciers and sea ice, changing ecosystems, sea-level rise, increase in the frequency and magnitude of severe weather events and impacts, species extinctions, and more frequent episodes of drought and flood, among others.[17] These changes in the Earth system will and are already starting to have profound effects on the global economy. Drought in Australia in 2006–2007 reduced overall grain production by 60% over the previous year's harvest,[18] and food prices are increasing worldwide.[19] Water scarcity is driving up the cost of water in many parts of the world.[20]

Increasingly, businesses will need to adapt to ensure security of supply of key commodities and resources, in terms of either quality or quantity, which may be affected by changing weather and climate.[21] Examples may include managing security of water supply, security of raw materials supply, labor, or even capital. Businesses with significant coastal or marine assets, including port facilities and offshore structures, will need to carefully consider the implications of climate change for design and operations. Industry also needs to consider the likely effects of a changing global climate on their existing and future operations. Investment decisions for long-term projects with expected life cycles of 20 years or more should consider how predicted changes in weather patterns, rainfall, wind and storm intensity, wave heights, rising sea levels, and warming air and sea temperatures might affect their designs and planning. Even access to insurance may be affected, with premiums rising for climate-exposed businesses. Table 5.3 summarizes some of the climate change adaptation risks that industry and business should consider and manage.

EXAMPLE

A new mining operation with an expected life of 100 years has limited choice of sites for the main crushing, processing, and stockpiling site because of extreme topography and the need to ship all product by sea. The main processing plant area is therefore planned to be built on a narrow strip of coastal plain with an average elevation of 0.2 m above sea level. Currently, there are no plans to consider climate change effects in the design of the facility.

However, best-available science now conservatively predicts that sea level is likely to rise up to 1 m before the end of this century.[22] On that basis, the risks to the planned $3.5 billion development are significant. As sea levels rise, the likelihood of storm surges inundating the site increases considerably. The consequences to business operations and safety from flooding, and the possible disruptions to production, are significant and could be catastrophic. The likelihood of such events occurring is high. Even if the likelihood of a rise of 0.5 m over the next 100 years is considered as moderate (50% likelihood), a typical risk analysis would suggest that mitigation is required (see discussion of risk assessment in Chapter 3).

One option is to raise the site elevation by 0.5 m by importing fill material, at a cost of over US$75 million. Other options include designing for future installation

TABLE 5.3

Climate Change Adaptation Risks for Industry

Risk	Likelihood	Consequence	Mitigation Measures
Rising sea levels, increased storm surges, increased wave heights, costal erosion damage	Increasing over next 10–20 years. Sea level already rising 3 mm/yr	Inundation of low-lying sites and structures, infrastructure damage, loss of production or capital, saline intrusion into coastal aquifers	*Protect or relocate existing facilities and population centers* Shift agricultural production from vulnerable coastal plains Select higher-elevation sites for new facilities Examine vulnerability of transport corridors Strengthen existing infrastructure for increased storm loads Adjust health and safety practices in coastal and offshore operations
Declining water security	High and increasing over 10–20 years Already a concern in many locations	Reduced availability and security of water; limitation on production and on expansion of existing facilities; limitations on new facilities; reduced growth; higher costs	*Develop long-term water management strategy* Retrofit existing facilities to improve water efficiency Design and build new facilities for water efficiency Secure access to water supplies less vulnerable to the effects of climate change Move production and population to areas where water supplies more plentiful and less vulnerable Examine potential for competition and conflict with other water users at key facilities
Declining raw materials and process input security	Rising over next few years	Greater difficulty in securing raw materials required for production	*Develop strategic plans to reduce dependence on raw material sources that are vulnerable to climate change (e.g., forest and agricultural products)* Secure access to supplies less affected by climate change
Rising global average temperatures	High and increasing over next 10–20 years, especially without meaningful progress on mitigation	Declining efficiency of process cooling systems, higher costs	*Take rising atmospheric and ocean temperatures into account in all process engineering designs* Examine cooling measures that will not depend on increased fossil fuel energy consumption

(continued)

TABLE 5.3 (CONTINUED)
Climate Change Adaptation Risks for Industry

Risk	Likelihood	Consequence	Mitigation Measures
More frequent and severe storms and weather anomalies	Moderate to high over next 10–20 years	Increased insurance costs, higher flood risks, disruption of operations, higher costs	*Adapt design and operational practice to account for higher frequency of weather events* Strengthen and upgrade infrastructure and flood defenses for key facilities Adapt heath and safety practices

Source: After Hardisty, P.E., Analysing the Role of Decision-Making Economics for Industry in the Climate Change Era. *Management of Environmental Quality*, 20(2), 205–218, 2009.

of raised protective sea walls and dewatering systems or choosing another location altogether. Note that this additional project expense, which exists because of global GHG emissions, is not directly related to the emissions of this facility but is rather a direct expression of the SCC.

A CAUTION TO DESIGNERS, ENGINEERS, AND MANAGERS

The previous simple example illustrates how climate change adaptation for industry is a key financial and economic concern that should be included in project decision making now. Planning and designing for adaptation now will reduce the risks of future impacts, reduce project life-cycle costs, and enhance business competitiveness. Furthermore, engineering designs completed today that do not take possible likely climate change into account could be subject to severe scrutiny in the future. Given our current state of knowledge and the best-available scientific information on the predicted effects of a changing climate, designers and managers who do not take this information into account could be exposed to accusations of professional negligence in the future.

EXAMPLE: GHG MANAGEMENT IN HEAVY OIL PRODUCTION

BACKGROUND

In situ extraction of heavy oil often involves the use of large quantities of steam to help mobilize viscous bitumen and drive it toward recovery wells. In some applications, this process is referred to as steam-assisted gravity drainage (SAGD). This is an energy-intensive operation, which typically can be justified financially only when oil prices are high. At many such operations in Canada, for instance, steam is generated using natural gas and sometimes power from the local electrical grid. If grid-supplied power is coal based and if gas is burned inefficiently, the carbon footprint of this type of oil production is large. This example examines options available for reducing the overall GHG emissions associated with extraction of heavy oil from a hypothetical SAGD facility (based on several real sites) while meeting the power and steam needs of the facility. The objective of the EESA in this example is the identification of the most economical and sustainable option for production of steam and power at the facility.

OPTIONS

The following options are considered:

- *Option 1A: Once-through steam generation (BAU)*: Use of once-through steam generation (OTSG) is common in the oil sands industry. It is a relatively unsophisticated and inefficient way to use natural gas to generate steam. Power requirements for the facility are met using electrical energy purchased from the local coal-fired grid. Source water is a combination of recycled process water and makeup water from local groundwater aquifers that provide a nonpotable water source. This option has the lowest capital cost of the options considered, is technically proven, and is widely used. However, the associated GHG emissions from coal-fired grid energy and combustion of gas are significant and represent a future liability to the operator.

- *Option 1B: OTSG with CCS*: Addition of a CCS system capturing the flue gas from the OTSGs combines the preferred technical option of postcombustion solvent-based technology for carbon capture with a CO_2 transport and sequestration scenario consisting of a merchant pipeline and storage partner. This option would reduce the carbon emitted by the facility by 90%. An ancillary benefit is a reduction in facility sulfur oxides (SOx) and nitrogen oxides (NOx) emissions. The additional energy required for CCS would be purchased from the grid. The postcombustion carbon capture process produces water as a by-product that is used to offset makeup water requirements. However, CCS implementation is expensive and energy intensive.

- *Option 2A: Complete cogeneration of facility power needs*: This involves installation of a cogeneration process consisting of a gas-fired turbine to generate power with hot flue gas fed into a heat recovery steam generation (HRSG) system to produce steam. This system uses natural gas more efficiently and would provide 100% of the power needs of the facility through the installation of a single Frame 5 gas turbine. The option, however, would not provide the necessary steam quantity; therefore, two additional OTSGs would still be installed to supplement steam production. This option would reduce the carbon footprint of the operation through substitution of coal-fired power (grid) by on-site gas-fired power and by using gas more efficiently. Natural gas produces significantly lower emissions than coal for an equivalent unit of energy produced. However, this option is more expensive than the BAU option and requires a larger facility footprint.

- *Option 2B: Complete cogeneration with CCS*: This option involves capturing the flue gas from the gas turbine and the remaining OTSGs using the same CCS design and operation as option 1B. No additional required energy for CCS would be required from the grid. This option would result in a significant reduction in carbon emissions, both from the substitution of gas for coal in energy production and from capturing CO_2 in flue gases on site. A reduction in SOx and NOx emissions would also be achieved. However, as with all of the CCS options (B series options discussed in the following), capital and operational costs are high.

- *Option 3A: Surplus power generation*: This option involves installation of a cogeneration process consisting of a gas-fired turbine to generate power with hot flue gas fed into an HRSG system to produce steam. This would provide more power than required for facility use but all of the steam necessary for the operation. A duct-fire process within the HRSG would be needed to ensure adequate steam quality. Surplus power would be exported and sold to the grid. This option completely removes the need to purchase external grid power and substantially reduces the carbon footprint of the operation through gas-for-coal substitution as discussed.

- *Option 3B: Surplus cogeneration power with CCS*: Option 3A is combined with a CCS system as described in option 2B.

- *Option 4A: Maximum power generation with full cogeneration*: Installation of cogeneration process consisting of a gas-fired turbine to generate power with hot flue gas fed into an HRSG system as discussed to produce steam. The option will generate all the steam necessary for the operation without the need for duct-fire, maximizing the power-to-steam ratio, producing significant surplus power that can be exported and sold to the grid. This is an expensive option; capital cost is about $260 million more than BAU and involves increased facility complexity.

- *Option 4B: Maximum cogeneration power with CCS*: Option 4A is combined with a CCS system as described in option 2B. This option reduces the CO_2, NOx, and SOx emissions of the facility by 90%. The additional energy required for the CCS scenario is compensated for by increased power generation. Surplus power is exported to the grid. The postcombustion carbon capture process produces water as a by-product that will be used to further offset makeup water requirements. This is the most expensive of all of the options but significantly reduces the risk to the operator of exposure to rising carbon taxes by significantly lowering the GHG footprint of the operation. The total CO_2 emissions produced by each option are presented in Figure 5.2. Since the cogeneration emits less carbon per kilowatt than the coal-fired grid power, any power exported to the grid would result in a net decrease in grid emissions.

COST ESTIMATE BASIS AND ASSUMPTIONS

The capital cost estimates for each option are shown in Table 5.4 and were developed for this conceptual high-level study on a ±40% basis. Costs were developed by scaling from historical cost estimates of similar facilities escalated to 2009 prices. Operating costs were based on estimates of consumable chemicals, power, and fuel gas from similar operations. Operation and maintenance costs were included based on an empirically derived percentage of capital expenditure (CAPEX) basis. Initial unit costs for key operating expenditure (OPEX) variables are provided in Table 5.5.

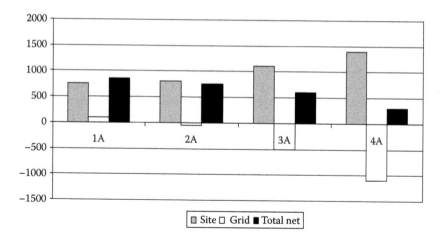

FIGURE 5.2   GHG emissions in kt/day (kiloton per day) for the four non-CCS options, showing the contributions from on-site gas combustion, the grid (positive if energy is purchased from the grid and negative if it is provided to the grid), and the net total emissions. Option 4A results in a 60% reduction in overall GHG emissions compared to BAU (option 1A).

TABLE 5.4

CAPEX Estimates for Each Option

Option	CAPEX (Millions of U.S. Dollars, 2009)
1A: Business as usual (buy electrical power from grid)	56
2A: Co-generation: on-site gas-fired power	104
3A: Co-generation with surplus power	158
4A: Complete co-generation with significant surplus power	260
1B – 1A + carbon capture and sequestration (CCS)	239
2B – 2A + CCS	305
3B – 3A + CCS	389
4B – 4A + CCS	532

TABLE 5.5

Base Case Values for Key Financial OPEX Inputs (2009 U.S. Dollars)

Parameter	Base Case Unit Value
Electricity from the coal fired grid: buy	$85/MWh
Sale price of electricity back to the grid (in the case of surplus)	$75/MWh
Fuel gas price	$0.25/sm^3 ($7.5/GJ)[a]
Operation and maintenance costs (general – nonpower)	3% of CAPEX annually

[a]   sm^3 = standard cubic meter

These values were also varied in the sensitivity analysis (discussed separately here) to examine their effect on the NPVs of each option.

BENEFITS ASSESSMENT AND VALUATION

The following external benefit categories have been considered in the analysis:

- The disbenefits (or external costs) associated with *GHG emissions*, based on an understanding of the current and likely direct cost of carbon and the SCC, discussed in Chapter 3. The commercial (internal financial) benefits of carbon capture can stem from reduction in payments of the current carbon tax in this jurisdiction, sales of emissions performance credits, and possible sales of carbon dioxide for use in enhanced oil recovery (EOR).
- The *total economic value of water (TEVW)*, which is discussed in Chapter 3. The majority of the water used in the process would be sourced from shallow aquifers close to the site. These aquifers are generally not used for potable or agricultural supply in the vicinity due to naturally elevated levels of total dissolved solids. Low quality and no competing current uses suggest a relatively low unit value for the water that would be used. The lowest commercial metered water rate in the area at the time of the study was on the order of US$0.10/m³. This is used as an estimate of the base case value of water. For sensitivity analysis purposes, the high value was set at the highest current commercial rate for water in the jurisdiction, on the order of US$1.40/m³ (a conservative overstatement of the value of this water). The combustion process creates water that would normally be emitted in the flue gas. Carbon capture from the flue gas recovers a portion of the water in the flue gas stream. This water can be fed into the SAGD water treatment facilities and can reduce or eliminate the need for makeup water from water wells for the SAGD facility. Thus, well water is left untouched or its use reduced.
- The *disbenefits associated with NOx and SOx emissions*, based on an understanding of the social costs of the emissions. NOx and SOx emissions result in external costs to society (disbenefits) such as respiratory illness and acid rain. Like markets for GHGs, markets for NOx and SOx emissions both limit the volume of these substances released and allocate the emissions in an economically efficient manner. The largest markets and auctions for these gases are in the United States. Chapter 3 provides ranges of market-based costs and social cost estimates for NOx and SOx.

Benefits Summary

Based on the information provided, the range of expected values for each of the major benefit categories is provided in Table 5.6. Each of the values in the table is based on a reference from Chapter 3. As can be seen, the unit values for benefits vary over a considerable range. Low estimates have been deliberately chosen to reflect the likely absolute minimum of the range and the high to bracket the likely uppermost value and to provide

TABLE 5.6

Monetized Unit Benefit Values (U.S. Dollars, 2009)

Benefit Category	Units	Low	Medium	High
CO_2	$/tCO$_2$e	0	15[a]	85
Water total economic value	$/m^3$	0.10[a]	1.10	1.40
NOx	$/t	750[a]	1,350	3,150
SOx	$/t	600[a]	1,200	2,200

[a] Unit benefit value used in base case.

an indication of the *likely future value trend*. The medium value is thought to represent the current best estimate of the unit value of the environmental benefit or cost.

Proportion of Benefits Realized by Each Protection Measure

Not all benefits are realized by each option, and in some cases benefits are only partially realized by a given option. Setting appropriate and consistent baselines for comparison is critical. For example, CCS provides 90% capture of CO_2 emissions from the flue gases; therefore, the remaining 10% is applied as GHG emission disbenefit. The baseline is set at zero GHG emission, therefore every tonne of CO_2e emitted is counted as an external cost (or disbenefit). Benefits (or disbenefits in some cases) are thus apportioned over time for each issue, based on the anticipated level of damage that will occur for the given option, based on the unit of valuation, and the physical quantity of the emission or production.

ECONOMIC SUSTAINABILITY ASSESSMENT RESULTS

Scope and Basis of the Analysis

This is a marginal economic analysis. It considers only the costs and benefits associated with the various options designed to manage power, steam, and carbon within the development and does not look at the wider processing facility development itself. The analysis does not consider the revenues generated by the project from the sale of oil or the wider economics of the project. The baseline for comparison and monetization is the current operating condition. If an external asset is damaged by implementation of a particular option, this damage appears as a disbenefit (negative benefit). If the value of the asset is maintained as it is (undamaged), then there is no effect, and no benefit or disbenefit is created. So, for example, if a water resource is left intact, in place, the current ecological support and option values of the water remain, and there is no benefit or disbenefit imputed into the analysis. If carbon dioxide, as another example, is emitted, a negative benefit (disbenefit) is imputed.

Base Case

For the base case, the following assumptions were made: (1) the benefit valuations described apply; (2) the planning horizon was assumed to be 20 years; and (3) the discount rate was 3.5% (which is the current U.K. Treasury rate for social discounting). The results of the base case assessment are presented in Table 5.7 and

TABLE 5.7

Base Case Results (Millions of 2009 U.S. NPV Dollars)

	Option 1A	Option 2A	Option 3A	Option 4A	Option 1B	Option 2B	Option 3B	Option 4B
Financial								
CAPEX	-56	-104	-158	-260	-239	-305	-389	-532
OPEX	-591	-590	-573	-551	-757	-778	-796	-833
Total financial NPV	-648	-695	-731	-811	-997	-1,084	-1,186	-1,367
Marginal Financial (Compared to BAU)								
CAPEX	—	-48	-102	-203	-182	-249	-333	-476
OPEX	—	0.8	18	40	-166	-187	-205	-243
Marginal financial NPV	—	-47	-83	-163	-349	-436	-538	-719
Externalities								
NOx	-2	-2	-4	-4	-2	-2	-4	-4
SOx	0	0	0	0	0	0	0	0
Water	-4	-4	-4	-4	-3	-3	-2	-2
CO_2	-195	-179	-140	-82	-72	-37	36	144
Total externalities	-200	-185	-147	-90	-76	-42	30	138
Total EcoNomics NPV	-1,940	-1,972	-1,937	-1,920	-2,473	-2,565	-2,627	-2,772
Marginal Economic (Compared to BAU)								
Marginal NPV (compared to BAU)	—	-31	3	21	-533	-624	-687	-832

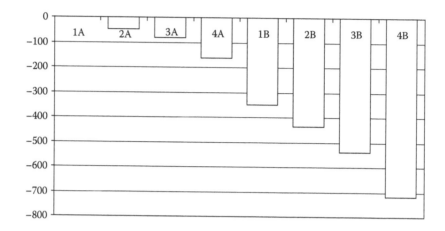

FIGURE 5.3 Base case 20-year financial NPVs compared to option 1A (BAU). All options are worse than BAU in a 20-year financial analysis, even at the low 3.5% discount rate used in the base case analysis.

in Figures 5.3 and 5.4. The results show the analysis using the central base case assumptions for values of all parameters, representing best estimates of currently reasonable and accepted values at a 3.5% discount rate. In the figures, the BAU case was selected as the baseline for comparison, so costs and benefits presented for each option are shown as they compare to BAU. In all of the figures presented, options are presented in order of increasing CAPEX, cheapest on the left (option 1A) and most expensive on the right (option 4B).

From a purely financial perspective, as expected, all of the options represent a net cost to the operator (Figure 5.3). BAU is financially superior because it is the

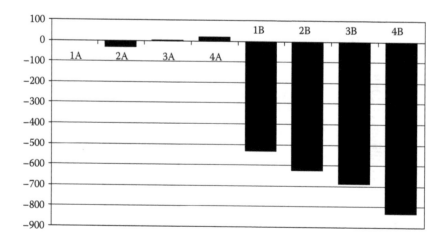

FIGURE 5.4 Base case 20-year full social NPVs compared to BAU. Options 3A and 4A emerge as just slightly more positive than BAU.

least-cost option under base case conditions and assumptions about future energy and carbon prices (low and constant). The analysis also reveals clearly that, under base case assumptions, CCS is not financially or economically viable. At the base case carbon value of US\$15/tCO_2e, none of the CCS options yields positive marginal financial or economic NPVs. However, given that the BAU option will produce over 0.9 mtCO_2e/year over the 20-year period and that option 4B, for example, results in elimination of a large fraction of these emissions and produces cleaner exportable power, resulting in a net decrease in emissions of about 1.5 mtCO_2e/year, carbon costs do not need to rise very far during the 20-year life of the project to make CCS more attractive from a social economic point of view. This is examined in more detail in the Sensitivity Analysis section.

As shown in Figure 5.4, when significant power generation is realized and exported, options have marginally positive economic NPVs compared to BAU, indicating that they are more economic choices for society as a whole. Since this analysis included all of the most important measures of sustainability for this project, options 3A and 4A can also be described as the most *sustainable* of the options examined; they generate the maximum benefit for all stakeholders over the 20-year life of the project. These options are more economic and sustainable than the BAU option in part because they both significantly reduce carbon emissions. Option 4A eliminates more than 0.5 mtCO_2e/year in GHG emissions compared to BAU. Figure 5.4 reveals that the relative economics of the options considered are dominated, unsurprisingly perhaps, by the OPEX (which notably includes costs of gas and grid power), CAPEX, and carbon costs. This is an important consideration when examining the sensitivity analysis results presented next.

SENSITIVITY ANALYSIS

Any analysis of this type is inherently subject to uncertainty. Cost estimates provided are ±30% accurate, and the valuation and estimation of benefits is subject to even larger changes, as discussed in this chapter and in Chapter 3. However, the key to the analysis is to reveal not absolutes in terms of dollars, but better and worse decisions overall compared to the range of possible decisions that could be made. This is where the EESA emerges from the monetary and economic focus and delves into decision making. From this perspective, sensitivity analysis is important because it allows the overall conclusions of the analysis to be tested across a wide range of parameter inputs. Table 5.8 shows the range of assumptions in financial parameters used in the sensitivity analysis.

Sensitivity to Energy Price Escalation

Figure 5.5 shows the sensitivity of the NPV (compared to BAU) of each option to variation in the value of all energy in the economy in terms of percentage per annum. The energy price consists of both fuel gas and power price combined. Due to the fact that the two factors actually work to drive NPVs in opposite directions, as shown in the previous sections, the combined energy price escalation analysis is critical. The analysis shows that as energy prices overall increase, the non-CCS options become increasingly advantageous while the CCS options become worse

TABLE 5.8

Ranges of Financial Assumptions Used in Sensitivity Analysis (2009 U.S. Dollars)

Parameter	Units	Low	Medium	High
Natural gas price	$/sm³	0.2	0.25[a]	0.4
Electrical energy purchased from grid	$/MWh	85	85[a]	185
Electrical energy sold to grid	$/MWh	75	75[a]	160
Discount rate	%	0	3.5[a]	14

[a] Unit benefit value used in base case.

in a marginal comparison. This is because CCS is a particularly energy-intensive process.

Considered separately, gas prices also affect the results. At low fuel gas prices, options 3A and 4A have a marginal advantage over the BAU case. As the fuel gas price increases, all options become undesirable compared to BAU since the cogeneration options all use more fuel gas than BAU. As the price of grid power increases from the base case assumption, the non-CCS options are all increasingly superior since these options replace coal-fired grid power with on-site gas-fired generation and sell surplus power. Under the high-end power price of US$185/MWh, option 4B yields a positive marginal NPV. Options 3A and 4A are superior to BAU over all values for power, and options 3B and 4B are inferior to BAU over all values for power. Under

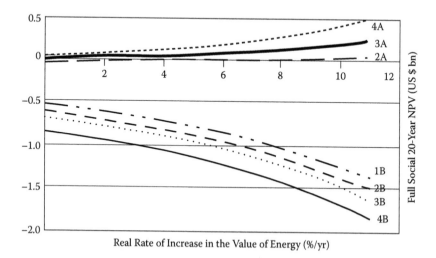

FIGURE 5.5 Sensitivity of NPV to annual rate of change in the value of energy (all types) with all other parameters fixed at base case values. NPVs are compared to BAU. Note that CCS and non-CCS options respond very differently as energy costs rise in real terms: CCS options fare increasingly worse, compared to BAU, because CCS requires significant amounts of energy to implement, and the non-CCS options provide increased energy efficiency.

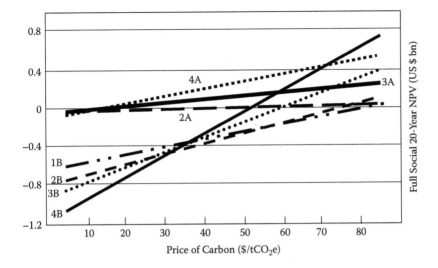

FIGURE 5.6 Sensitivity of NPV for each option to change in the value of CO_2e with all other values fixed at base case levels. (NPVs are compared to BAU in 2009 U.S. dollars.)

these conditions, CCS appears to be too expensive to provide a socially optimal and sustainable solution except when coupled with the most efficient of the options overall (4A) and only if the value of grid-fired power and carbon are high. CCS at the scale being considered at this facility would become economic and sustainable only in a future dominated by high power costs and high carbon costs (whether direct or social/external). Whether this is a likely scenario is left to the reader to decide.

Sensitivity to CO_2 Value

Figure 5.6 shows the sensitivity of the NPV of each option to variation in the value of the carbon dioxide (with all other values fixed at base case levels). The sensitivity of the options with respect to carbon value was one of the critical objectives of the study due to the potential risk of future liability associated with changes in legislation, the strong current worldwide impetus to significantly reduce GHG emissions, and the operator's desire to significantly reduce GHG emissions overall. The analysis shows an improvement in marginal NPV with an increase in the value of CO_2 emissions. For the non-CCS options, gas-fired power generation contributes fewer carbon emissions per unit of power produced than coal-fired grid power. Therefore, for the non-CCS options that export power, that replacement within the grid of a lower emission power unit provides a net carbon reduction to the option. The value of carbon is a dominant driver of the analysis. Option 4B, the most economic of the CCS options, becomes economic, and therefore sustainable (using our definition), with all other parameters set at base case values, if the value of CO_2 is greater than US$52/$tCO_2$e.

These results are particularly interesting in light of current worldwide carbon price trends. The current E.U. phase 2 ETS (Emissions Trading Scheme) average price is on the order of US$25/$tCO_2$e and has been higher in the recent past. At that value, using the economic analysis shown here (at the social discount rate), options 3A and 4A become even more strongly economic and sustainable, and even option 2A begins to approach a social break-even point. It is also worth noting that if the currently estimated SCC (or the true value of the damage of each additional tonne of GHG to the planet) of about US$85/tCO$_2$e were used, that all of the options examined would prove superior to BAU *even without any energy price changes.*

Sensitivity to NOx and SOx Emissions

The choice of option is insensitive to NOx values. The analysis shows a minor decrease in marginal NPVs with an increase in the social cost of NOx emissions, although under the values used, there is no threshold by which any option becomes either more advantageous or disadvantageous. The implication is that NOx emissions are not a significant decision driver for this particular project. SOx behaves in an almost identical fashion. The quantities produced, even at the highest unit value used in the assessment, are not sufficient to alter the relative economic and sustainability performance of the options considered.

Sensitivity to Water TEV

The base case for water TEV was evaluated at the relatively low value of US$0.10/m^3 due to the fact that all makeup water is supplied by nonpotable groundwater aquifers. The analysis shows an improvement in marginal NPV for the CCS cases with an increase in the TEV of water because CCS produces water as a by-product. However, none of the CCS options becomes superior to the BAU case under any water values. Even at the highest value of water used in the sensitivity analysis, the relative ranking of options does not change; the TEV of water is not a significant decision driver in this example.

Sensitivity to Discount Rate

Figure 5.7 shows the sensitivity of the NPV of each option to variation in the applied discount rate using the base case assumptions for all other parameters. The analysis shows an interesting change in marginal NPV with an increase in the discount rate. The non-CCS options become less economic, all other factors remaining equal, while the CCS options show an improvement (although not enough to surpass the marginal threshold). Discount rate variation provides some insight on the relative influence of OPEX and CAPEX on overall decision making. The higher the discount rate, the less future flows of costs and benefits impact the decision today. Higher discount rates effectively devalue the future.

Note how the NPVs of options converge as discount rate increases (Figure 5.7), reflecting that future flows of cost and benefit of all kinds become less important at higher discount rates, and CAPEX emerges as the dominant factor. These results reinforce the original base case findings that BAU is the most financially advantageous option when current prices are considered, where carbon costs to the operator are low and will remain so, and where discount rates (IRR hurdle rates) and payback expectations are high. However, as has been shown, even a modest increase over time in real energy prices and carbon values will shift this view in favor of lower carbon emissions options.

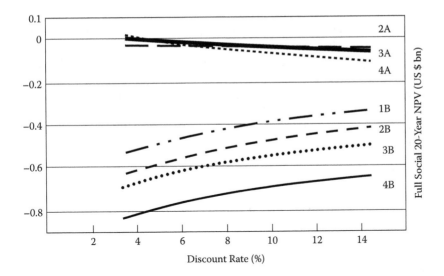

FIGURE 5.7   Sensitivity of NPV for each option to change in the applied discount rate with all other values fixed at base case levels. (NPVs are compared to BAU in 2009 U.S. dollars.)

OPTION SELECTION

Using the method described in Chapter 3, a complete equiprobable database can be compiled of NPV outcomes for each option using the full range of values for each parameter being investigated. This way of presenting the results was discussed in Chapter 3 and in examples in Chapter 4, and this example can be examined in exactly the same way. What this reveals is that option 4A is the most environmentally, socially, and economically sustainable option under about 90% of all the conditions examined (Figure 5.8).

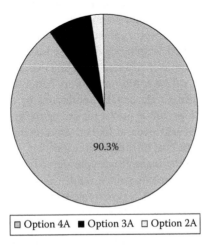

FIGURE 5.8   Makeup of the most economic and sustainable option under the full range of parameter values considered.

Option 2B is the best choice only under a narrow range of conditions (7.6% of conditions examined). Option 4B (4A + CCS) is the next-best choice under only a narrow range of conditions (16% of the conditions examined), basically when carbon costs are high and energy prices low. What is more, option 4A is robustly the most NPV positive of all the options over the widest range of conditions. In fact, under more than half of all possible futures examined, it provides a total NPV of at least US$1 billion more than BAU. There is also significant upside potential for option 4A. As carbon prices rise, and as energy prices rise, option 4A becomes quite rapidly more NPV positive.

This reveals the main outcome of the environmental and economic sustainability assessment—not a pronouncement that a particular option provides a specific number of dollars in net value (the future is so inherently uncertain that this is patently not possible)—but rather a decision-focused outcome that effectively leaves money behind. It identifies that one option is robustly superior to the others examined over the widest reasonable range of values considered for the key determining parameters. Option 4A is superior for a wide range of reasons despite its higher capital cost. The identification of option 4A, in this case, is the output of the EESA; it is identified to decision makers as the best option when all stakeholders' views are considered in a fair, objective, and quantitative sense. The EESA also provides decision makers with a simple and clear way of communicating that decision to stakeholders.

IMPLICATIONS

This example examined a typical power generation option assessment for a supposed oil heavy development. Under a conventional financial analysis, using a 15% discount rate, including current prices for energy and carbon, and assuming that carbon and energy costs will remain essentially stable in real terms over the next 20 years, the BAU power and steam option (option 1A) is most financially advantageous. Under this scenario, total CO_2 emissions are estimated at about 0.9 $mtCO_2e$/year or 18 mt over the 20-year planning horizon used in this analysis.

But, as has been discussed at length in previous chapters, sustainability by definition requires a longer-term perspective, and thus on the margin, a lower, more socially geared discount rate is applied. Employing the base case values for externalities, BAU is no longer the most economic or sustainable option over the 20 years. In fact, using the social cost for carbon, *all other alternatives* are better than BAU. Option 4A performs best, largely because it reduces GHG emissions by over 530,000 tCO_2e/year and produces additional low-carbon power for use on the grid, despite a high capital cost. Option 4A performs best even at higher, more commercial discount rates if reasonable estimates for the cost of carbon are used.

However, at the scale of the example site considered here, the CCS options appear to be too expensive for the benefits gained (under base case assumptions) to make economic sense, partly because CCS is extremely energy intensive in its own right, and as energy costs rise over time, CCS becomes less and less attractive at this scale. In this example, driving toward increasing GHG abatement results in diminishing returns for the company and society. Only if carbon is priced at or near the SCC, now or in the future, does CCS start to make sense.

SUMMARY

Managing GHG emissions and understanding the economics of achieving sustainability objectives will become increasingly important for industry and business as the world moves to tackle climate change. Many companies are already establishing their own internal emission reduction targets and are planning for a carbon-constrained and carbon-impacted future. Significant emission reductions can be achieved now at negative or low cost, in many cases actually reducing overall costs to operators and improving profitability.

But, unlocking these opportunities and moving away from BAU requires that we examine the overall benefits that emissions reduction can achieve, not only in terms of carbon itself but also for all of the associated issues that may be affected. How industry responds to these challenges will be an important factor in its future success. In both mitigation and adaptation, there is evidence that the risks of inaction far outweigh the costs of well-considered, economically viable action using all of the tools, expertise, and market mechanisms currently available to industry. Companies that wait to take action run increasing risks of higher costs, disrupted operations, and mounting stakeholder scrutiny. Climate change carries with it a clear procrastination penalty for industry and the planet.

NOTES

1. Intergovernmental Panel on Climate Change (IPCC). 2007. *Fourth Assessment Report. The Physical Science Basis.* Cambridge University Press, Cambridge, UK.
2. Esty, D.C. 2007. Transparency: What Stakeholders Demand. *Harvard Business Review*, October: 5–7.
3. Porter, M.E., and F.L. Reinhardt. 2007. Grist: A Strategic Approach to Climate. *Harvard Business Review*, October: 1–3.
4. Citibank. 2007. *Investment Implications of a Change in Climate.* Thematic investing global research report. Citigroup Global Markets, New York.
5. World Business Council on Sustainable Development (WBCSD). 2004. *Energy and Climate Change: Facts and Trends to 2050.* WBCSD, Geneva, Switzerland.
6. Organization for Economic Cooperation and Development/International Energy Agency (OECD/IEA). 2008. *World Energy Outlook, 2008.* OECD/IEA, Paris, France.
7. Department for Environment, Food, and Rural Affairs (DEFRA). 2003. *The Scientific Case for Setting a Long Term Emissions Reduction Target.* United Kingdom DEFRA, London. http://www.defra.gov.uk/environment/climatechange/pubs.
8. Stern, N. 2006. *The Economics of Climate Change—The Stern Review.* Cambridge University Press, Cambridge, UK.
9. Ibid.
10. Hardisty, P.E. 2007. A Climate Change Risk-Assessment for Business. *Middle East Economic Survey*, 41(12), 30–32.
11. Hardisty, P.E. 2007. The economics of climate change management in the petroleum industry. *Middle East Economic Survey*, 50(33), 31–33.
12. Hoffman, N., and J. Twining. 2009. Profiting from the Low Carbon Economy. *McKinsey on Corporate and Investment Banking*.
13. Ibid.
14. Hardisty, P.E. 2009. Analysing the Role of Decision-Making Economics for Industry in the Climate Change Era. *Management of Environmental Quality*, 20(2), 205–218.

15. Pearce, D., and D. Warford. 2001. *World Without End*. World Bank, Washington, DC.
16. Hardisty, P.E. 2009. Analysing the Role of Decision-Making Economics for Industry in the Climate Change Era. *Management of Environmental Quality*, 20(2), 205–218.
17. Intergovernmental Panel on Climate Change (IPCC). 2007. *Fourth Assessment Report. Impacts, Adaptation and Vulnerability*. Cambridge University Press, Cambridge, UK.
18. Australian Agricultural Stats ABARE. 2006. *Australian Crop and Livestock Report, Drought Update*. Australian Bureau of Agriculture and Resource Economics, Canberra.
19. World Food Programme. 2008. *WFP and Global Food Price Rises*. United Nations, Rome.
20. Pacific Institute. 2009. *Water Scarcity and Climate Change: Growing Business Risks for Business and Investors*. Ceres report, February, Ceres, Boston, MA.
21. Ibid.
22. IPCC, *Fourth Assessment Report. Impacts, Adaptation and Vulnerability*.

6 Energy

INTRODUCTION

CREATING A SUSTAINABLE FUTURE

The population of the world is growing rapidly, particularly in the less-developed world, where hundreds of millions of people still live without access to basic amenities that those in developed countries take entirely for granted: access to clean water, a basic level of sanitation, electricity in their homes. Perhaps more than any other factor, it is the access to safe, reliable, and affordable energy that allows real improvement in standards of living. Even basic electrification brings a multitude of self-reinforcing benefits: improved security and health, access to refrigeration for food storage, pumping of water to replace hand carrying, heating, and the potential to access electronic communications and information technology. In effect, access to electricity is one of the first major steps on the road to escaping poverty.

Meeting the legitimate aspirations of billions of people for a better life, free from poverty, disease, and the eternal grind of manual labor, depends in significant part on our ability to provide affordable energy. But, these hopes are starting to collide head-on with the costs of providing electrical power: Conventional least-cost power generation using coal and other fossil fuels carries with it a tremendous burden—the not-so-well-hidden external costs of air pollution and climate change. Added to this are the sometimes staggering social and environmental damage created during the exploitation of these fossil fuels—coal mining deaths in China, for instance, and the massive environmental destruction caused in the United States, where the tops of entire mountains are removed to expose coal seams for opencast mining.[1] Our current global energy system is now described by many as being in crisis, patently unsustainable.[2]

Among the most pressing challenges of the twenty-first century is to provide affordable electrical power to all without causing irreparable damage to the planet. This chapter provides examples of how the environmental and economic sustainability assessment (EESA) process can reveal optimum energy solutions and help to put the life-cycle costs and benefits of electrical power production into context. In particular, the real overall environmental, social, and economic benefits of various forms of renewable energy are examined in the context of our current fossil-fuel-powered world.

SUSTAINABILITY AND ENERGY

To become more sustainable and give ourselves a reasonable chance of avoiding the worst effects of climate change, by 2050 we need to decarbonize the economy of the world by as much as 60% to 80%.[3] The scale of this change is daunting and represents one of the biggest and most important challenges humankind has yet faced. Reaching this goal will require a wide range of technological developments, policy changes, and behavioral shifts on a massive scale. Appropriate price signals will have to be put into place to progressively drive up the cost of carbon and thus push individuals and businesses to seek lower carbon power generation and mobility solutions. Regulatory instruments, including mandatory renewable energy targets (MRETs), can also play an important role in achieving this extremely challenging goal. Carbon-intensive operations will need to make profound changes to avoid large cost increases and subsequent effects on profitability, competitiveness, and organizational sustainability.

Emissions trading systems, such as the one operating now in Europe (EU Emissions Trading Scheme [ETS]) and the scheme planned for Australia,[4] are intended to provide a price signal on carbon, accelerating the development and introduction of new technologies and operational techniques within industry, and providing an impetus to move away from carbon-intensive practices. Over the medium term, it is likely that this will also improve the competitiveness and profitability of those industries by driving them to become significantly more efficient, particularly with energy.[5] Such changes will not only reduce carbon emissions and the costs that industries have to bear under an ETS but also will promote longer-term energy security for those businesses, insulate them further against conventional energy price escalations over time, and improve their overall operational efficiency in other areas, such as water and air quality management, and even the protection of biodiversity.

However, it is clear that these changes will require significant innovation, not only in terms of technology development and deployment but also in how energy and resource projects are conceived, evaluated, and designed. In the short term, costs will rise. However, the transition to a lower-carbon economy carries with it significant procrastination penalties; the longer we take to make the required changes, the more expensive the overall changes will become. In fact, the latest studies completed by the International Energy Agency and the Organization for Economic Cooperation and Development (OECD) indicated that the savings in fuel costs alone that will arise from achieving significant reductions in greenhouse gas (GHG) emissions will exceed the costs of the changes themselves.[6] Long-term investment decisions in these sectors need to look beyond short- or medium-term carbon costs under an ETS and realize that the social cost of carbon (SCC) will become the eventual benchmark for the cost of GHG emissions.[7]

AN ENERGY MIX FOR THE FUTURE

Providing reliable and reasonably priced power, while reaching necessary GHG reduction targets, is a massive challenge. The energy mix for electrical power generation in many countries, including the United States, Australia, and China, for instance, is currently dominated by coal, which is among the most carbon-intensive ways of producing energy.[8] Over the past several decades, various renewable energy technologies have emerged as legitimate alternatives to fossil-fuel-based power, despite chronic underfunding (between 2002 and 2008, the U.S. government alone spent over US$72 billion on subsidies to the fossil fuel industries and only US$12 billion supporting the development of wind, solar, and geothermal renewable energy).[9]

FIGURE 6.1 Wind power is rapidly moving toward cost parity with conventional fossil fuel electricity generation in many parts of the world. This is despite the fact that global subsidies to the fossil fuel industry continue to dwarf those provided to promote the introduction of renewables.

In many parts of the world, renewable energy costs (measured in the conventional sense, without considering externalities) are moving ever closer to parity with conventional energy sources (Figure 6.1). But, there remains a strongly held conventional view that renewable energy requires subsidy; therefore, it is often associated with a socialist political perspective—it can only survive with government support. However, if every energy producer were made to pay the *real value* of the damage they inflict on society and the environment (through emissions to atmosphere, water used throughout the life cycle of production, and through the ecological damage created during the exploration and extraction of the fuel), and if the lopsided government subsidies paid in many countries to *support the burning of fossil fuels* were to be withdrawn, we would find that most forms of renewable energy are actually less costly overall, right now, than coal or oil.

On a level playing field, with everyone fully responsible for all of the implications of their actions, many renewable energy alternatives are economic and sustainable now and require no subsidy to compete. Currently, it is fossil fuels that are most subsidized. By accepting the damages caused by fossil fuels, without compensation, we are effectively providing a massive subsidy, which renewable do not receive (or need).

EXAMPLE: THE EXTERNAL COSTS OF POWER PRODUCTION

A simple example of the importance of examining the external costs in decision making comes in the electrical power generation sector. It is a commonly held fact that generating electricity from coal is cheap, typically cheaper than using most

Cheap

Expensive

FIGURE 6.2 Conventional wisdom (a term coined by the economist John Kenneth Galbraith) has it that coal-fired electrical power generation is cheap, while renewable energy in most forms is comparatively expensive.

other fuels and far cheaper than using renewable energy (Figure 6.2). That is why coal remains one of the dominant fuels for power generation on the planet and continues to grow in importance. China, for instance, has plans to add another terawatt (TW) of coal-fired power generation capacity over the next 20 years.[10] That is the equivalent of a new 500-MW plant every 5 days, much of it using technology that is less than optimally efficient.

Data on the global costs of power production seem to bear out the view that coal is cheap. However, data on the external costs of air emissions from various types of power production in Europe show that in fact coal has the highest average external cost; in comparison, renewable energy is relatively benign. Table 6.1 provides a summary of internal and external costs for four types of power production and sums them to reveal the total true cost to society. On average, coal is one of the most expensive ways to produce power, mainly because the social and environmental damage resulting from emissions of nitrogen oxides (NOx), sulfur oxides (SOx), carbon dioxide (CO_2), and fine particulates are considerably larger than for other forms of power generation.

Other, sometimes considerable, life-cycle external costs of power generation are not included in the data. For coal, these include the damage caused to the environment from coal mining (Figure 6.3), the often-significant methane emissions from mining activities, and in the case of China (in particular, but not exclusively), the significant loss of life in mine-related accidents.[11] For nuclear power, the considerable

TABLE 6.1

Estimates of the Total Costs of Electrical Power Generation (U.S. Dollars per kilowatt hour, 2005)

Power Generation Method	Average Tariff Cost[12]	External Cost[13] Low Estimate	External Cost[14] High Estimate	Total Cost to Society (Average Values)	Other Life-Cycle External Costs
Coal	0.045	0.02	0.23	0.17	Methane emissions from coal mines, environmental damage from strip mining, social costs (death of miners)
Natural gas combined cycle	0.03	0.01	0.04	0.055	Methane leakage from gas distribution systems, estimated at over 85 Mm^3/yr, with Russia, the United States, and the Ukraine the largest contributors
Nuclear	0.035	0.002	0.006	0.039	GHG emissions throughout the uranium mining life cycle, environmental and water resource impacts of uranium mining, long-term costs of managing nuclear waste, risks of nuclear weapons proliferation
Wind	0.10	0.001	0.002	0.01	Life-cycle GHG footprint of turbine production, noise, and visual disamenity
Solar PV	0.36[a]	0.005[15]	0.01	0.37	Life-cycle GHG footprint of solar panel production, visual disamenity, physical footprint

[a] Costs of PV power have dropped substantially due to breakthroughs in technology, providing increased efficiency. The value for PV cost shown here is from a 2008 report by McKinsey, who also suggest solar PV power costs could reach $0.05/KWh by 2030.

possible external costs associated with damage from long-term impacts from radioactive waste are also not included.

Nevertheless, this simple comparison reveals that natural gas, for instance, is actually a much more economically sustainable choice than most other forms of power generation in the near term.* Some forms of renewable energy, such as wind, geothermal, and concentrating thermal solar, so consistently pilloried for being too expensive, are all actually far cheaper than coal in real terms under most circumstances.

---

* This statement assumes that gas is transported from field to power station in the gas phase. Liquefied natural gas (LNG) can be transported over great distances but requires large amounts of energy for the compression and liquefaction of the gas, significantly increasing its effective GHG footprint per unit of energy produced.

FIGURE 6.3 The external environmental costs of power production also include the damage associated with mining coal, extracting oil, and producing the metals and component materials that go into the manufacture of wind turbines and solar cells.

EXAMPLE: COMMERCIAL-SCALE SOLAR THERMAL POWER IN AUSTRALIA

INTRODUCTION

Australia is particularly blessed with a range of plentiful sources of renewable energy. Studies of concentrating solar power (CSP) thermal technology have shown that Australia has a significant and exploitable solar resource that can be used to provide utility-scale power at a price competitive with fossil fuel power sources when even a relatively low cost of carbon is included.[16] CSP technology harnesses the thermal energy of the sun to produce steam, which is then used to drive conventional power generation turbines. It is estimated that by 2020 CSP technology alone could provide 40% of Australia's renewable energy needs.[17] As discussed, if the market provided a truly level playing field, with all industries paying the true value of the external damage they caused, no subsidies would be required for many renewable energy technologies. The best technologies would be adopted by the market.

This example considers the feasibility of a commercial utility-scale solar thermal power plant in Australia.[18] Other renewable energy sources that could be developed successfully in Australia include wind power (already being rapidly deployed in many other parts of the world) and geothermal energy. MRET legislation, enacted in 2009 in Australia, will help considerably in bringing renewable energy meaningfully into the mix in Australia. Setting strong carbon price signals will also encourage new renewable energy deployment.

CSP TECHNOLOGY OVERVIEW

The development of CSP technology has been dormant since the early 1990s but has recently undergone a renaissance in countries with good solar resources. Within the renewable energy sector, wind power has to date occupied the leading position in terms of both cost and overall deployment. However, large-scale continuing deployment of wind generation is hampered by factors such as the scarcity of development sites, delays in obtaining development approvals, and the inherent problems of constancy: The wind does not always blow, and accurate predictions of the reliability of wind resources are difficult to make.

CSP development began in earnest in the 1980s. Between 1985 and 1990, several plants were built in various parts of the world, but since 1991 the new Nevada Solar One (64-MW) and Abengoa Solar's central receiver PS10 (11-MW) projects built in 2007 were the first CSP projects commissioned.[19] New projects are now being considered or are under development in the United States, Spain, and China. Spain has emerged as a world leader in renewable energy development, driven in part by aggressive government incentives and a stable regulatory environment. CSP incentives of €0.26/kWh (US$0.38/kWh) for 25 years for plants up to 50 MW (with a 500-MW total limit) are currently being offered. As a result, developers are now planning over 2,900 MW of CSP projects in Spain. In the United States, various tax credit schemes and state-administered renewable portfolio standards (RPSs) are also igniting interest and development in CSP. California, for example, has set RPS targets of 20% by 2015 and 33% by 2020. Globally, CSP capacity is expected to increase from about 350 MW in 2008 to as much as 26,000 MW by 2020 (a 75-fold increase).[20]

A number of CSP technologies have been developed, each with its own merits and weaknesses. Some of these are listed in Table 6.2. Among the most widely used, simple, and inexpensive is parabolic trough (PT) technology. Figure 6.4 shows the PT arrangement at a power plant in California. Heat reflected from the curved mirrors is focused on an insulated tube system that runs the length of the rows of mirrors, through which flows thermal oil. The oil carries the thermal energy to a series of heat exchangers (Figure 6.5). The heat is used to make steam, which drives turbines.

PT systems have been operating for over 17 years in different parts of the world, including in Israel and California. This experience has meant that operating costs have dropped considerably, and efficiency has risen as operators have gained knowledge on how best to operate, maintain, and upgrade systems. For these reasons, many consider that PT technology currently enjoys a lead over other technologies and thus carries less risk for investors.[21] In addition, it is expected that the use of thermal storage using molten salt will stretch the life of the PT technology. It is likely to be several years before other CSP technologies reach the general level of acceptance now enjoyed by PT systems.[22]

FACILITY DESCRIPTION AND COSTING

The feasibility of developing a commercial utility-scale thermal solar power plant in Western Australia was examined.[23] The study assumed that a 250-MW peak conventional PT facility utilizing a single 250-MW turbine/generator unit would be

TABLE 6.2
CSP Technology Comparison

Technology	Basic Description	Efficiency	Demonstrated Capability	Advantages	Weaknesses
Parabolic trough	Parabolic mirrors concentrate thermal energy of the sun to generate steam.	Practical experience has greatly improved overall efficiency over the last 15 years. Demonstrated performance.	Several major facilities in operation worldwide. Plants in United States and Israel have been operating for over 15 years without major problems.	Low risk; most highly developed and tested CSP technology; proven at utility scale.	Requires large areas of flat ground.
Stirling dish engine	Individual parabolic dishes up to 10 m diameter, up to 25 kW per dish.	Theoretically can achieve the highest efficiency of CSP systems (up to 29% at 750C).	Has not developed beyond demonstration scale.	Possibility of significantly reduced capital costs and good scalability.	Technical and manufacturing challenges.
Linear Fresnel	Variant on parabolic trough but uses flat mirror.	Lower efficiency than PT.	Pilot plant deployment in Spain and Australia.	Cheap to manufacture; no requirement for complex tracking systems.	Lower output and lower efficiency than PT.
Central receivers	Mirrors concentrate the energy of the sun on a central tower.	Higher temperatures may provide for improved efficiency over PT but remains to be proven at scale.	Small plants have been commissioned in Spain, United States, and Israel.	High-temperature output and increased storage capacity; does not require flat land.	Yet to be proven at commercial utility scale; investment risk remains.

constructed. This was based on initial studies that revealed the significant impact of facility scale on energy cost. Economies of scale are evident: A 10-MW CSP system would provide power at over twice the cost of a 250-MW CSP system. The facility was conceptualized with a 1,000-MWht (megawatts of heat) molten salt thermal energy storage (TES) system, which considerably lengthens the daily operational time of the facility (Figure 6.6). The solar field design consisted of single-axis-tracking PT solar collector assemblies, which track the sun as it moves during the day. The facility would be located in a part of Australia that enjoys good insolation and where access to the network is available. Figure 6.7 shows the considerable seasonal variation in overall output resulting from changes in insolation from winter to summer.

FIGURE 6.4 A thermal solar parabolic trough power plant.

The capital cost for the facility, which includes the solar field, land purchase, all power systems, and the molten salt TES system, was estimated as US$1.1 billion.[24] The facility is expected to produce 574 MWh of power annually over a 40-year operating life.

FINANCIAL ANALYSIS

A conventional financial analysis of the project, based on expected power tariffs in Australia (whose grid is currently dominated by coal-fired power), revealed that the thermal solar project would currently require subsidy to compete within the major electricity markets of the country. The financial analysis for the notional CSP facility can be examined in three distinct parts: (1) modeling of the capital expenditure (CAPEX) and operating expenditure (OPEX) over the life of the plant to determine the cost of electricity (the "production cost"); (2) estimation of total revenue (including

FIGURE 6.5 Oil–water heat exchangers at a solar thermal power station.

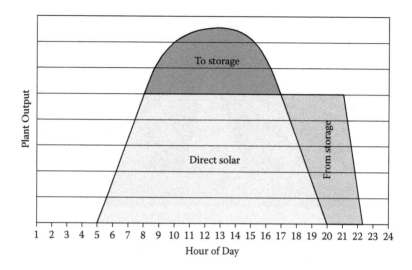

FIGURE 6.6 Expected daily power output for rated 250-MW solar thermal power plant in Australia. Note the contribution of molten salt storage, which allows peak daytime energy to be delivered later in the day, after the sun has set. (Courtesy WorleyParsons, *Advanced Solar Thermal Implementation Plan, Final Report—Vol. 1,* 2008, WorleyParsons, Sydney, Australia.)

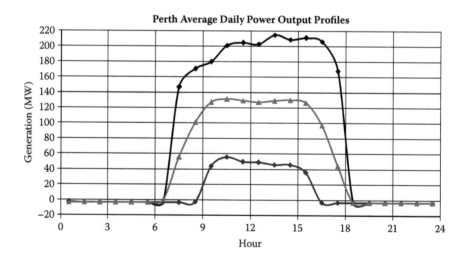

FIGURE 6.7 Expected average daily power output profiles for a notional 250-MW power plant located in Perth, Western Australia (courtesy WorleyParsons, *Advanced Solar Thermal Implementation Plan, Final Report—Vol. 1,* 2008, WorleyParsons, Sydney Australia). The top curve is the summer profile, the lower curve is for winter.

capacity payments and renewable energy credits [RECs]) generated on the current local electricity market in 2008; and (3) projecting revenue forward 20 years based on historical data (to be examined in more detail in the sensitivity analysis).

The break-even cost per megawatt hour, estimated assuming a debt-financed project (5.07% for 20 years), a 40-year planning horizon, straight-line depreciation over 12 years, an effective tax rate of 22.5%, a 80–20 debt-to-equity ratio, and a 10% real rate of return, was US$165/MWh.[25] Average revenues over the same period were expected to be in the range of US$135/MWh, a US$30/MWh initial shortfall.

These estimates were based on a detailed analysis of temporal variations in power production across daily, yearly, and life-cycle scales. On this basis, a CSP plant at this location is not financially viable. Given conventional return-on-investment expectations and a conventional view of how power costs are expected to rise in future, the facility would lose money every year of its operation. Interestingly, the same analysis, conducted for a plant that starts operation in 2020 (assuming that current industry expectations for a 30% decrease in capital costs are realized), reveals that CSP would reach financial cost parity in the market. Nevertheless, the clear conclusion is that a CSP project of this scale would require significant subsidy to be workable, from a conventional financial perspective, with no carbon cost in the economy (the current situation).

However, remote off-grid energy users in the region, including many large industrial facilities, currently pay over $170/MWh for their power and are exposed to potentially significant fuel price rises over the coming decades. For these remote users who require power at scale, CSP is in fact cost competitive today. Its implementation for remote users is hindered at present, however, by the large capital outlays required to launch a project at this scale.

As the Australian economy grows, additional power generation capacity will have to be brought online. All new-build power stations (renewable or conventional) will face the same challenges in terms of rising costs of construction and materials, land and water availability, and permitting. The financial analysis compared new thermal solar to existing power generation through the current tariff structure. However, comparing new CSP to the existing stock of power generation assets, much of it older than 10 years and already significantly depreciated and paid for, is not a valid like-for-like comparison. CSP should be compared to other new-build power options going forward. This will have the effect of lessening the financial gap between CSP and conventional generators.

ENVIRONMENTAL AND ECONOMIC SUSTAINABILITY ASSESSMENT

Using the basic methodology outlined in Chapter 3, the overall environmental, social, and economic sustainability of CSP was examined and compared to conventional coal-based electrical energy production in Australia. For this simple exercise, only the direct external costs associated with emissions to atmosphere were included. Table 6.3 shows the estimated impact on the dollars per megawatt hour tariff associated with various assumptions for the external costs of NOx, SOx, and CO_2. For midrange assumptions of the values of these externalities, the average power tariff for coal-fired electricity would increase by almost $40/MWh.

TABLE 6.3

Impact on Power Tariff from Air Emissions for Conventional Coal-Fired Power

Externality	Midrange External Costs (US$/t)	Midrange Tariff Effect (US$/MWh)	High External Costs (US$/t)	High Tariff Effect (US$/MWh)
CO_2	18.75 (with 2%/yr escalation)	18.25	63.75 (with 2%/yr escalation)	62.00
NOx	870	1.95	1,562	3.50
SOx	1,012	0.90	2,232	2.00
Sum	—	39.35	—	67.50

On this basis, CSP becomes significantly more attractive. From a social economic perspective, eliminating transfer payments from the equation and using a social discount rate of 3.5% as the base case for assessment, the unit cost of coal-fired power generation at the proposed location would be on the order of US$35/MWh. Using the CSP facility design discussed, the cost of power from solar thermal would be about US$120/MWh, a gap of about US$85/MWh. However, adding midrange air emission externalities as discussed, the gap closes to about US$35/MWh, and at the high estimates (with carbon close to the SCC), to about US$17/MWh. Solar thermal also gets closer to parity with coal if future fuel cost escalation is considered (CSP enjoys free fuel for the life of operation, while coal prices are likely to rise over the next few decades).

The argument here is clear: Who is subsidizing who? In the conventional financial analysis, solar needs subsidy to compete with coal on the grid (but not off-grid). But, by not paying for the real value of the damage done by its emissions, coal-fired power enjoys an apparent cost advantage over solar thermal power. The higher the value of the damage, the closer CSP comes to parity with coal. In this example, society is subsidizing coal-fired power by accepting (willingly or unwillingly) the damage associated with the air emissions without compensation. This in turn makes solar power appear as if it needs subsidy to be economic, when in effect it needs (almost) no subsidy of its own to be competitive, simply the removal of the subsidy enjoyed by conventional fossil fuel power.

A true level playing field, which would allow different forms of energy production to compete in a free and open market, would require simply that all forms of energy pay for the true costs of their production. Consumers would be presented with a clear choice based on the actual market price of the power. This would lead to optimal choices that balanced environmental, social, and economic considerations and a truly sustainable energy system.

Of course, there is far more complexity to this issue than has been represented here. Deployment of renewable energy on a large scale requires major and complex changes to the entire energy infrastructure, which can only be accomplished through integrated planning and investment. In addition, a wider range of external costs and benefits needs to be captured, across the entire energy generation life cycle, to provide a complete picture. Solar thermal, for example, also has other distinct overall advantages compared to traditional fossil-fuel-based power generation that should be considered, including energy security, GHG emission mitigation, and other air quality improvements (reduction in emissions of particulates to the atmosphere).

Energy Security

Once constructed and operational, a CSP facility delivers dependable power over the long term independent of an external fuel supply. The energy of the sun is free, plentiful, and dependable. This provides a high degree of security to CSP power supply, which will not be subject to the vagaries of market changes in fuel prices and availability or disruptions in supply. CSP also provides a tangible fuel price risk hedging strategy. Since ongoing fuel costs are essentially zero, CSP is insulated from any perturbations in traditional fuel prices that could substantially affect the profitability of other forms of power generation. In fact, if fossil fuel costs rise at 10% per annum or more, in real terms, CSP becomes, over a 20-year planning horizon, the *financially* superior power generation choice for grid or remote applications. Given the fuel price changes over the past decade, this level of fuel cost rise would seem a distinct possibility. As part of a wider power generation portfolio, CSP could play an important fuel price risk management role, blending down the overall impact of fuel price impacts on the rest of the portfolio. CSP is not only immune to the fuel price rises, but also, as fuel prices rise, so do the costs of power generation by other means, thus increasing the market value of the CSP power.

Carbon Cost Reduction

With the impending introduction of the ETS in Australia, carbon may shortly come to be priced. This will impose a cost impost on CO_2 emissions from fossil-fuel-based power generation. CSP produces almost no GHG emissions and would therefore not incur carbon costs. As carbon costs through tax or cap-and-trade mechanisms rise over time, which they almost certainly will, the cost advantage of CSP over other forms of power generation will increase. As carbon costs paid by industry rise, CSP is expected to become progressively more competitive. Again, as part of a power generation portfolio, CSP can be seen as part of a carbon cost risk hedging strategy. As carbon costs rise, CSP would help to buffer the financial effects to the rest of the portfolio.

Carbon Emission Reductions

Because CSP produces no GHGs, it provides value to society by helping combat the onset of global climate change, which will cause substantial damage to Australia and the economy of the world if left unchecked.[26] Using an SCC estimate rising at about 2% per year (as discussed), CSP is today very close (within the error inherent in the analysis) to economic parity with coal for the Australian grid. The difference between the SCC and what we pay for carbon today (zero) represents the unpriced social cost of GHG emissions. The ETS is designed to provide a price signal for decision makers to reduce this social damage. CSP is expected to become more economic as time goes on by virtue of the continually rising SCC. The carbon emission reduction benefits of CSP are a key driver for government assistance in developing solar power generation in Australia. The societal benefits of carbon elimination of CSP are significant, on the order of $150 million a year for a 250-MW power station.

Early Mover Advantages

Solar thermal power at scale can provide Australian businesses with the opportunity to develop early mover advantages in a world that is increasingly concerned about climate change and looking for ways to move toward a lower carbon economy. The institutional, regulatory, engineering, and technological expertise required to develop and deploy commercial utility-scale solar power has the potential to rapidly become a source of economic prosperity for Australia and Australian businesses. Exploiting the tremendous potential of large-scale solar power is a significant commercial opportunity for Australia.

Public Relations and Corporate Responsibility Benefits

CSP and renewable energy in general enjoy significant support with the public, not only in Australia but also around the globe. For example, the *Greener Times*, published by the West Australia Conservation Council (which represents over 90 environmental organizations in the state) called an initiative to deploy CSP in Australia "rare, heartening news."[27] Involvement in a project that has clear social value (which can be explicitly monetized as discussed) generates significant goodwill, both from the public and from within the proponent's own organization.

EXAMPLE: COMPARING RENEWABLE ENERGY OPTIONS

INTRODUCTION

There is a wide variety of renewable energy technologies available on the market today. Some are well-developed and widely used, like wind power and biomass, while others are highly experimental, such as wave and tidal systems. Different renewable energy technologies produce power in different ways, using different media, require vastly different CAPEX, and produce power under varying circumstances—wind only when the wind blows, solar only when the sun shines. What is more, as with every other industrial activity, renewable energy has its own life-cycle internal and external costs that must be examined if a full appreciation of the true relative merits of various systems can be determined.

This example examines a range of renewable energy alternatives that a power company might use to develop its renewable portfolio in the face of ever-broadening MRETs around the world. In this case, an EESA is used to examine a range of relatively small renewable power opportunities and determine which optimizes environmental, social, and economic benefits.

OPTIONS DESCRIPTION AND COSTING

A range of renewable energy options, suitable for deployment in an agricultural region in Australia, was compared. This particular region of the country has been badly affected by soil salinization, which has been caused by extensive clearing of native vegetation. Removal of up to 95% of the deep-rooted native trees over vast areas has caused water tables to rise, introducing salt into the shallow soils. This phenomenon has rendered large areas of land unable to support agriculture. To arrest the impacts,

TABLE 6.4

Renewable Energy Option Comparison (2008 U.S. Dollars)

Option	Description and Costing Basis	Capacity (MW)	CAPEX (Millions of Dollars)
1. Cofiring at existing coal-fired power plant	Mallee oil trees are cultivated in salt-impacted areas with high water table and copsed, and the biomass is shipped to a nearby coal-fired power plant for cofiring. CAPEX includes land purchase costs.	5	10.0
2. Biomass energy facility	Fuel feedstock assumed 30 km from facility. Land leasing costs included in OPEX.	7	30.0
3. Solar photovoltaic (PV)	Flat-plate solar PV facility. Land purchased for $1,000/ha and is suitable for solar array in high-insolation area.	5	37.2
4. Integrated wood plant (IWP) biomass	Mallee oil trees cultivated in salt-impacted areas with high water table and copsed, and the biomass is used to generate power in a biomass power plant. High-value eucalyptus oils are also recovered in the facility and sold. Combustion also yields high-quality activated carbon sold on the open market for additional revenue. Feedstock assumed average of 30 km from facility.	5	41.2
5. Solar photovoltaic (PV)	Flat-plate solar PV facility. Land purchased for $1,000/ha and is suitable for solar array in high-insolation area.	20	49.2
6. IWP biomass	IWP facility assumes a simple fourfold scaling up from the 5-MW facility using the same assumptions. Feedstock is now on average 100 km from facility.	20	121.5
7. Wind	Wind power costs based on existing projects in the region.	100	300.0
8. Concentrating thermal solar	CSP project cost based on land purchased at average $1,000/ha in high-insolation area.	250	1,890

farmers and the government have started to plant oil mallee eucalyptus trees, which drive down the water table and reverse the effects of soil salinization. A number of the renewable energy options evaluated in this example involve planting and copsing mallee trees for use as feedstock for energy production, either in purpose-built biomass plants or for cofiring in the existing coal-fired facility in the area. Wind and various solar possibilities were also examined. Table 6.4 shows the options being considered, their expected power production and CAPEX.

BENEFITS ASSESSMENT

The financial benefits produced by each option include revenues to the power company from the sale of energy to the grid and sale of by-products such as eucalyptus

TABLE 6.5

Financial Benefit Assumptions (2008 U.S. Dollars)

Option	Energy Revenue ($/MWh)	REC ($/ MWh)	Capacity Factor	Eucalyptus Oil Revenue Calculation Inputs	Granular Activated Carbon Revenue Calculation Inputs
1. Cofiring at existing coal-fired power plant (5 MW)	75	57	1	n/a*	n/a
2. Biomass energy facility (7 MW)	85	57	1	n/a	n/a
3. Solar photovoltaic (PV) (5 MW)	75	57	0.2	n/a	n/a
4. IWP biomass (5 MW)	75	57	1	864 t/yr at $3,000/t	6,456 t/yr at $3,000/t
5. Solar photovoltaic (PV) (20 MW)	75	57	0.2	n/a	n/a
6. IWP biomass (20 MW)	75	57	1	3,456 t/yr at $1,000/t	25,824 t/yr at $1,000/t
7. Wind (100 MW)	75	57	0.35	n/a	n/a
8. Concentrating thermal solar (250 MW)	100 (peak)	57	0.15	n/a	n/a

*n/a = not applicable to this option.

oil and activated carbon (AC) to the market (in the case of the integrated wood processing [IWP] facilities). These options also attract RECs in this jurisdiction. Table 6.5 summarizes the value of key financial benefits used in the assessment.

Note that the 5-MW IWP facility enjoys relatively high prices for its by-products compared to the 20-MW IWP facility. Market research has shown that the national market for these products is small, and that beyond the output from the 5-MW facility, oversupply would most likely depress prices considerably. In this example, the energy market regulator in the state pays a capacity credit of $0.13 million/MW. Each of the technologies considered in this analysis produces power with differing levels of certainty and at different times of the day; as a result, they are subject to differing capacity credit payment factors reflective of this. The factors adopted for each option are also presented in Table 6.5.

The following external benefit categories have been considered in the analysis:

- The disbenefits associated with *GHG emissions*, based on an understanding of the social cost of carbon dioxide equivalents (SCC). Disbenefits may also be considered as external costs.
- The disbenefits associated with *NOx, SOx and particulate emissions*, based on an understanding of the social costs and market prices discussed in more detail in this chapter.

TABLE 6.6
Unit Values for Externalities Considered (2008 U.S. Dollars)

Benefit Category	Units	Low	Median	High
GHG	$/t CO$_2$e	0	25	85
NOx	$/t	0	770	2,850
SOx	$/t	0	630	2,250
PM$_{10}$	$/t	0	24,084	48,168
TEV of water	$/kL*	0	1.65	3
Regional benefits from salinity amelioration	$	See valuing salinity amelioration benefits section		

*kL = kiloliters

- The *total economic value of water (TEVW)*, broken down into three components: the direct use value (used or potentially usable by humans), the ecological support value, and the option value (value to society from having the resource available at some time in the future to be used).
- *Regional economic and social benefits arising from the amelioration of salinity*, specifically benefits from the prevention of salinity accruing to the regional economy, local and state governments, and the regional ecosystem.

Table 6.6 summarizes the base case and high and low unit values for each of these externalities.

Valuing Salinity Amelioration Benefits

For the IWP options, establishment of mallee plantations as a source of biomass and the expected additional benefits from salinity reduction are inherently long-term propositions.[28] The effects of salinity on agricultural productivity can only be reversed over long periods of time. The benefits associated with attenuation of salinity damage can accrue to agricultural producers, households, commercial and retail businesses, industry, and government, and these benefits grow over time as the plantations grow and begin to lower water tables.[29] As such, salinity amelioration benefits were estimated as a time function, using data for two regions chosen as the notional sites for the 5-MW and 20-MW IWP plants. Benefits were estimated directly as the elimination of costs associated with salinity based on available studies in the literature from similar areas, scaled for land area.[30] Tables 6.7 and 6.8 provide salinity benefit estimates over time for each of the facilities.

COST–BENEFIT ANALYSIS

The data were used as input to the cost–benefit analysis (CBA) part of the assessment. For a notional 2009 start, a planning horizon of 20 years was used. For the base case analysis, a social discount rate of 5.5% was applied. As each option has a different operating capacity, option net present values (NPVs) were normalized based on capacity. Results are thus provided on a basis of dollars net present value per megawatt hour.

TABLE 6.7

Salinity Amelioration Benefit Estimates (Million U.S. Dollars, 2008): 5-MW IWP Plant

Beneficiary	Year 1	Year 10	Year 20
Agriculture[31]	0	2.87	10.40
Households[32]	0	0.25	0.97
Government	0	1.26	5.47
Biodiversity	0	3.94	11.24
Total	0	8.32	28.16

Base Case Results

The results of the base case assessment are presented in Figures 6.8 and 6.9. The results show the analysis using the base case assumptions for values of all parameters. Figure 6.8 provides the unit NPV results in U.S. dollars per megawatt hour for each of the options being considered, broken down by each component of internal and external value (cost and benefit bar pairs for each option). This analysis examined the costs and benefits associated with the production of renewable energy only. It did not examine the more detailed economics of grid connectivity issues, transmission infrastructure, and electricity market conditions.

As shown in Figure 6.9, a number of options have a socially positive NPV and are economic; thus, according to our definition, they are also sustainable (society receives more ongoing benefit from the action than cost and therefore is likely to want to keep supporting the action on an ongoing basis).

Under base case conditions, option 4 (the 5-MW IWP) is the most economic at $220 NPV/MWh, followed by option 1 (cofiring at an existing coal-fired power plant) at US$91/MWh, and option 2 (bioenergy) at US$83/MWh. The solar options are uneconomic under the base case conditions, as is option 6 (20-MW IWP). Option 7 (100-MW wind) is economic and is the next-best nonbiomass energy option after options 4, 1, and 2, with an NPV per megawatt hour of US$52. The majority of the benefits of option 4 are derived from the value of its nonenergy products, AC and eucalyptus oil, with US$260/MWh for the entire option derived from the sale of both

TABLE 6.8

Salinity Amelioration Benefit Estimates (Million U.S. Dollars, 2008): 20-MW IWP Plant

Beneficiary	Year 1	Year 10	Year 20
Agriculture	0	2.35	8.38
Households	0	0.77	3.16
Government	0	1.92	8.10
Biodiversity	0	3.79	10.88
Total	0	8.83	30.52

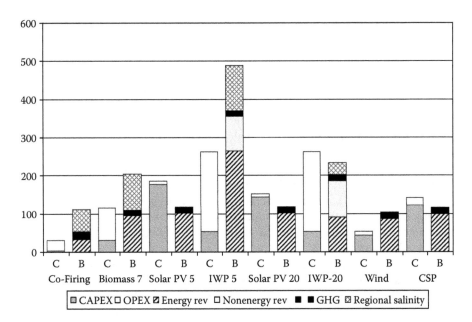

FIGURE 6.8 Breakdown of 20-year base case unit present value costs (C) and benefits (B) per megawatt hour (US$/MWh) for each option at base case conditions. Options are presented from cheapest overall CAPEX on the left (co-firing) to most expensive on the right (CSP). External costs for NOx, SOx, and particulates emissions and the effect of the TEV of water are negligible (<US$2/MWh) and are therefore not included.

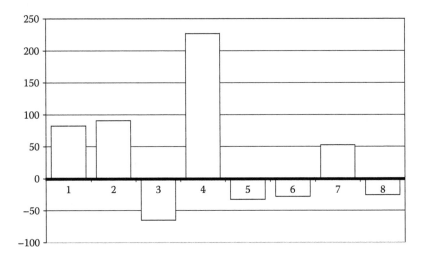

FIGURE 6.9 Twenty-year base case NPV (US$/MWh) comparison. The options are ranked from cheapest overall CAPEX at left to most expensive overall on the right. The small 5-MW IWP facility is the most economic and sustainable option (option 4) on a basis of dollars per megawatt hour but cannot contribute appreciably to an overall renewable energy portfolio. Expanding the IWP concept to a 20-MW facility (option 6) cannot achieve the same level of benefits.

products in present value terms (Figure 6.8). AC makes the most significant contribution of the two products, with eight times the volume of eucalyptus oil produced as AC and sold at the same price under base conditions ($3,000/tonne). The regional benefits of salinity reduction are valued at approximately US$115/MWh in present value terms. The value of the energy from the plant is only US$90/MWh, which reinforces the value proposition of the project from its nonenergy contributions. While the capital and operating costs and energy production revenues are similar for both the 5-MW and 20-MW IWP options (in terms of present value dollars per megawatt hour), the value of the nonenergy products and the regional benefits are both significantly less for the 20-MW IWP option. The 5-MW facility would, on its own, saturate the market for the carbon and eucalyptus oil products. Additional supply from a larger facility would depress overall prices in the market considerably, making it a significantly less-economic proposition. In addition, the larger facility is assumed to be located in a region where the salinity effects are not as serious, reducing the overall benefits of salinity amelioration. As a result, the benefits of the 20-MW IWP option are not sufficient to justify the capital and operating costs.

The solar options all have high capital costs relative to the other options, and the value of the energy revenue as well as the benefits from the displacement of carbon and other atmospheric pollutants are not sufficient to overcome these capital costs in this example. Wind is an attractive option under the base case condition; the value of the energy revenue is sufficient to justify the CAPEX and OPEX.

SENSITIVITY ANALYSIS

The sensitivity analysis provides a view of how each option performs across a wide range of conditions. In this hypothetical example, cost estimates are indicative (±25% accuracy), and the valuation and estimation of benefits are subject to even larger changes, as discussed in Chapter 3. However, as discussed, the key to the analysis is to reveal not absolutes in terms of monetary units, but better and worse decisions overall compared to the range of possible decisions that could be made. From this perspective, sensitivity analysis is important because it allows the overall conclusions of the analysis to be tested across a wide range of parameter inputs. If a decision is favorable or economic over a wide range of parameter inputs, compared to other possible decisions, then despite the overall uncertainty in the actual dollar figures, the decision can be identified as superior to its competitors. This is particularly useful when considering sustainability of options. By definition, sustainability is concerned with the future, which is inherently uncertain. By varying key input parameters over a wide but reasonable range, the implications of a range of possible futures can be examined.

Examination of the sensitivity of results to all parameters shows that the changes in NPV dollars per megawatt hour resulting from changes in the value of NOx, SOx, PM_{10}, and TEVW were insignificant under all conditions. These parameters were therefore excluded from further analysis. The parameters that were included in more detailed sensitivity analysis are GHG value, salinity benefits, AC revenue, energy price, eucalyptus oil revenue, and social discount rate. Figure 6.8 shows

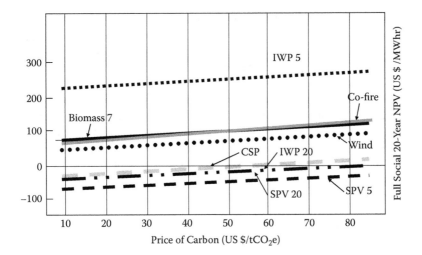

FIGURE 6.10 Sensitivity of each option (NPV US$/MWh) to change in GHG value ($/tCO$_2$e) with all other parameters set to base case values. Note that the 20-MW solar present value and the 20-MW IWP curves are almost identical. SPV denotes solar PV.

their relative contributions to costs and benefits under base case conditions. Each is discussed next.

Sensitivity to GHG Value

Figure 6.10 shows the impact of variation of the value of GHG emissions, with all other parameters fixed at base case values. Each of the options becomes more economic as the value of GHG emissions increases—reflecting the value from the displacement of current GHG-producing energy on the grid. The increase is modest when compared to the impact of changes to other variables (at most US$40/MWh), and the increase is consistent for all options. Other than option 2 (7-MW biomass) improving from third-best option to second-best option with increasing GHG value, there is no change to the ranking of the options. Option 4 (5-MW IWP) remains the most economic option under all values of GHG considered in the analysis, with all other values set to base case conditions. Note that the large-scale thermal solar option becomes economic and sustainable at carbon values above about US$70/tCO$_2$e (tonnes CO$_2$ equivalent). The current SCC is likely already higher than this.[33] When higher GHG values are coupled with rising energy prices (discussed below), all of the renewable energy options yield net positive environmental, social, and economic benefit overall.

Sensitivity to Salinity Amelioration Benefits

Salinity benefits arising from the beneficial effects of the mallee tree feedstock were varied from zero to twice the base case value. Under the range of values examined, the 5-MW IWP option remains the most economic.

Sensitivity to Revenue from Activated Carbon

The economic value of the 5-MW IWP option, the most economic option under base case conditions and the sensitivity parameters analyzed thus far, is most dependent on the value of AC. AC provides the largest revenue source for this option, and as a result, its NPV (US$/MWh) is extremely sensitive to changes in the sale value of the product. For values of AC above US$1,200/tonne, the 5-MW IWP option is the most economic and sustainable. However, for values less than this amount, the 7-MW biomass energy and 5-MW cofiring options are most economic. At extremely low values for AC (~US$500/tonne), the 20-MW IWP becomes the least economic of the options; however, with increasing AC price, this option rapidly improves such that at a moderately high value of US$4,000/tonne, it is the fourth-most economic option behind the 5-MW IWP, 7-MW biomass energy, and 100-MW wind options.

Sensitivity to Energy Price

Each of the options exhibits a similar increase in economic NPV (US$/MWh) with rising energy prices, except the 5-MW cofiring option, which assumes no additional revenue generation as a result of the displacement of coal at the notional power station. The 5-MW IWP option remains the best option under the entire range of values considered.

Sensitivity to Social Discount Rate

The 5-MW IWP option is the most economic option under a wide range of discount rates. While all option NPVs (US$/MWh) vary significantly with varying discount rates, for discount rates lower than the base case of 5.5%, the 7-MW biomass energy option is the next-best option, while for values above this base case, the 5-MW cofiring option is the next-best option.

Cumulative Probability

An ordered plot of all the possible NPV (US$/MWh) values for every combination of the input parameters under the full ranges described is provided in Figure 6.11. Each option is described by a curve following the methodology described in Chapter 3. If an option is consistently to the right of the plot (more positive NPV), without being intersected by any others, then that option is on average the most sustainable and economic of those being assessed under the range of conditions considered.

Figure 6.11 shows that the 5-MW IWP option is the furthest to the right under approximately 60% of all the conditions analyzed. In addition, 80% of all the possible values of NPV per megawatt hour for the 5-MW IWP option are positive. It is also the flattest line, reflecting its strong sensitivity to the price of AC.

The rest of the curves are relatively vertical compared to the 5-MW IWP option, signifying that their NPVs are less sensitive to variations in the various input assumptions. A group of three options (7-MW bioenergy, 5-MW cofiring, and 100-MW wind) cluster together as the next-best options. Across 90% of all the possible values of NPV per megawatt hour examined, these three options cluster as the second-most economic and sustainable choice and are NPV positive. In fact, these three options are NPV positive across a larger range of conditions than is the 5-MW IWP option. The economic performance of the 5-MW IWP option, as discussed, is extremely

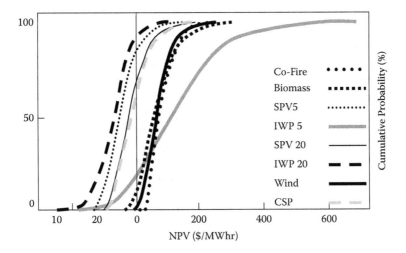

FIGURE 6.11 Cumulative probability distribution for each option.

sensitive to the price of AC. The long positive tail to the SMW IWP curve exhibits significant "up-side" potential should the market price of AC rise. However, should the price of AC fall, the option rapidly becomes uneconomical.

DECISION-MAKING IMPLICATIONS

Figure 6.12 shows the option that is most economic and sustainable under the full range of conditions examined. The 5-MW IWP option is the most economic and sustainable option in approximately 57% of the total cases assessed, followed by the 5-MW cofiring option (25% of conditions), the 100-MW wind (10% of conditions), and 7-MW biomass energy (8% of conditions). The ranking of the 5-MW IWP as most economic and sustainable depends heavily on achieving strong returns from the sale of nonenergy by-products and, given the market conditions for these by-products, cannot be replicated for any other facilities. In other words, although this

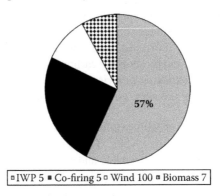

FIGURE 6.12 Most economic and sustainable decision over the full range of conditions examined, percentage of 1,000 cases calculated.

option performs well, it is literally a "one-off," and 5 MW alone does not represent a significant contribution to renewable energy in the area. If these by-products are removed from the equation, the other biomass options (to the degree that they realize appreciable salinity amelioration benefits) will continue to fare well in relative terms. All of the renewable energy options produce benefits in terms of GHG reduction that increase as the value of carbon rises toward the SCC.

Examining the issue from a renewable energy portfolio perspective, an environmentally, socially, and economically optimum approach overall might be to proceed with construction of a single 5-MW IWP or a properly situated 20-MW IWP facility (geared to produce only the economically optimum amount of by-products while maximizing salinity amelioration benefits). Clearly, biomass makes sense when coupled with the benefits of salinity management in this part of the world. For sheer capacity, wind power seems to be a strong candidate under the conditions examined. With rising energy costs and the increasing likelihood of a rising cost of carbon, all of the renewable energy options examined here deliver economic benefits to society. An optimal mix of options would involve deploying the most economic and sustainable options first and moving to less-economic options next, as demand requires, over time.

SUMMARY

A more sustainable and economic energy future depends on a mix of solutions, including reducing demand by improving efficiency, reducing waste, and simply using less energy. Renewable energy can play a much more extensive role in our future energy mix than conventional wisdom suggests.[34] While different types of renewable energy generation fare quite differently when examined through the lens of an EESA, it is clear that rising fuel costs and carbon prices, combined with the other external costs of conventional fossil fuel generation, make a wide variety of renewable energy technologies economic and sustainable.

Coal, plentiful and cheap, is revealed as one of the least-economic and least-sustainable energy choices available (if burned using conventional power generation technology). The combination of high carbon emissions, production of a range of other air pollutants, and other significant life-cycle external costs (including the mining footprint) severely disadvantages coal compared to most other forms of energy generation. Coal can be made into a sustainable fuel for the future through significant improvements in efficiency and by adopting carbon capture and sequestration (CCS) on a significant scale, but achieving this vision will require massive investment and strong collaboration among industry, government, nongovernmental organizations, and the public.

Just because something is cheaply available on the market, does not mean that its real cost is low or that it delivers real value. This is particularly the case in the energy sector. Providing plentiful, truly affordable and secure energy for all of the world's people is a defining challenge of the twenty-first century. Accomplishing this objective without the sustainability of the planet's climate and ecosystems will require a new, more complete, and balanced examination of the real costs and benefits of power generation options.

NOTES

1. Mitchell, J.G. 2006. When Mountains Move. *National Geographic*, March, 1–7.
2. International Energy Agency and Organization for Economic Cooperation and Development (IEA/OECD). 2008. *Energy Technology Perspectives 2008. Scenarios and Strategies to 2050*. IEA/OECD, Paris.
3. Intergovernmental Panel on Climate Change (IPCC). 2007. *Fourth Assessment Report. Mitigation of Climate Change*. Cambridge University Press, Cambridge, UK.
4. Commonwealth of Australia. 2008. *Green Paper on Carbon Pollution Emissions Reduction Scheme*. Government of Australia, Canberra.
5. International Energy Agency and Organization for Economic Cooperation and Development (IEA/OECD). 2008. *Energy Technology Perspectives 2008. Scenarios and Strategies to 2050*. IEA/OECD, Paris.
6. Ibid.
7. Stern, N. 2006. *The Economics of Climate Change—The Stern Review*. Cambridge University Press, Cambridge, UK.
8. International Energy Agency and Organization for Economic Cooperation and Development (IEA/OECD). 2008. *Energy Technology Perspectives 2008. Scenarios and Strategies to 2050*. IEA/OECD, Paris.
9. Environmental Law Institute (ELI). 2009. *Estimating US Government Subsidies to Energy Sources, 2002–2008*. September, ELI, Washington, D.C.
10. BBC. 2006. *China Building More Power Plants*. June 19. BBC News, London, U.K.
11. Xiaohiu, Z., and J. Xueli. 2007. Coal Mining: Most Deadly Job in China. *China Daily*, Beijing, China. http://www.chinadaily.com.
12. Voss, A. 2002. Life Cycle Analysis and External Costs in Comparative Assessment of Electricity Chains: Decision-Making Support for Sustainable Electricity Provision. In *Externalities and Energy Policy: The Life-Cycle Approach* (ed. D. Pearce). Nuclear Energy Agency, OECD, Paris. pp. 2–18.
13. European Commission Directorate General Environment. 2005. *Damages per Tonne of Emissions of PM2.5, NH3, SO2, NOx and VOC from Each EU25 Member State and Surrounding Seas*. CAFÉ Programme. AEA Technology, European Commission, Bruxelles.
14. Ibid.
15. Lorenz, P., D. Pinner, and T. Seitz, 2008. The Economics of Solar Power. McKinsey & Company *McKinsey Quarterly*, June. pp, 1–10.
16. Meurs, P., J. Gill, D. Aberle, and S. Int'Veld. 2008. A Water-Secure and Sustainable Australia—Providing Solar Thermal Power and Desalination in Western Australia. *Proceedings of the Enviro08 Conference*, Melbourne. pp. 152–161.
17. Hardisty, P.E., P. Meurs, D. Mofflin, J. Bowen, and A. Kirvan. 2008. *Submission to the Government of the Commonwealth of Australia's Green Paper on a Carbon Pollution Reduction Scheme*. Government of Australia, Canberra.
18. WorleyParsons, 2008. *Advanced Solar Thermal Implementation Plan. Final Report—Vol. 1*. WorleyParsons, Sydney, Australia.
19. Emerging Energy Research (EER). 2007. *Global Concentrated Solar Power Markets and Strategies, 2007–2002*. EER, Cambridge, MA.
20. Ibid.
21. Ibid.
22. Ibid.
23. WorleyParsons, 2008. *Advanced Solar Thermal Implementation Plan. Final Report—Vol. 1*. WorleyParsons, Sydney, Australia.
24. Ibid.
25. Ibid.

26. Stern, N. 2006. *The Economics of Climate Change—The Stern Review.* Cambridge University Press, Cambridge, UK.
27. WA Conservation Council. 2008. *Greener Times*, 2. WA Conservation Council, Perth, WA, Australia.
28. Cacho, O. 2001. An Analysis of Externalities in Agroforestry Systems in the Presence of Land Degradation. *Ecological Economics,* 39(1), 131–143.
29. Van Bueren, M., and Bennett, J. 2000. *Estimating Community Values for Land and Water Degradation Estimates.* Final Report prepared for the National Land and Water Resources Audit, Project 6.1.4. CSIRO, Canberra, Australia.
30. Wilson, S. 2002. *Cost of Salinity to the Glenelg-Hopkins Region.* A report to the Glenelg-Hopkins Catchment Management Authority. Prepared by Wilson Land Management Services and Ivey ATP, May. Glenelg-Hopkins Catchment Authority, Hamilton, Victoria, Australia.
31. Wilson, S.M. 2004. *Dryland and Urban Salinity Costs Across the Murray-Darling Basin: An Overview and Guidelines for Identifying and Valuing the Impacts.* MDBC Publication 34/04. Murray-Darling Basin Commission, Canberra.
32. Ibid.
33. Stern, N. 2006. *The Economics of Climate Change—The Stern Review.* Cambridge University Press, Cambridge, UK.
34. Jacobsen, M.Z., and M.A. Delucchi. 2009. A Path to Sustainable Energy by 2030. *Scientific American*, November: 58–65.

7 Contaminated Sites and Waste

INTRODUCTION

As the world's population grows, industrial activities continue to degrade land and water at a faster pace. But, reclaiming contaminated land and repairing polluted aquifers can be expensive, technically difficult, and time consuming.[1,2] Deciding if and when to remediate, and to what degree, can be regarded in the context of alternative environmentally and socially beneficial actions. What else could be done with the money required to restore a site or aquifer? Could we purchase and preserve several acres of rain forest or other valuable natural habitat? Which would provide the greatest benefit to society? And then, what are the commercial realities facing those who are called on to pay for the restoration of contaminated sites?

Under the "polluter pays" principle, increasingly adopted as the fundamental ethical precept for remediation policy, the responsibility for planning, funding, and executing remediation lies with the polluter. This could be a government, a municipality, or a private sector enterprise. In the background, ever present and increasingly vocal and powerful, are the public, the neighbors, the inhabitants of the planet, demanding that their interests be served also, and that the dwindling resources of the planet be protected for their future and the future of their children and grandchildren. Combining and prioritizing these diverse interests into a decision-making process, using a common unit of value, is essential if equitable, practical, and rational economic decisions are to be made.

Significant amounts of time, effort, and money have already been devoted to remediation of contaminated sites and aquifers worldwide. A tremendous diversity of methods and technologies has been applied in conditions as variable as the individual sites themselves. Along the way, consultants, problem holders, individual professionals, and government institutions have accumulated wide knowledge of the costs of remediation. Until very recently, selecting the least-cost remedial option passed for "economic" analysis.[3,4] The benefits to the problem holder were sometimes considered; the wider benefits to other parts of society rarely were.[5]

Borrowing from the wider environmental economics literature,[6] the costs and wider economic benefits of remedial alternatives can be compared within an EESA to select optimal remediation approaches.[7] A critical part of this equation, rarely considered, is the cost of secondary effects or by-products of the remedial action, including the external costs of many common remedial practices, such as excavation and landfilling contaminated soils and materials. This chapter expands on previous work and presents the case for applying the environmental and economic sustainability assessment (EESA) to remedial decision making.

CONCEPTUAL FRAMEWORK

At the outset, it is important to distinguish between the different levels at which remedial decisions can be made. The current literature makes reference to "remedial approaches," "remedial options," and "remedial technologies," sometimes interchangeably and often without clear definition. For contaminated sites and groundwater, the distinctions between objectives, approaches (or strategies), and technologies is important. These are formally defined as follows:[8]

- *Remedial objective* is the overall intent of the remediation. Objectives could include restoration of a parcel of land to productive use, the protection of specific receptors, or the elimination or reduction of certain unacceptable risks.
- *Remedial strategy* (or approach) is the way in which the objective is to be reached and is defined specifically in the risk assessment context by identifying the pollutant linkage component it addresses: source removal, pathway elimination, source protection/isolation, or a combination of these.
- *Remedial technologies* are the specific tools that form the components of the strategy. For example, physical containment (a pathway elimination approach) can be achieved through use of slurry walls, sheet pile walls, or liners, often in conjunction with groundwater pumping and treatment. Source removal can be achieved through excavation and on-site treatment of contaminated soils (by a variety of techniques) or through many available in situ mass destruction techniques. A remedial strategy will very often involve the use of several different remedial technologies.

These levels are all interlinked. The remedial objective should be known before detailed design (technology selection) occurs. The choice of a remedial approach is a critical intermediate step that can be used both to help set objectives (by considering and comparing various approaches at the conceptual level) and to guide the selection of the technological components that will make up the final design. Each of these three levels of analysis is discussed.

SPACE AND TIME

Contamination issues must be seen in the context of time and space and are inherently dynamic in nature. This presents a number of challenges for the setting of remedial objectives and assessing the most economic remediation alternative: (1) objectives must be framed in a temporal context, (2) technology changes with time, and (3) regulations change with time. In the same way, the *scale* of a contamination problem is not necessarily fixed. A spill that is initially concentrated in a small area may over time expand and affect a considerable volume as contaminants migrate laterally and vertically, bringing them into contact with other media and receptors. The scale of a contamination problem may have significant impact on how it is valued by society.

REMEDIAL OBJECTIVE

The remedial objective is the level at which the benefits of remediation are most readily and fundamentally determined. If a valuable receptor is protected, a benefit to society accrues. If a receptor is not protected, damage results. Benefits are tied clearly to the fundamental objective and the basic approach used to achieve it. For a groundwater contamination problem, for instance, the choice of whether to achieve the objective using pump and treat, a biobarrier, or natural attenuation has a direct impact on costs (including any external costs associated with the method, such as release of off-gases to the atmosphere, for instance), but benefits remain essentially constant.

Choosing a remedial objective can become quite complex when mobile groundwater plumes are involved. Figure 7.1 provides a simple visual schema for considering the overall consequences of various remedial objective options under such conditions. A fixed point source actively introduces contaminants into groundwater at a mass rate. Sodium chloride contamination, for instance, will behave as a conservative solute, moving at the linear advective groundwater velocity. Many organic contaminants, such as benzene, will biologically degrade over time and are also subject to adsorption onto matrix material. As time passes, the plume migrates in groundwater, dispersing laterally and transversely due to the effects of chemical diffusion and mechanical mixing. At time 1 (Figure 7.1), for instance, the plume has migrated only a short distance and is relatively highly concentrated. Only a relatively small volume of aquifer has

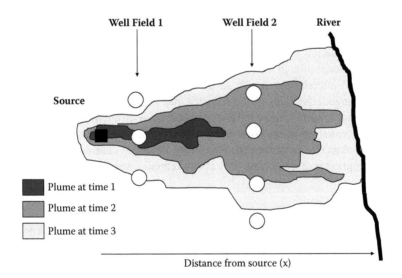

FIGURE 7.1 A plume of contamination in a groundwater aquifer is subject to lateral and transverse dispersion as it migrates over time away from the source. As it moves, if not intercepted, first well field 1, then well field 2, and finally the river will be impacted. Remedial decisions need to be made in context of where the remediation will take place (x) and when (t). These decisions will profoundly affect the economic outcome of the remediation. Various choices of the timing and location of remediation can be tested and compared using EESA.

been impacted, including well field 1. As plume migration continues unabated, other receptors will become impacted, first well field 2 and finally the river. The result is that the number of receptors impacted increases with time. Depending on the behavior of the contaminant solute (degree of attenuation by adsorption, dispersion, biological and chemical degradation), impacts could also vary with time at a given receptor. In such a situation, we need to describe risk as a function of time and space.

As the plume migrates and disperses with time, various receptors are impacted at different times. For interventions that take place at a given point in time and space ((x,t) coordinates), a remedial cost C_i and benefit B_i (equal to damage avoided if the remediation takes place) would be realized. So for the generic case, we see that the costs and benefits of remediation vary not only with time but also with the location in space at which we decide to implement our remedial action. The various remedial objective options can be evaluated within this context.

The remedial objective in the case of Figure 7.1 could be to prevent impact to the river. A secondary objective could be to prevent contamination from surpassing a given concentration in the wells of well field 2. Wells in well field 1 may already be impacted by the time we discover the problem. Note that remedial objectives can be interdependent: achieving one objective might be a prerequisite for achieving another or might help substantially in achieving it. Achieving one objective might automatically mean another is achieved at the same time and so on. This is a result of the spatial and temporal dynamics of plume movement.

Naturally, this leads to situations involving remedial objectives that change with time. For example, an initial evaluation could indicate that net benefits are maximized if a certain noncritical receptor could be sacrificed (well field 1, for instance) and that the situation with respect to a more distant receptor (the river) could be reevaluated at some time in the future (given that migration times would be expected to be long and attenuation active). At the reevaluation point, remedial technology may have changed, the value of the receptor may have increased, and the degree of attenuation and migration may have turned out to be different from that originally predicted. A reanalysis of the costs and benefits at that point could then indicate that a change of objective is warranted. Flexibility based on continuous monitoring of remedial progress is an important part of any remedial strategy.

REMEDIAL STRATEGY

There exist a staggering number of different remediation technologies available to achieve any specific technical outcome. For example, dissolved volatile organic compounds (VOCs) can be removed from pumped groundwater by air stripping (packed tower, shallow tray, or other configurations); advanced oxidation methods (ultraviolet [UV]-ozone, hydrogen peroxide, TiO_2-UV, and other systems); granular activated carbon (GAC); and biological reactors (many available configurations), among others. Many other means of controlling the migration of contaminants in groundwater exist, including physical barrier systems, funnel-and-gate technology, and permeable reactive barriers. PAHs (polyaromatic hydrocarbons) can be removed from contaminated soil by excavation and treatment by soil washing, enhanced biological treatment, or chemical oxidation. Often, several different technologies, each designed to achieve a specific technical outcome, will be required to create a system that can accomplish the remedial objective. Because so many technologies exist that achieve such different technical outcomes at such widely varying costs, direct comparison of technologies using EESA is often not practical or useful without first framing them within a set of remedial strategies.[9]

The remedial strategy does not focus on technology per se but on ways of breaking the pollutant risk linkage that causes (or will cause) damage. The list of possible remedial strategies is relatively short: remove the source, eliminate the pathway, or protect, move, or manage the receptor. Consideration of remedial strategy can be useful in aligning the EESA with current risk-based guidance and in streamlining the EESA process since only a limited number of strategies need be considered.

Remedial strategy provides a link between remedial objectives and the hundreds of cleanup technologies available. Also, the degree to which the risk linkage is broken, the timing of the action, and the spatial location at which the action is taken are all variables that must be considered when choosing a strategy. Constraints analysis can be undertaken to help assess which strategies are realistically achievable. Preliminary high-level costs can be assigned to each strategy that can feasibly achieve the desired objective and compared to the benefits of achieving the objective. This provides a relatively quick strategic analysis of the costs and benefits of remediation and a basis for selection of an optimal remedial strategy before proceeding to detailed technology evaluation and cost analysis.

REMEDIAL TECHNOLOGY

The remedial technology selection level involves choosing the most cost-effective way of implementing a remedial strategy. This requires detailed comparison of capital and operation and maintenance (O&M) costs for technologies over a projected project life span. The external costs of remediation over the life cycle should also be incorporated into the cost analysis. Application of constraints to remediation helps to reduce the number of viable technologies that can be feasibly applied to the problem. In some cases, different technologies may also realize different benefits (within a given strategy). If these are significant, they should be included in the assessment.

THE ECONOMICS OF REMEDIATION

An economic model describing the full costs and benefits of site remediation was presented by Hardisty and Ozdemiroglu (2005).[10] In this analysis, it is assumed that there is full knowledge that the contamination has occurred, and that damage is occurring. Situations in which damage is occurring without the knowledge of the public or regulators are not considered, although the same method is applicable with added risk or uncertainty. The main variables in the economic analysis are the timing of remedial action and the spatial context and scope of the action. So, preventive action could be taken now, thus avoiding future damage; can be postponed, allowing existing damages to continue; and possibly also allowing future damages to occur. The other variable is spatial: the location at which the avoidance or remediation takes place.

For each option considered, at whatever level of interest, the environmental and economic sustainability assessment (EESA) examines the sum of the benefits of the action over the planning horizon and compares them with the full life-cycle environmental, social, and financial costs of remediation. As discussed in Chapter 3, varying the values of key parameters over the planning horizon in a sensitivity analysis then allows selection of the most robustly superior options. The goal is to determine the option that provides an environmental, social, and economic optimum for all stakeholders, over the widest range of possible future conditions.

The EESA can be used to determine an environmentally, socially, and economically sustainable remedial objective by considering the costs of remedial alternatives, the benefits of remediation, and the costs of secondary impacts (sometimes called disbenefits, but referred to here as external costs). Benefits of remediation can be expressed as the value of avoided damage or prevention of future damages that would have occurred if the remedial action had not been taken.[11] Action that eliminates or reduces damage already incurred is also considered. Each of these elements is discussed in more detail.

Financial Costs of Remediation

The costs of undertaking remediation of contaminated sites and groundwater are relatively well-known given the decades of activity in this area, particularly in the United States and Europe. Remediation costs in practice will vary with size of plume, type of contaminants, and the nature of the geologic and aquifer material and properties. So, at least in some cases, the later we intervene, the higher the costs of remediation are likely to be. If the prevention or remedial actions taken produce a secondary impact, it should be included in the analysis as an external cost of remediation.

External Costs of Remediation

External costs of remediation are conceptually similar to disbenefits in that they are damages that accrue over time as impacts on stakeholders or resources. Thus, as long as they continue, they will accumulate. External costs reflect the damages that occur as secondary impacts of the main remediation, after the application of available mitigation measures. External costs of remediation can be divided into planned or process-related external costs that cannot or will not be mitigated against and unplanned or unforeseen external costs (to which a probability of occurrence can be attached).

Planned external costs may include the landfilling wastes excavated from contaminated sites, which may result in secondary damage at the new location and will generate costs to society associated with transporting waste to landfill using heavy goods vehicles (HGVs) by way of increased congestion, impacts on health from emissions, noise impacts, and increased probability of accidents.[12,13] For instance, the external costs of transporting contaminated materials by road in HGVs have been estimated at US\$0.78/vehicle-kilometer (v-km).[14] This can add up quickly. For 1,000 vehicle movements, each of a 500-km round-trip, an external cost of US\$390,000 is added to the overall cost of remediation. The relevance of this impact can be seen by considering typical private remediation costs for excavation and landfilling of 10,000 tonnes of contaminated soil. A typical remediation program of this size would cost on the order of \$2 million to \$5 million, depending on location, contaminant type,

TABLE 7.1
Examples of Planned External Costs of Remediation

Activity	Secondary Effect	Comments
Air stripping of volatile compounds from groundwater without off-gas treatment	Release of volatile compounds to atmosphere	Still occurs in many jurisdictions; can be mitigated against
Thermal treatment of contaminated soils	Release of CO_2 and other gases to atmosphere	Greenhouse gas emissions
Permanent geosequestering of contaminated groundwater (deep-well disposal)	Permanent loss of injected groundwater as a resource	Widely used for difficult and recalcitrant contaminants
Excavation of concentrated source of contamination to protect underlying groundwater results in habitat destruction	Habitat in excavated area destroyed	Mitigation "banking" approaches can be used to offset

tipping fees, and the complexity of the dig. In this example, the expected private or internal cost for remediation using "dig and dump" was expected to be approximately $3.2 million. Adding $0.4 million to reflect the real cost of the remedy represents a 12% overall increase in cost. Note also that if clean fill has to be imported to the site to fill in the excavation, additional vehicle movements will be required, further boosting the external cost of transport. Furthermore, the other possible external costs of landfilling have not yet been added (see the discussion on the external costs of landfill in this chapter).

Examples of other types of planned external costs of remediation are listed in Table 7.1. In general, planned external costs are increasingly being mitigated against. In many jurisdictions, specific regulatory measures are being put in place to ensure that remediation methods that deliberately shift costs from the problem holder to society are reduced or eliminated.

Accounting for unplanned or unforeseen costs of remediation is of course problematic: We may not know they are going to happen, or we may have discounted them as only a remote possibility. Sometimes, despite the best planning and care, remediation activities result in the creation of a secondary impact to the environment or to other stakeholders. If the impact is an unplanned or unforeseen result of remediation for which mitigation measures have not been provided or have not been successful in countering, then the value of this damage is included as an external cost of remediation. Table 7.2 provides a list of examples of unplanned external costs.

Accounting for unplanned external costs within an economic evaluation of remedial alternatives is not straightforward. For any given remedial approach considered, the possibility that its implementation may cause additional external damages must be carefully evaluated. In most situations, experienced remediation engineers and specialists should be able to identify possible secondary damages. In all cases, mitigation measures should be put into place to deal with these possibilities. Whatever probability remains of that damage occurring should be applied to the value of the

TABLE 7.2

Examples of Unplanned External Costs of Remediation

Activity	Secondary Effect	Example
Remediation causes LNAPL to revert to DNAPL due to preferential removal of lighter compounds.	NAPL sinks, contaminating a new volume of aquifer, worsening dissolved-phase problem.	SVE (soil vapor extraction) preferentially removes volatile aromatics from an LNAPL containing less-volatile dense compounds.
Bioremediation results in creation of daughter products that are more toxic than parent.	Toxicity to receptors increases.	TCE (trichloroethylene) degrades to VC (vinyl chloride), and VC persists in aquifer.
Remediation inadvertently increases mobility of contaminant within the aquifer through alteration of physiochemical properties.	Impact on receptors worsens due to further spreading of plume, increased mass flux, or more rapid breakthrough.	Surfactant flush greatly increases dissolution and mobility of NAPL, which migrated into previously uncontaminated rock.
Remediation inadvertently increases mobility of contaminant within the aquifer through alteration of properties of the aquifer itself.	Impact on receptors worsens due to further spreading of plume, increased mass flux, or more rapid breakthrough.	In situ fracturing of aquifer to enhance NAPL recovery inadvertently allows increased NAPL mobility toward receptors.
Remediation compromises adjacent confining layers or geological features.	Contaminant is introduced into a hitherto uncontaminated geologic unit.	Pumping wells completed across a confining layer, cross connecting two groundwater-bearing zones.

damage anticipated in case the event does occur (the cost of implementation of mitigation measures should also be added to the overall cost of remediation). Assigning a probability to an eventuality that is being mitigated against is a matter of professional judgment and should be based on experience, knowledge of the limitations of remedial technologies, and the mitigation measures themselves.

BENEFITS OF REMEDIATION

Traditionally, when examining the "economics" of site remediation, the focus has been placed on cost. This has led to a fixation, in many parts of the industry, at many levels, on least-cost solutions. However, there has been little consideration of whether the lowest-cost solutions actually yield commensurate benefit for society. Rational, economically balanced decision making requires that some consideration be given to understanding whether the sum that is to be spent is actually "worth it." If the lowest-cost remedial solution that can be devised to meet a specific cleanup target (of say a certain concentration of a contaminant in soil) is far greater than the value of achieving the cleanup goal, society is worse off if the remediation is undertaken. Conversely, the opposite also holds: If there is significant value in achieving a remedial objective, then any method that achieves that cleanup at a cost lower than the benefit that will be realized is a good deal for society. But, without being able to express remedial benefits in the same unit as the costs (e.g., dollars, euros, pounds, francs, yuan), society cannot know definitively if the remediation is actually achieving some real good.

In the author's experience, much of the contaminated site remediation done around the world over the last 20 years has likely not produced an increase in overall human welfare—it has not been economic or sustainable. Conversely, much of the remediation that has been undertaken to attempt to deal with significant problems has been underdone, often because insufficient funds were devoted to the effort in relation to the significant social value that could have been realized by remediation. Understanding and quantifying the benefits of remediation are critical in allowing a more balanced allocation of resources to remediation and selection of objectives, strategies, and technologies that optimize outcomes.

PRIVATE BENEFITS

If the analysis is undertaken at the company (or problem-holder) level, at which only the costs and benefits that will accrue to the problem holder are considered, then the analysis is a *financial* analysis. When estimating the financial costs and benefits, market prices are used, including the subsidies or taxes that are included in the market price. Financial analysis does not deal with environmental or other social impacts of an investment unless these have a direct implication for the costs and benefits of the problem holder. In essence, financial analysis is what is traditionally done when evaluating remediation. Table 7.3 presents a selection of benefit categories that can be used in a financial analysis.

EXTERNAL BENEFITS

If the analysis is undertaken for the whole of society, then the analysis used will be an economic (or social) analysis. The EESA method focuses on the wider social analysis (which of course includes the financial) to examine the costs and benefits that accrue to society as a whole. External benefits of remediation are those that accrue to the rest of society when a problem holder undertakes remediation. If contaminated sites create damage, either because they are not remediated or because only some of the effects of the contamination are dealt with, then this damage is an external disbenefit. Table 7.4 presents a selection of benefit categories that can be used in an economic (social) analysis.

In practice, only some of the external benefits of remediation can be readily quantified and monetized. The degree to which monetization of benefits is taken depends on the circumstances of the analysis. For more complex, high-profile, and serious problems, a greater degree of analysis would be warranted. When benefits can be reliably monetized, they should be. Benefits valuation methods are discussed in Chapter 3. Some of the more important benefit categories listed in Tables 7.3 and 7.4 that are particularly important for contaminated sites are discussed in more detail next.

TABLE 7.3
Private Benefits of Remediation

Private Benefit	Comment
Increase in property value	Applies to increase in the value of the property owned by the proponent that results from the cleanup action. Benefit is the net increase in value over the preremediation value. In many instances, this can be a major internal driver to remediate contaminated sites on high-value urban properties.
Elimination or reduction of corporate liability	Remediation (and possible sale) of a contaminated site may allow an owner to eliminate a financial liability or provision currently affecting its balance sheet. This can be seen as a direct financial benefit to the owner or company. Contaminated site liability provisions are, depending on the jurisdiction, often based either on guesswork or on an estimate of remediation cost. On that basis, the net position can often be neutral.
Public relations value	Remediation of a contaminated site can result in a reduction in ongoing negative public relations, which may result in improved stakeholder relations, lower cost of capital, or perhaps improved financial performance through customer attraction. In practice, it can be difficult to quantify this benefit.
Avoidance of prosecution or fines	If remedial action avoids fines or legal action against the company, the costs avoided are a direct financial benefit to the company. However, when a complete economic analysis is done, these costs are not included as benefits as they are simply transfer payments (the payments are a cost to the company but a benefit to society [the government], so they cancel each other out).
Health and safety benefits	If remediation of contamination reduces health and safety impacts on workers of the company, then this will be a direct benefit to the firm in terms of reduced expenditure on ongoing protective measures, increased workforce productivity, lower absenteeism, and lower medical costs borne by the company. Improvements in the workers' own health and the benefits the workers themselves realize are counted as external benefits (see Table 7.4).
Protection of resources used as production inputs	If remedial action protects the quality or quantity of a resource that the company uses in its production, financial benefits may also be realized. For example, if contamination from the facility is making its way into an aquifer that the company uses as a source of production water and the company then has to treat that water to allow it to be used, remediation may result in the avoidance of some or all of those ongoing treatment costs.

Remediation of Brownfield Sites: Unlocking Private Benefit

As discussed in Tables 7.3 and 7.4, one of the most robust ways of examining the economic impact of site contamination is to consider its effects on property value. In many places, the value of property drives efforts to remediate and then sell the land.

First, there is the value of the site itself. A contaminated property will almost always sell at a discount over the price that could have been fetched if the site were fit for purpose (assuming the purchaser has knowledge of the contamination on site). In many cases, one of the key benefits of remediation to the owner of a contaminated site is the increase in property value achieved. If the increase in value is greater than the costs of remediation, a net profit is realized. This can be a powerful impetus to clean up sites. Throughout the developed economies, a growing number of

TABLE 7.4
External Benefits of Remediation

Private Benefit	Comment
Increase in value of neighboring properties not directly affected by contamination	A cleanup action will often cause an increase in the value of neighboring properties that are not physically or directly impacted by the contamination but are simply affected "by association" with the site, typically through the effects of odor, visual, or aesthetic concerns, or worries over possible health impacts. Benefit is the net increase in value over the preremediation value. This effect is widely observed in the literature and in empirical studies of property value.[15] Reduction in blight is typically valued through hedonic pricing, which captures a bundle of benefits accruing to those in the affected properties.[16]
Increase in value of neighboring properties directly affected by contamination	If nearby properties are physically affected by the contamination—contaminants have been deposited or have migrated from the site to neighboring properties—then remediation of the site itself may also improve the value of the adjacent properties. If the remediation extends to the nearby properties themselves (the company cleans up the neighboring sites also), then the increase in property value of the nearby properties is also an external benefit of remedial action.
Health benefits to neighboring residents	If remedial action improves the health of residents living near the site, an external benefit would result. These benefits are real and can be measured through reduced medical expenditure, improved productivity, or increased income due to improved work attendance. They are not typically captured in a traditional financial analysis. Care must be taken to ensure that a health component is not also included in the property value benefit (double counting).
Health and recreational benefits to visitors of the area	Remediation of a contaminated site may reduce the real or perceived health impacts on visitors to an area. This can be valued in the same way as the health benefits to residents of the neighboring area. If site contamination has been affecting the perceived enjoyment of visitors to a nearby recreational area, either by direct contamination or by "association," then remedial action can also trigger an increase in wider benefit experienced by the user group.
Reduction in ecological damage	In many situations, contaminated land may act as a long-term source of deleterious impact on ecological resources, either within or outside the site. Wetlands, forests, aquatic ecosystems, coastlines, marine habitat, and individual species may be adversely affected by contamination. Elimination of these impacts, and the improvement in the quality and health of the affected ecosystems, can be an important external benefit of remediation and one that is rarely, if ever, captured in conventional financial analysis.
Protection of resources used by others	If remedial action protects the quality or quantity of a resource that is being used (or may be used in the future) by other parties, this would accrue as an external benefit to society. For example, if contamination from the facility is making its way to an aquifer that is used as a source of water for a community and the community is being affected by that contamination, directly or indirectly, remediation will result in benefits to society. In this example, the benefit could be the elimination of the need for end-of-pipe treatment for the contaminants before distribution of the water to users or elimination of the need to develop an alternative source of water. For water resources, all elements of the total economic value (TEV) may benefit from remedial action, including nonuse and option values.

"brownfield" developers are seeking to capitalize on the often-considerable price margin between "dirty" and "clean" sites.

Brownfield projects typically involve purchasing a contaminated property at a substantial discount, remediating the site, and then selling the property for more than the sum of the original purchase price and the remedial cost. Proponents of these projects take on the liability for the contamination under the assumption that they can remediate the site, eliminate or manage the liability, and produce a site that can be sold at or near the undiscounted market value. Different organizations and individuals have different levels of risk tolerance, so their willingness to pay (WTP) for liability reduction can be markedly different. Firms or groups that understand the technical, legal, and financial complexities of the contaminated land business are more likely to be tolerant of environmental risk and liability and are better placed to execute a brownfield project profitably. Since market prices are often driven at least partially by perceptions of risk, and these perceptions can vary considerably, regulators have a key role to play in brownfield transactions. Regulatory approval of remedial designs and results can be instrumental in creating comfort in the market that remediation has been successful, thus unlocking site value.

Revenue realized by cleaning up a site accrues to the site owner as an internal or "private" benefit. The benefit of remediation is the increase in site value that is attributable to the remediation being completed. In other words, the benefit is the difference between the value of the site before remediation and the value after remediation. It can be readily and accurately measured in most countries by market techniques. Property agents maintain detailed and thorough listings of property values and selling prices. This allows ready monetization of this private benefit of remediation. A case history of brownfield redevelopment in Canada is provided later in this chapter to illustrate these points.

Blight Reduction: External Benefit

When a contaminated site is remediated, it is not only the site owner who benefits. By removing what might have been an eyesore at best or a potentially dangerous or hazardous condition at worst, the whole neighborhood benefits. Several recent economic studies have shown that people and businesses experience real economic benefit when a neighboring waste site or polluted site is remediated. This is due to the removal of blight (or disamenity) from the properties in the vicinity of the remediated site. This effect is intuitive—people would rather live in an area without contamination and waste if they had the choice.

Recent research has found substantial negative effects on property values in areas subjected to transshipment of radioactive wastes in the United States.[17] Another study showed a 35% difference between average selling price of homes located within a 2-mile radius of a low-level radioactive waste site in the United States compared to those outside a 2-mile radius.[18] Proximity to landfills, active or closed, has also been shown to have a significant depressive effect on property values. In the United States, decreases in property values between 12% and 25% have been recorded, depending on the distance from a hazardous waste landfill.[19] A study considered the disamenity costs of landfill in Great Britain.[20] The study considered 11,300 landfill sites and over half a million residential property transactions within 2 miles of those landfills over the period 1991–2000 (inclusive). Residential property prices were found to be negatively affected within 2 miles of landfills. Across Great Britain, property values were found to suffer a 7% reduction within 0.25 miles, decreasing with distance to a 1% reduction between 0.5 and 1 mile from the landfill and 0.7% between 1 and 2 miles from the landfill. In Scotland, however, impacts on property values were greater, decreasing 41% within 0.25 mile, 3% between 0.5 and 1 mile, and 2.67% between 1 and 2 miles away.

Removal of disamenity by remediation will cause average property prices in the affected area to rise. This increase, multiplied by the number of properties affected, can be used as a direct market valuation of disamenity or blight reduction and as an estimate of the economic benefit that accrues to those stakeholders involved. This benefit will be greatest in dense urban areas with many neighbors and higher property values.

ECONOMIC SUSTAINABILITY ANALYSIS FOR CONTAMINATED LAND

OVERVIEW

Management of contaminated sites can benefit from application of the EESA process. The EESA process starts with a clear understanding of the contamination problem and the risks associated with the site to all receptors; this is achieved through detailed site investigation and risk assessment.[21] This should include identification of a clear remedial objective at the appropriate level of consideration, listing of a range of viable options that can meet that objective, analysis of the constraints to remediation that may limit the application of certain remedial approaches or techniques,[22] and development of designs that comprise each option. At that stage, each option is physically described in terms of its life-cycle inputs and outputs and the effects it is expected to have on its surrounding environment and society. Then, full life-cycle capital and operational cost estimates can be developed, and remedial benefits assigned to each approach.

Different approaches will often accrue different levels and categories of benefits at different times. Calculation and comparison of the discounted present value (PV) of all the costs and benefits of applying each remedial approach yield full social net present value (NPV) estimates for each option. Then, a detailed sensitivity analysis, using the approach discussed in Chapter 3, allows an overarching view on decision making to take shape (the NPV itself is not of interest). The future is uncertain, so any longer-term analysis cannot realistically hope to come up with "the answer." Rather, we seek an indication of which options are clearly superior over a wide range of possible future conditions and which are clearly and consistently worse. The results of the EESA can then help to inform decision makers regarding the balanced, economic, and sustainable course of action.

In some cases, more than one objective can be selected for the site. For example, we could choose to both protect an identified off-site receptor from future harm and clean up the site to an appropriate standard for sale for residential redevelopment.

In a risk-based environment, objective selection can also be based on identification of the remedial strategies that manage identified risks most economically.

APPLICATION

EESA is perhaps most useful at the objective-setting and strategy levels. Here, the benefits of achieving a certain objective are assessed in a high-level preliminary analysis. To the extent that benefits can be monetized, they are. If benefits cannot be monetized but clearly exist, a benefits threshold approach is applied: "We know that achieving this objective will result in benefits of at least this much" (monetized) and that several key additional benefits will be accrued (nonmonetized; can be described qualitatively). Technology selection is usually (but not always) a least-cost analysis exercise (what is the least-cost way of implementing the strategy within the identified constraints?).

One of the main advantages of broad economic analysis is that all stakeholders can have input into the analysis and have their concerns measured in the same units (money). This is where wider economic issues (external or social benefits and costs) are considered and the fundamental direction of the remediation is decided. Once the objective has been identified and its economic viability demonstrated, the details of implementation, such as which remedial technologies to deploy, can be developed. EESA can be applied at this next level also, if warranted, but the real value comes from ensuring that the designed-for objective makes overall sense to society. There is not much use in working hard to develop a least-cost technical solution that will achieve a socially net-negative outcome. After all, remediation is undertaken, in large part, because the status quo is unacceptable to society. It follows then that the way in which remediation is conducted, and what it is designed to achieve, should actually yield some net benefit to society. The rest of this chapter presents a series of case histories in which EESA, or elements of the process, is used at various levels of sophistication to identify more socially optimal positions.

THE SOCIAL COST OF WASTE MANAGEMENT USING LANDFILL

OVERVIEW

Worldwide, dumping rubbish into a hole in the ground (more commonly known as landfilling) remains the most common and widely used way of dealing with the ever-increasing amounts of solid waste generated by society. In addition, it remains one of the most popular methods of remediating contaminated sites; contaminated soil is excavated and removed by truck to a secure landfill. Figure 7.2 shows an industrial landfill constructed specifically to contain the waste generated at a major gas-processing facility.

The reason for the widespread popularity of landfilling is its low cost. Plainly put, it is the cheapest way to isolate waste (at least temporarily) from society. Growing material consumption associated with rising global product (GP) has led to a significant surge in the amount of solid waste generated worldwide (including municipal solid waste [MSW], construction and demolition waste, and special and hazardous wastes).[23] Because landfill is cheap, the majority of this waste is directed to landfill. Depending on the jurisdiction, a landfill might be simply a hole in the ground, an abandoned quarry, or, where legislation is in place and is properly enforced, a fully lined and monitored secure facility. But, landfilling is not as cheap as we might think. There are myriad hidden costs associated with the practice.[24]

FIGURE 7.2   A newly constructed secure industrial waste landfill at a gas-processing plant. The landfill is lined and capped to prevent infiltration and generation of leachate. The landfill contains a variety of wastes generated by the facility over several decades, including contaminated soil generated during a major remediation effort at the site.

An Overview of Waste Trends

In the United States, total MSW generation rose from about 80 million tonnes (mt)/year in 1960 to over 250 mt/year in 2007, driven by increasing affluence and a population that grew from 180 to 300 million. During the same period, per capita waste generation in the United States almost doubled.[25] In Australia, total waste generated rose from 22.7 mt in 1997 to over 32 mt in 2003, partly due to a 32% increase in the amount of waste generated per person.[26] In 2008, Australia produced 41 mt of solid waste,[27] with each Australian responsible for over 350 kg of waste sent to landfill every year (Americans and Cypriots landfill most, 625 kg and 653 kg per person, respectively).[28]

More people generating more waste is a trend duplicated across the developed world over the last decades. And now increasingly, the same behavior is being seen in the developing countries. Asia is the most populous and rapidly growing region on the planet. The economies of India and China are now major engines of economic growth, and increasing prosperity throughout the region is driving booming consumption. Waste management is now an exponentially growing problem in Asia-Pacific, with projected per capita waste generation in countries such as India, the Philippines, and Thailand expected to grow by more than 50% over the next 15 years.[29]

However, in many developing countries, recycling has begun to make an impact on the amount of waste that ends up in landfill. As the value of recovered materials such as plastics, aluminum, and paper has increased, recovery and recycling of these materials has begun to make financial sense. In the United States, materials recovered for recycling and composting increased from a mere 5.6 mt in 1960 to 85 mt in 2007, by which time a third of MSW nationally was recycled.[30] In Australia, 14.8 mt of waste were recycled in 2003 compared to only 1.5 mt in 1997, almost a ninefold increase in just 6 years, which has contributed to an overall 24% drop in the amount of waste per person directed to landfill.[31] Recycling efforts in Europe have been heterogeneous. The Netherlands, for instance, landfills less than 5% of its MSW and recycles over 65% (the remainder is incinerated in waste-to-energy plants). Germany, Sweden, and Belgium perform similarly. At the other end of the spectrum, Greece landfills over 90% of its waste and the United Kingdom over 75%. Portugal recycles less than 5% of the solid waste it produces.[32]

SOCIAL COSTS

There are hidden costs to landfill, costs that typically are not paid for by the waste generator but are borne by society as a whole. Depending on the types of wastes in the landfill and the construction and operation of the facility, these social costs may include nuisance and odor from operating sites that may have an impact on local residents, generation of greenhouse gases (GHGs) and other air pollutants from decomposing organic materials, and impacts on groundwater and surface water from the generation of contaminated leachate. A number of studies have estimated the social costs of landfill. Table 7.5 provides a range of estimates broken down into major contributing components.

Other estimates of the external costs of landfill have been developed using other economic methods. In the United Kingdom, for instance, empirical data have shown

TABLE 7.5

Estimates of the Social Cost of Landfill (2008 U.S. Dollars/tonne)

Study	Air Pollution (Including GHG)	Leachate	Disamenity (Including Noise and Odor)	Total External Cost	Notes
European Commission (2000): new facility[33]	9.20	0	18.50	27.60	Methane and CO_2 emissions priced at 1999 estimates for cost of carbon emissions
European Commission (2000): old facility[34]	14.70	2.76	18.50	36.80	Methane and CO_2 emissions priced at 1999 estimates for cost of carbon emissions
U.K. Department of Environment (1993): existing landfill, no energy recovery[35]	8.0	1.1	0.6	10.83	Disamenity includes only the cost of increased road accidents
Estimate based on social cost of carbon (Stern, 2006)[36]	106	—	—	106	United States Energy Information Administration estimate of GHG generation from MSW,[37] assuming no gas capture

that people living within half a kilometer of an existing landfill suffer, on average, a decline in their property value of US$10,400 (2008 dollars) or approximately 5% of mean property value.[38] Another study from the United States revealed that people were willing to pay on average US$352 per household per year (2008 dollars) to avoid living in proximity to a landfill.[39] These studies reflect the fact that impacts such as noise, odor, and the risk of possible effects from vapors, gases, and other releases are of real concern to people—they are willing to pay to avoid them, and if they are already experiencing them, they would benefit directly from the removal of these conditions.

In addition, the loss of valuable materials to landfill that might otherwise be recycled represents an opportunity cost to society, driven by the fact that recycled materials may eliminate external costs associated with production from primary resources. Recycled materials typically produce benefits to society through the elimination of GHGs, energy and water savings, and reduced requirements for landfill space.

Social Costs and Landfill Taxes

Landfilling, then, produces significant costs that society must bear. For this reason, governments in some countries, notably in Europe, have introduced taxes on landfilling to curb disposal and encourage waste reduction, recycling, and materials recovery. In the United Kingdom, a landfill tax has been in place since 1996, during which time the amount of waste sent to landfill dropped by 26%. Table 7.6 provides

TABLE 7.6
Landfill Taxes in Selected Jurisdictions

Jurisdiction	Landfill Tax (2008 US$/ tonne)	History	Waste to Landfill (% of total)
Netherlands	122.40	First introduced in 1995 at $18.72/tonne and increased in three steps to the present level. Netherlands has one of the lowest levels of landfilling in the Organization for Economic Cooperation and Development and one of the longest-running and highest landfill taxation programs.	5
Belgium	92.16	Varies by jurisdiction within the country. Introduced in 2004 with annual increases to 2010.	12
Sweden	59.04	Introduced in 2000. Since 2005 illegal to landfill organic waste. Extended producer responsibility enacted in 1994; now includes vehicles, tires, packaging.	15
France	13.18	Introduced 2002. Hazardous waste tax $26.35/tonne in licensed facilities only.	38
United Kingdom	52.16	First introduced in 1996 at a low rate, increased to current level in 2008, and will increase at $13/tonne/ year over the next several years.	75
Greece	0	No landfill tax.	95

data on current landfill tax rates in selected countries and their respective current landfilling rates.

Table 7.6 shows a clear correlation: Landfill taxes reduce initial waste generation and divert waste from landfill and into recycling and reuse. The social costs of landfill are a useful benchmark for determining optimal taxation levels. If taxes reflect social costs, society can see the real price of landfilling and make choices about how it wants to dispose of waste within a market that adequately reflects the true environmental, social, and economic implications of those decisions. As shown in Table 7.6, tax rates need to be high enough to effect measurable change. Set too far below the social cost, they do not reflect the true burden placed on society. Experience also shows that it takes time for the price signals to effect change in the economy. The longer-standing levies have shown more effect.

There is appreciable elasticity of demand for landfill: As tax rates rise, the volume going to landfill falls. However, the European experience also suggests that there is typically a threshold value below which there is little real effect on behavior. The additional cost is simply not enough to promote change. Beyond that threshold, however, a real difference in volumes to landfill may begin to be seen. From a policy perspective, landfill tax revenues are best used to reinforce the purpose of applying the tax in the first place: waste reduction and landfill minimization. Because of the elasticity of demand, relying on landfill taxes as a source of general revenue for government is self-defeating. Combining tax with legislative measures and regulations, such as the introduction of mandatory extended producer responsibility (EPR) schemes, accelerates and augments diversion of waste away from landfill and toward reuse and recovery.

SUMMARY

Disposing of waste to landfill involves significant external social costs that are often hidden from consumers and industry. When landfilling is cheap, there is little incentive to recycle or reduce the amount of waste that society produces. Ample evidence has shown that as landfill taxes approach the social costs of landfilling, recycling and waste recovery become steadily more viable, and landfill disposal declines as a percentage of total waste generated. Understanding the social costs of landfill allow individuals, companies, and society to understand the real costs and benefits of waste management.

CASE HISTORY: BROWNFIELD REDEVELOPMENT IN CANADA

This example of brownfield redevelopment in Canada is intended to illustrate how the purely financial perspective can often trigger remedial action without the consideration of any externalities.

BACKGROUND AND SETTING

A consortium of partners with expertise in contaminated land approached the owners of a 22-hectare contaminated site in Canada. The property was first developed in the early 1960s as a small petroleum refinery that manufactured heating oil, gasoline, and diesel. In the early 1970s, the refinery was decommissioned, and the property was given to the local government, which operated a highway maintenance yard on the site. In 2000, the owner agreed to sell the site to the consortium under the condition that it remediate and redevelop the property for subsequent industrial or commercial use. This program was completed in 2001. Redevelopment of the property ultimately helped to conserve farmland and native uncleared forest that surrounds the town.

Site Description and Contaminant Distribution

The site was contaminated by crude oil and refined products associated with the original refinery and later by salt storage and underground fuel tanks associated with the highway maintenance yard. The surface soils are primarily composed of fill material and native soil (sandy clay). Previously developed areas of the site have been covered with up to 2 meters of fill material. Lagoons and pits associated with the former refinery had been filled in at the time of purchase. Glaciolacustrine clay underlies the fill material and native soils to a depth of approximately 6 meters. Two main groundwater-bearing intervals exist deeper beneath the site, with the clay material acting as an effective barrier to deeper contaminant migration, protecting deeper groundwater-bearing intervals.

Five main areas of hydrocarbon contamination were identified, associated with lagoons and foundations from the former refinery and underground storage tanks associated with the maintenance yard. During site investigation, approximately 25,000 m^3 of hydrocarbon-contaminated soil were identified at the site. Total soil BTEX (benzene, toluene, ethylbenzene, and xylene) concentrations were relatively low, generally less than 100 mg/kg, total phenol concentrations were generally less than 0.5 mg/kg, and PAHs were present in the former bunker and lagoon areas at concentrations up to 10 mg/kg. Elevated nickel and arsenic concentrations were observed in soil but were determined to be of natural original. All contaminated soils on site were classified as nonhazardous in accordance with local regulatory authority guidelines and hence were suitable for disposal at conventional industrial landfills. Salt-contaminated zones were present in the northwest corner of the site where salt had been historically handled and stockpiled and in the south near a former salt shed. The salt-impacted area exceeded 2 hectares but was limited to the upper 1 to 3 meters below the surface.

Hydrocarbon impact on groundwater was restricted to regions of historic use at the site, and in no cases did hydrocarbon concentrations in groundwater exceed the criteria agreed with the regulator. Significant salt impact to the shallow groundwater was observed. Nevertheless, the slow groundwater flow velocities, lack of receptors of significance, and relatively low concentrations of risk-driving compounds meant that no off-site risks resulting from the contamination on site were identified. The risk assessment results were shared with and approved by the regulator.

Remedial Approach

A remediation plan was submitted to the regulatory authority, the town government, and other local stakeholders; the objective was to create a parcel of land fit for commercial or light industrial use and to eliminate the risk that substances on site might cause adverse effects to the environment or human health. A least-cost remedial approach designed to meet the remedial objectives included: (1) excavating heavy-end hydrocarbon-contaminated soils that could not be practically treated on site, with disposal at an approved industrial landfill; (2) excavating light-end hydrocarbon-contaminated soils and biologically treating these soils in an on-site treatment area; (3) managing salt-impacted areas on site by constructing groundwater collection systems; and (4) dismantling and removing facilities and buildings that were no longer required or useful and disposing of the demolition debris in the local landfill. This remedial approach selection method was conventional and did not, at the time, employ the EESA methodology. The following discussion is presented as a retrospective view.

COST–BENEFIT ANALYSIS

The primary driver for remediation in this case, as with most brownfield sites, was the anticipated private benefit resulting from the sale of remediated land, fit for use. From this perspective, a financial cost-benefit analysis revealed that the total private benefits were expected to exceed the total costs of remediation. So, while other social benefits would clearly result from the cleanup and redevelopment of the formerly derelict site, they were not needed to justify the project from an economic perspective. Furthermore, the partners in the redevelopment, having taken on the financial burden of remediating the site at their own cost, were expecting a significant return on investment. Typically, social discount rates range from as low as 2% to as high as 6%. However, the private sector typically expects returns on risk capital of 10% or higher.

Table 7.7 shows the discounted costs of remediating the site and compares them with the discounted total revenue realized through sale of the reclaimed land. Costs are quoted as percentages of the initial land purchase price that the brownfield consortium had to pay to acquire the contaminated property. What is readily apparent is that the brownfield redevelopment team had to be prepared to accept a period of over a year of significant negative cash flow. At one point, total cash outflow reached over seven times the purchase price of the contaminated property. As parcels of remediated land were sold over time, the initial investment was recouped, and then the project gradually began realizing profits. Clearly, significant financial risk is taken on by the developer in these situations. Risks included higher-than-expected remediation costs, change in market conditions leading to a drop in land value, change in

TABLE 7.7
Cost–Benefit Analysis (Private)

Cost–Benefit Category	PV Cost (−)/Benefit (+) as % of Initial Land Purchase Price
Purchase of contaminated property	−100
Remediation cost	−300
Development of property (roads, access, services)	−333
Revenues realized from sale of developed lots	+1200
Net benefit (private)	+467

regulatory climate, cash flow limitations, higher-than-anticipated borrowing costs, and inability to sell land after cleanup. For all of these reasons, brownfield redevelopers will typically look for a significant private net benefit to defray the many risks involved.

IMPLEMENTATION AND OUTCOMES

The remediation was completed over 1 year. A verification program was implemented during both soil and groundwater remediation to confirm that remedial objectives were being met. The results were submitted to the regulator. The site was successfully redeveloped for industrial use in accordance with the remediation criteria and remediation plan submitted to stakeholders and the regulatory authority. The local community has welcomed the efficient use of space within the existing area of industrial development. The majority of purchasers and interested parties are businesses that need to expand or augment their existing facilities to meet the increasing demand for services in the area. Further, the former site owner has been able to sell and redevelop an unused facility in an efficient, reliable, and responsible manner.

A WIDER PERSPECTIVE

The brownfields project was clearly a success for all parties: the community, the regulator, and the developer. The financial analysis in Table 7.7 shows a healthy return on investment for the developers. However, in undertaking the remedial works, there were uncosted external damages that society had to bear, including GHG emissions from the remedial activities and the external costs of road transport and landfilling of contaminated soils. On the positive side, society accrued some significant benefits from the remediation, including the protection of groundwater and the elimination of blight on surrounding properties. Table 7.8 shows how the inclusion of external effects in this case demonstrates an ever more economic and sustainable outcome overall for society than is revealed by the purely financial.

However, the question remains: Did the objective selected at the outset using traditional decision making result in the maximum possible overall net benefit to all stakeholders? The values in Table 7.8 show the economics of only one of a range of

TABLE 7.8
Cost–Benefit Analysis (Social)

Cost–Benefit Category	PV Cost (–)/Benefit (+) as % of Initial Land Purchase Price
Total financial costs	−733
Total financial benefits	+1200
Net benefit (private/financial)	+467
Total external costs	−12
Total external benefits	+120
Net benefit (overall social)	+575

possible remedial options. There is no way to tell whether this is a comparatively optimal solution. The application of EESA can help to reveal if such optima exist and how to achieve them, as shown in the following examples.

CASE HISTORY: GROUNDWATER REMEDIATION AT A REFINERY IN EUROPE

OVERVIEW

A spill of organic liquid chemicals at an operational refinery in Europe led to the site owner undertaking a significant program of site investigation works and remediation pilot trials to determine the levels of contamination present and the feasibility of potential remedial solutions. This work revealed that the extent of the contamination due to the spill was limited, and that there were no current impacts to identified receptors. The investigation also confirmed what had been revealed in previous studies at the site: Other contamination existed at various locations and depths across the complex. However, the nature of the contamination in question—light nonaqueous phase liquids (LNAPLs) within fractured sandstone at depth—meant that remediation was likely to be technically challenging and costly.[40] The regulator, however, was of the view that significant remediation should take place to attempt to remove at least the bulk of the separate-phase LNAPL that had reached the aquifer at depth beneath the site.

Site Conditions

The refinery site comprises an area of over 600 hectares. One boundary is adjacent to a major estuary. The LNAPL spill in question occurred almost in the center of the complex, far from any site boundary. The spill site area itself was approximately 3 hectares and formerly housed a chemical production plant constructed in the mid-1970s. The plant was decommissioned in 1990 and was subsequently demolished to ground-floor slab level. The area is now vacant and unused. The LNAPLs were released at some point before the facility was decommissioned. Figure 7.3 shows the basic layout of the site.

The groundwater surface occurs at a depth of 5.0 meters, and regional groundwater flow is to the north and the estuary. Various areas of LNAPL contamination were identified at and above the groundwater table, and the hydrocarbons were generally in the C6 to C28 range. The volume of LNAPLs in this area was estimated to be 16–64 m^3, predominantly present in fractures with limited penetration into the sandstone matrix. A licensed on-site water abstraction well, BHA (borehole-A), pumps water from the regional aquifer and uses it for industrial applications on site. So far, there was no evidence that contaminants from the spill had reached the well, which is located approximately 1 km from the spill source.

Risk Assessment

The chemical compounds relating to the former chemical manufacturing plant were of low volatility, and there were no buildings above the area of contamination. The minimal vapor risks were addressed by on-site health and safety management

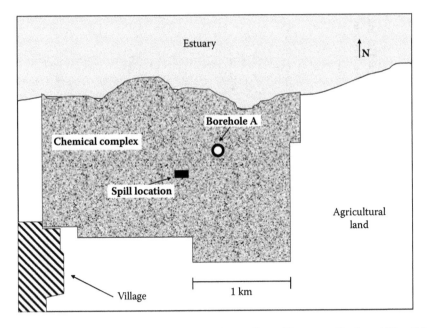

FIGURE 7.3   Map of the facility area and environs. The spill occurred in the middle of the operational complex. Regional groundwater flow is toward the estuary.

procedures, so were not considered in detail in this assessment. Direct human contact, migration via airborne dust, and migration via pipes and service trenches were not interpreted to be active pathways for the contamination to have an impact on on-site workers, and no pathways existed to any off-site receptors. The key pollutant linkages (hazards) relating to the site are summarized in Table 7.9. Despite the overall low risks associated at the site, regulatory pressure to remove liquid organics from the groundwater was intense, driven in part by an informal policy position that concentrated "subsurface sources" of contamination (such as nonaqueous phase liquids [NAPLs]) should be removed from groundwater-bearing zones wherever feasible.

Constraints to Remediation

Constraints to remediation can take many different forms. Time pressures, physical limitation on placement of equipment, technological limitations, and land ownership or property access constraints are just some of the issues that could affect the selection of remediation objectives. Also, some objectives, such as elimination of all contamination from the subsurface and groundwater, may be technologically infeasible even if unlimited funds are available. Constraints can be generally categorized as temporal, physical, and technological. The only significant constraint to remediation at the site was the technological difficulty associated with remediation of LNAPLs in fractured rock, which is generally considered very difficult.[41] Few remediation methods have been successfully demonstrated in these conditions. Indeed, the remediation pilot trials conducted at the site confirmed this.

TABLE 7.9
Hazards to be Managed

Source	Pathway	Receptor	Magnitude of Impact	Likelihood of Impact	Hazard ID and Overall Hazard
LNAPLs at and above the water table beneath old chemical facility compound	Free phase migration within fractures in the aquifer and vadose zone	Underlying aquifer	Low: Volume of impacted aquifer is small. Plume is stable, trapped as residual saturation, and extends less than 100 m from source.	High: Impact has occurred to small volume of aquifer. Likelihood of impact on wider regional aquifer: Low.	1: Low
		Abstraction borehole BHA	High: If LNAPLs did reach BHA, it would effectively remove it from production completely. Site would need to find an alternative water supply.	None: LNAPLs have not reached BHA and are not expected to in future based on modeling studies.	2: None to low
	Dissolution and migration via groundwater flow	Underlying aquifer	Low: The LNAPL spill is degraded, and the compounds within it have relatively low solubility. Site investigation shows low concentrations of dissolved organics. There are no licensed off-site boreholes down-gradient of the refinery (aquifer discharges to estuary).	Moderate within 100 m of spill; biological attenuation is occurring. Low beyond, where after 10 years since the spill, no dissolved-phase contamination from the spill has been identified. Note that other contaminants from other on-site sources have been detected throughout the aquifer across much of the site.	3: LOW
		Abstraction borehole BHA	Low: No impact from dissolved-phase contamination from the spill is detected.	Low: Low levels of organic contamination are already present in pumped water and are currently being removed by end-of-pipe treatment to a level suitable for industrial use of the water.	4: Low
		Estuary	Low: The low levels of organic compounds discharging to the estuary from the sitewide contamination are not affecting the already degraded water quality or ecology of the estuary.	Low: The net effect from the spill, in the context of the overall situation, is negligible.	5: Low

TABLE 7.10

Remediation Objectives and Hazards Managed

Remedial Objective	Hazards Managed
Make complex suitable for redevelopment to agricultural-residential standard	1, 2, 3, 4, and 5
Protect BHA abstraction well	2 and 4
Remediate underlying aquifer already impacted	1, 2, 3, 4, and 5
Protect estuary	5
No action/maintain status quo	None

Level of Analysis

The decision to remediate a site, the remediation objective selected, and the methods used to achieve that objective lie at the core of sustainably and economically implementing remediation. For the spill in question, a number of possible remedial objectives could be defined, as listed in Table 7.10. Most are capable of managing one or more of the hazards identified in Table 7.9.

Each remediation objective can be achieved in a number of different ways, using a number of different technologies, applied in various combinations, at different places and times. Together, these variables are used to describe the remediation strategy. For example, the objective of protecting the groundwater abstraction well from future contamination could theoretically be achieved by source removal, a hydraulic barrier between the source and the abstraction, treating the groundwater abstracted, or a replacement water supply. All of these approaches achieve the objective, but each focuses on a different part of the risk linkage (source, pathway, or receptor).

Remediation strategies focus on how the source-pathway-receptor (SPR) linkage is to be broken to achieve the specified objective. Benefits could vary depending on the remediation strategy selected. Thus, the benefits part of the economic analysis requires that an objective be coupled with an economically optimal remediation approach. Approach selection will also have a direct impact on cost because different approaches will typically involve different technologies and thus varying costs. Determining which objective is best involves identifying high-level alternative approaches for each objective and evaluating the costs and benefits of each approach. Only then can the most sustainable and economic objective for a given pollutant linkage be determined.

Potential remediation approaches for the spill are presented in Table 7.11 along with relative overall costs, advantages, and disadvantages. Consideration of relative costs, at this point in the analysis, provides a precursor to development of high-level cost estimates for each approach. Benefits vary depending on which remediation objective is achieved. The costs and benefits of each approach–objective combination are then assessed. This allows each to be compared to the others on the basis of the benefits that will result and the cost to implement.

TABLE 7.11
Remediation Strategy Alternatives Preliminary Screening

Designation	Approach Description	Relative Cost	Advantages/Disadvantages
		Minimal Intervention	
N1	Monitored natural attenuation (MNA)	Low	Low monitoring cost. Unlikely to protect receptors. Extent of contamination may worsen with time.
N2	Do nothing	Low	No capital cost. Does not address liability or risk of action from regulators. Extent of contamination may worsen with time.
		Source Methods	
S1	Readily mobile LNAPL removal	Moderate	Removes main source in the vicinity of the site area. Likely to leave a significant mass of residual contamination in place. Unlikely to completely protect receptors. No improvement to surrounding areas of the complex. Potential for recontamination of chemical manufacture area from NAPL/dissolved-phase migration from surrounding areas.
S2	Full NAPL removal (mobile and residual)	High	Prevents NAPL spill from manufacture area from migrating laterally or vertically. High relative costs and technically difficult. No improvement to surrounding areas of the complex. Potential for recontamination of site area from NAPL/dissolved-phase migration from surrounding areas.
S3	Full NAPL removal and site remediation, as part of complex decommissioning at end of refinery life	Very High	Removes liability from whole complex. Very high relative costs. Remedial technology likely to improve in effectiveness over time and drop in cost. Remediation of all contamination at once removes the very real issue of recontamination after cleanup from other remaining contamination and ongoing sources and future spills over the life of the refinery operation.
		Pathway Methods	
P1	Hydraulic containment at chemical manufacture area boundary	Moderate	Prevents migration of contaminants laterally toward BHA. NAPL would be contained, but contaminated groundwater is likely to be able to migrate off site. Technically very difficult to prevent NAPL and contaminated groundwater bypassing the barrier at the sides or base.
		Receptor Methods	
R1	Divert groundwater pumping to other abstraction wells	Low	Only addresses the risk to the abstracted groundwater from BHA. Does not protect the underlying aquifer.
R2	Treat BHA groundwater abstraction	Moderate	Only addresses the risk to the abstracted groundwater from BHA. Does not protect the underlying aquifer.
R3	Replace extracted groundwater with main water	High	Only addresses the risk to the abstracted groundwater from BHA. Does not protect the underlying aquifer.

REMEDIAL COSTS

Remedial costs were estimated based on available data from similar remediation projects undertaken over many years in Europe and North America. In all cases, for both costs and benefits, the base case assessment used a discount rate of 5% and examined a 20-year planning horizon.

N1: Monitored Natural Attenuation

Monitored natural attenuation (MNA) involves further site investigation, field and laboratory analysis, and ongoing monitoring to demonstrate that natural attenuation processes are operating and residual contaminant concentrations are reducing as a result. The program was anticipated be undertaken for the duration of the 20-year planning horizon considered. The capital cost to set up the monitoring network was estimated to be US$1.5 million with an annual operating cost of US$0.12 million/year over a period of 20 years.

S1: Readily Mobile NAPL Removal Plus MNA

This approach involves the removal of readily mobile NAPL from the chemical manufacture-related NAPL plume located in fractures in the sandstone bedrock, coupled with MNA for dissolved-phase contamination. The plume was estimated to cover an area of approximately 0.35 hectares. The remediation would be carried out using multiphase high-vacuum extraction (HVE). This would be achieved via wells installed in a 20 × 20 meter grid system (assumes 10-meter effective radius of influence per well) across the NAPL plume area over a 3-year period. The capital cost for the remediation was estimated at US$0.2 million with an annual operating cost of US$0.7 million/year.

S2: Full NAPL Removal Plus MNA

This approach involves the removal of as much of the NAPL from fractures in the sandstone bedrock as practicable. The remediation technique employed would be that used in option S1 (HVE), with the addition of steam or surfactant injection to assist in removal of NAPL that was not readily mobile, and MNA for dissolved-phase contamination. As in option S1, NAPL removal would be achieved via wells installed in a grid system across the site over a 3-year period. The assumptions for the spacing and number of the wells were the same as those for option S1. Capital cost for the NAPL removal was estimated as US$0.27 million plus US$1.2 m/year operating costs.

S3: Full NAPL Removal and Site Remediation as Part of Site Decommissioning Plus MNA

This remediation approach (full NAPL removal and site remediation as part of site decommissioning plus MNA) assumes that the facility has reached the end of its life and the decision is made to close the refinery permanently and decommission the entire complex. The objective would be to render the land fit for redevelopment purposes. Under this hypothetical scenario, it is assumed that the complex would be decommissioned in 20 years. An overall estimate of decommissioning costs was developed by comparing costs for similar facilities in other parts of Europe and North America. A typical per hectare remediation and reclamation cost for several other facilities was used and the per area costs translated to this facility after some crude adjustments for the relative areas of process and tank farm facilities. After discounting, and including 20 years of complexwide MNA for the facility, the cost of S3 was estimated to be US$250 million.

P1: Hydraulic Containment at Chemical Manufacture Area Boundary

This approach involves the installation of a hydraulic barrier at the chemical manufacture area boundary to prevent further migration of contaminants toward BHA. The remediation technique adopted would be HVE along the western and northern site boundaries, operating over the 20-year planning horizon. The capital cost of installing the hydraulic containment system would be US$0.35 million plus US$0.70 million in annual operating costs.

R1: Divert Groundwater Pumping to Other Abstraction Wells

This option simply diverts groundwater pumping from BHA to other wells that already exist on the complex and are already equipped to provide the necessary water to the facility. Shutting down BHA would reduce the possibility of contaminants from the spill being drawn toward the well. There would be very little additional cost for this option.

R2: Treat BHA Groundwater Abstraction

An option for protection of the groundwater abstraction is to treat water at the abstraction point. This would be achieved by GAC filtration. The estimated cost for this was estimated to be on the order of US$0.20 million capital costs and US$0.90 million in annual operating costs over the 20-year planning horizon. This option was not carried forward in the analysis as it achieved the same objective as R1 but with a greater cost.

R3: Replace Extracted Groundwater with Main Water

This option was simply to cease groundwater pumping from BHA and use main water supply. This would have negligible capital costs and an annual operating cost of US$1.10 million (based on a supply rate of 1,440 m^3/day and a water supply cost of US$0.30/m^3) for a total 20-year period. This option was not carried forward in the analysis as it achieves the same objective as R1 but with a greater cost.

External Costs

The only external costs considered as part of this assessment were those of transporting waste to landfill using HGVs. These were estimated based on the assumption that increased HGV traffic on the road network will result in a number of costs, including increased road congestion, impacts on health from emissions, noise, and increased probability of accidents. A unit cost of US$0.78/v-km was used.

BENEFITS

The possible benefits of remediation at the site include increased property value, both on site and in neighboring areas; improvement in aquifer quality; and reduction

TABLE 7.12
Monetizable Benefits (Base Case), (Millions of 2009 U.S. Dollars)

Benefit Category	One-Time Benefit If Remediated Now	Annual Damage Avoided ($/Year)	20-Year PV Sum of Benefits	Valuation Method
Recovery of land value, site area	2.65	—	1.0[a]	Current market value of land minus current book value. Note that this parcel of land could not realistically be sold now as the complex is still operational, and recontamination would no doubt occur.
Recovery of land value, whole complex	531	—	200	Current market value of land minus current book value. This value could be realized now (on completion) if the whole complex were remediated and sold.
Removal of blight on property value surrounding site	20.9	—	7.90	Current book value of land outside complex but within 2 km of chemical manufacture area, with 5% blight factor applied.
Improvement of aquifer water quality	—	0.0004	0.005	Calculated groundwater production impacted; base case TEVW applied.
Avoidance of BHA water replacement	—	0.16	2.0	Base case value of water applied to current BHA production.
Avoidance of impacts to estuary	—	0.02	0.28	Estimated value of stretch of estuary impacted, assuming 100% damage attributed to the spill in question for 20 years.

[a] Since land value cannot be unlocked until the complex is decommissioned, value is assumed to flow as a one-off benefit in year 20. The same applies for recovery of land value for the whole complex and the elimination of neighboring blight.

in effects on the estuary. Each of the benefits of remediation is discussed next and is summarized in Table 7.12.

Property Value

The chemical manufacture complex area occupies approximately 3 hectares of land. It lies entirely within the refinery complex, which covers approximately 600 hectares. In terms of this assessment, one remedial benefit is the value of the property that is unlocked as a result of remediation. The book value of the complex used in the site owner's corporate asset valuation was US$15,000/hectare. However, this value reflects the state of the land at that time (an occupied refinery process facility). The average market value for fit-for-purpose commercial-industrial land in the area was

US$900,000/hectare. The incremental value of the facility land that would result from remediation was estimated as the average market value minus the book value. It is important to note that the incremental value realized for the whole complex was only assumed to be realized at the end of the life of the facility, at which time a complete decommissioning could be undertaken.

At the time of the assessment, the site owner had no plans to decommission the facility. Nevertheless, for the purposes of this analysis, it was assumed that complete facility remediation would occur in 20 years. In the same way, even if the chemical manufacture area were completely remediated, it could not feasibly be sold for commercial-industrial use to another party until the end of life of the facility, since it would remain within the core of the complex. The PV of the 3-hectare site in 20 years, if remediated, was estimated at US$1.0 million, and of the entire complex was US$200 million (all values in 2009 dollars).

Water Abstracted from BHA

The commercial market value of water at the time of analysis was approximately US$0.30/m³. This value was chosen as the base case value of water for this example. This is a conservative assumption; it understates the full total economic value of water (TEVW) since the commercial use value is only one component of total economic value (TEV), as discussed in Chapter 3. For the purposes of a preliminary, high-level analysis, this value serves as a threshold value for TEVW; we know that water is worth at least this much, but without significant additional study, we cannot determine exactly how much more. Using the sensitivity analysis, the effects of higher values of TEVW can be examined.

Along with a network of other on-site water supply wells, the on-site BHA abstraction well is used to provide water to the various facility processes. The well has a production capacity of 525,600 m³/year. The value of the water produced, using the base case water value of US$0.30/m³, is approximately US$0.16 million/year. At present, the complex does not pay this market rate for water. If the presence of significant contamination in BHA water rendered the water unusable in the refinery process and the other wells on site were also unavailable, the complex owner could be forced to replace this supply by purchasing water at the market rate.

Benefits to Neighbors

The potential benefits of remediation to off-site residents were valued by the increase in property value that could be realized by remediation (from the elimination of negative effects on property values). For valuation purposes, the perception of damage to the surrounding people and lands, and to public health (which is not in reality affected), can be assumed to be reflected by a blight effect on property values in the neighboring community, as discussed in the blight reduction section of this chapter. In the vicinity of the complex, it is conceivable that properties were suffering some loss in value by virtue of their location near an active petrochemical facility. This loss of value could conceivably be the result of aesthetic impacts, concerns over air or water quality, concerns over the presence of subsurface contamination at the complex, and the possibility of contaminant migration off site. Such concerns would likely not exist if the area occupied by the facility were a greenfield or occupied by residential housing or other more benign uses. A conservative analysis (worst-case for site owner) attempts to identify the largest possible benefits accruing from remediation. In a methodology similar to that used commonly in risk assessment, such a conservative approach means that assessment results are worst case. In other words, if the worst-case assessment shows that remediation is not warranted on economic grounds, then the more likely case will definitely not be economic.

Conservatively, the radius of the area surrounding the spill site that could conceivably be affected by blight was taken at 2 km. Within 2 km of the site, only 682 hectares lay outside the boundaries of the complex. Of this, 125 hectares are residential land (occupied by houses and gardens), 125 hectares are occupied by commercial-industrial properties, and 436 hectares are agricultural land and public open space. Market values in this area for property not adjacent to landfills or major industrial complexes (unblighted) were US$200,000/hectare for agricultural land, US$900,000/hectare for commercial and light industrial, and US$1.75 million/ hectare for residential land. Blight factors of between 5% (base case) and 10% (high case) were applied as per the findings of research discussed in this chapter.

Therefore, the total PV of completely removing blight on surrounding property (assuming that this occurred in year 20 at the end of the life of the facility) was estimated to be approximately US$7.9 million for the base case or US$15.8 million for the high case. It is important to reemphasize that it is not implied that blight actually exists, but rather that in constructing a conservative analysis, it is useful to purposefully overestimate the economic value of perceived risks. In this way, we are biasing the analysis toward justification of higher expenditure on remediation.

Aquifer Value

The value of the aquifer being damaged due to the presence of the LNAPL in the subsurface at the chemical manufacture area can be represented by the additional lost potential water production from the portion of aquifer rendered unusable by contamination. The volume of abstractable groundwater rendered unusable by the presence of the spill was estimated using simple volumetric calculations at 1,278 m³/year. Using the base case value for water, the potential benefit of remediating the LNAPL to restore the damaged part of the aquifer to a standard suitable for drinking water was estimated at US$383/year using the same base case value for water.

Estuary

The risk assessment suggested that it was unlikely that contamination from the chemical manufacture area would have an impact on the local estuary to any significant degree. Nevertheless, as part of the conservative approach to benefits analysis, it was assumed that some impact does occur. Thus, depending on the approach selected, remediation would result in a benefit to the estuary. In this highly industrialized area, there are numerous heavy industries clustered around and adjacent to the estuary. The overall environmental quality of the estuary has been graded as poor by regulatory authorities and is currently not considered to be a high-value coastline in terms of recreational or ecological value. Nevertheless, it was assumed that the spill in question impacted 4 hectares of the estuary at its discharge point to the degree that 100% of the value of that part of the estuary was lost over the full 20-year planning horizon. A conservatively high unit value for the estuary was assumed by using the coastal wetland value estimate of US$5,517/hectare/year provided in Table 3.7. Therefore, a nominal annual value for the damage avoided to the coastline by eliminating the notional impacts from contamination was set at US$0.022 million/year.

Benefit Apportionment

Not all benefits are realized by each remediation approach, and in some cases benefits are partially realized by an approach. The proportion of each benefit that is assumed to be realized by each remediation approach is summarized in Table 7.13. Many of the remediation approaches take a number of years to complete. Realization of the value of property, for instance, would not be achieved until the remediation has reached closure.

BASE CASE ASSESSMENT RESULTS

Table 7.14 provides the net benefit and benefit cost ratio (BCR) for each of the shortlisted approaches under base case conditions (base case value assumptions, 5% discount rate, and 20-year planning horizon). What is readily apparent is that all of the approach options that are being considered for remediation of the spill are NPV negative, except the full decommissioning of the facility in 20 years at its assumed end of life (option S3) and approach R1 (shifting pumping to other wells).

Option S3 was the most economical overall. Option R1 was the next most economic, with a slightly positive NPV by virtue of its low cost, and was also the cheapest of the options considered under base case conditions. R1 is an immediate response and delivers more benefits than it costs, but it does not deal with the contamination issue in any physical way. From this perspective, R1 does not satisfy the overall objective of dealing with the contamination. It can, however, be used in combination with S3 to provide a complete solution.

For all of the other options, the benefits are too small overall to justify their implementation. This is a particularly strong result given that all efforts have been made to bias the analysis in favor of remediation. Only the option that deals with the spill as part of a wider end-of-life decommissioning strategy makes sense from an overall environmental, social, and economic perspective—it is the only sustainable option. Attempting to remediate an isolated spill that sits within a large operational site that also has other contamination issues makes little economic sense. The economies of scale associated with a complete end-of-life decommissioning effort make this a much more effective way to proceed.

SENSITIVITY ANALYSIS

Given the base case results, a highly conservative case (from the perspective of society) was considered in parallel to the base case to provide a basic sensitivity analysis. A high case was developed to examine the results with higher benefits to remediation, maintaining the same level of costs as the base case. A higher value for water (US$1.20/m^3) was used, which arguably better reflects the TEV of water. In addition, it was assumed that ten times as much abstraction potential from the aquifer is compromised by the

TABLE 7.13
Benefits Apportionment by Remediation Approach (Percentage of Full Value)

		Benefits					
Code	Approach	Increase in Site Land Value	Increase in Whole-Complex Land Value	Avoidance of Damage to BHA Supply	Blight Avoided	Improved Aquifer Quality	Avoidance of Damage to Estuary
N1	Monitored natural attenuation	0	0	5	0	5	5
S1	Readily mobile NAPL removal	0	0	10	0	10	10
S2	Full NAPL removal	0	0	100	0	100	100
S3	Full NAPL removal and site remediation as part of complex decommissioning	100	100	100	100	0	0
P1	Hydraulic containment at site area boundary	0	0	100	0	100	100
R1	Divert groundwater pumping to other abstraction wells	0	0	100	0	0	0

TABLE 7.14

Net Present Values and Benefit–Cost Ratios for the Base Case (Millions of 2009 U.S. Dollars)

Code	Remediation Approach	Present Value Cost	Present Value Benefit	NPV (Net Benefit)	NPV Ranking (1 = best)	Benefit–Cost Ratio
N1	Monitored natural attenuation	3.0	0.11	−2.89	3	0.04
S1	Readily mobile NAPL removal (site NAPL plume area only)	8.9	0.22	−8.78	5	0.03
S2	Full NAPL removal (site NAPL plume area only)	15.2	2.29	−12.9	6	0.15
S3	Full NAPL removal and site remediation as part of complex decommissioning	94.25	209.9	115.7	1	2.23
P1	Hydraulic containment at chemical manufacture area boundary	9.1	2.29	−6.81	4	0.25
R1	Divert groundwater pumping to other abstraction wells	1.1	2.0	0.9	2	1.81

presence of contamination. The high case also assumed a 10% blight factor on property. The results of the high-case analysis are presented in Table 7.15.

It can be clearly seen from the sensitivity analysis that while the absolute numbers change, the relative ranking of options remains virtually the same (except that MNA (option N1) is replaced by hydraulic containment (option P1) as the second-best option). Higher values for all benefits (except the value of the site itself) mean that NPVs and BCRs for all options increase. However, remediating at the end of facility life (S3) remains the most economic and sustainable course of action, for all of the reasons discussed, by a considerable margin. Option R1 is even more beneficial but remains incapable of dealing with the central issue. In other words, R1 does not achieve a meaningful remedial objective. Considerably higher benefit values do not alter the choice of the most optimal remedial approach, which again is a combination of R1 to deal with the immediate concerns over site water supply and a comprehensive end-of-life decommissioning program designed to deal with all of the contamination at the site at once.

Discussion

The most environmentally and economically sustainable remedial approach under base case assumptions was S3, the remediation of the entire refinery complex,

TABLE 7.15

Net Present Values and Benefit–Cost Ratios for the High Case (Millions of 2009 U.S. Dollars)

Code	Remediation Approach	Present Value Cost	Present Value Benefit	NPV (Net Benefit)	NPV Ranking (1 = best)	Benefit–Cost Ratio
N1	Monitored natural attenuation	3.0	0.43	−2.57	4	0.14
S1	Readily mobile NAPL removal (site NAPL plume only)	8.9	0.86	−8.04	5	0.10
S2	Full NAPL removal (sites are NAPL plume only)	15.2	8.76	−6.44	6	0.58
S3	Full NAPL removal and site remediation as part of complex decommissioning	94.25	223.8	129.6	1	2.37
P1	Hydraulic containment at site area boundary	9.1	8.76	−0.34	3	0.96
R1	Divert groundwater pumping to other abstraction wells	1.1	8.0	6.90	2	7.27

assumed to occur at the time of facility decommissioning in 20 years. Despite being by far the most expensive option, the costs are deferred to 20 years in the future. This opens up the possibility that remedial technology would have progressed, and costs may have thus dropped overall. In addition, it is very likely that land values will have risen in real terms, making the benefits of remediation at that time commensurately greater. Option R1 proposes management of the pumping at BHA in combination with other supply wells available to the site owner to ensure that contamination does not reach BHA and does not spread. The additional cost for this approach was negligible. Even though approach R1 benefits only one stakeholder (the site owner), the result reflects the fact that the contamination does not significantly affect outside receptors, which in turn means that the benefits of remediation are small if undertaken now, within an operating site.

MNA (option N1) is also quite inexpensive, but its very low BCR reveals that it provides very little value for money; the plume is being monitored, not remediated, so society benefits little over the longer term given that the plume is migrating within a large operating site subject to ongoing other impacts on groundwater. Option S1, the program of remediating mobile LNAPL that was originally favored by the regulator, is expensive and yields negligible benefits. As a result, this approach is highly uneconomic and, by definition, highly unsustainable for society. In fact, it is the least economic of all remedial actions considered. This makes intuitive sense. Removing only part of the contamination that damages water resources or nearby ecologically important receptors means that benefits of remediation are very small. In addition, remediation of this area alone, in the center of an active large industrial complex, will do little or nothing in terms of removing blight to surrounding properties and will not by itself result in an increase in refinery complex land values—the site cannot reasonably be used for other purposes until the whole complex is decommissioned.

The high-case analysis did little to change the overall ranking, although BCRs and NPVs did increase.

This analysis implies that the site owner may seek, with agreement from the regulator (which in a similar case was reached), to divert the funds budgeted for remediation (to accomplish S1) into another part of its environmental compliance program that is expected to yield significant overall benefits. This is clearly a more sustainable outcome than the initial position of the regulator, which would have required significant expenditure on a remediation program with benefits that did not justify the costs.

IMPLICATIONS

This case study illustrates that the EESA approach can be a useful tool in helping decision makers assess what is an appropriate level of remediation for a given site. In this case, the regulator's initial preference was to undertake a level of remediation that was revealed to be uneconomic and unsustainable. Communication of the EESA results allowed the problem holder and the regulator to reach agreement on a mutually acceptable course of action.

Proper understanding of the risks and benefits of remediation can reveal a different outcome from that expected by the problem holder, regulator, or community, sometimes justifying higher spending on remediation, sometimes lower. The goal, as discussed throughout this book, is an optimal solution that balances the needs of society, the environment, and industry.

REMEDIATING NAPLs IN FRACTURED AQUIFERS

INTRODUCTION

The problems associated with NAPLs in fractured aquifers have received significant attention in the technical literature over the last decade.[42–44] Concerns over the impacts of chlorinated solvents on groundwater have led to a significant body of work examining DNAPLs (dense non-aqueous phase liquids) in the subsurface, including in fractured rocks.[45,46] More recently, the unique behavior and problems associated with LNAPLs in fractured aquifers have been studied.[47–49] The highly heterogeneous nature of fractured systems combined with the inherent complexity of multiphase flow typically make characterization and remediation of NAPL in fractured systems difficult and expensive.

Because this type of remediation can be so challenging and costly, there is a need to understand clearly the justification for spending potentially large amounts of money. Indeed, the remedial methods that tend to be employed to remove or immobilize NAPLs in fractured aquifers have tended to be complex, intrusive, and energy intensive. This in turn brings the sustainability of such efforts into question. Is that level of effort always warranted, given the value of the damage to society, the economy, and the environment caused by the presence of the contamination? This section examines the application of EESA in setting an optimal remedial objective and strategy for this difficult problem.

TECHNICAL CONSIDERATIONS

Remediation of LNAPL and DNAPL in fractured aquifers is a complex undertaking. DNAPLs may migrate to significant depths via fractures, and if spill volumes are large and fracture interconnectivity high, DNAPL may invade progressively smaller aperture fractures with depth. As NAPL fluid pressures increase, matrix invasion may also occur. The vertical migration of LNAPL in fractured aquifers is constrained by the water table, but despite this, significant penetration beneath the water table may occur, and lateral migration may occur in directions independent of the hydraulic gradient.[50] Within fractured aquifers, NAPL movement is governed by the geometry of the fracture network (including fracture orientations, densities, interconnectivity, apertures, and wall roughness); capillary pressure and fluid saturation relationships; and the properties of the NAPL (density, interfacial tension, viscosity).

Whether dealing with LNAPL, N-NAPL (neutral-buoyancy NAPLs), or DNAPL, significant challenges exist when contemplating remediation. First, characterization of the distribution and behavior of NAPLs in fractured rock is notoriously difficult.[51] In a deterministic approach, fracture networks need to be characterized, major fracture sets identified in the field, and representative fracture parameters determined. The occurrence of NAPL within these fractures then needs to be ascertained, areally and vertically. For DNAPLs, definitive characterization to depth may be problematic.[52] Rarely in practice is a complete characterization feasible.[53]

Next, proven techniques for NAPL removal from fractures are few. Pump-and-treat methods, while effective for containment, have proven disappointing for NAPL removal, even when coupled with targeted NAPL recovery pumping and skimming.[54] More aggressive in situ NAPL removal methods have been field tested, including HVE, thermal heating, and surfactant-assisted aquifer remediation.[55] These relatively expensive methods have shown good results in some cases but have not yet been widely applied in fractured rock environments.

Finally, when the understanding of contaminant distribution is sketchy, even the simplest remediation techniques can prove unsuccessful. The combination of new or unproven remedial techniques, incomplete characterization, and complex aquifer and contaminant distribution conditions makes remediation success uncertain. Within this context, a clear understanding of the financial and broader social, environmental, and economic implications of remediation provides decision makers with the means to select achievable and ultimately beneficial remedial objectives.

EXAMPLE: DNAPL IN A FRACTURED CARBONATE AQUIFER, UNITED STATES

BACKGROUND

A small site in the United States was used since the 1950s in the recycling of transformers. Until the 1980s when the site was closed, various chlorinated solvents were disposed of at the site in shallow unlined trenches. The site is situated on a small hill on the outskirts of a small rural town. Down-gradient of the site are fields, a small wetland, and a creek.

Over the years, DNAPL soaked into the 8-meter thick layer of fine-grained sediment that covers the site and in places reached the highly weathered top of the fractured carbonate aquifer below. The groundwater surface at the site is within the bedrock aquifer, at about 12 meters below ground. The aquifer itself is characterized by low yields and marginal quality from a drinking water perspective.

An extensive remediation program resulted in the on-site thermal treatment of over 20,000 m³ of NAPL-contaminated soil, removing the vast majority of the NAPL on site. Groundwater monitoring revealed low concentrations (in most cases below current regulatory limits but occasionally slightly above in two of fifteen monitoring wells off site) of certain chlorinated solvents moving off site within the fractured bedrock aquifer, toward the wetland and creek. Small amounts of DNAPL may still be present in selected fractures within the bedrock at the site, perhaps as residual ganglia or as adsorbed phase within sediment-filled fractures; however, this is inferred only from dissolved-phase concentrations measured in three on-site monitoring wells (Figure 7.4).

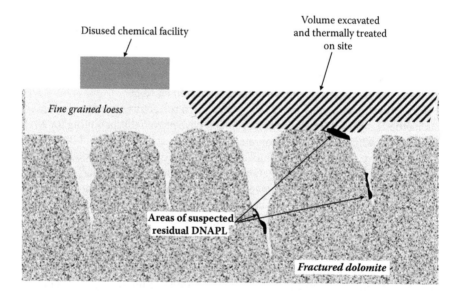

FIGURE 7.4 Conceptual cross section of disused site. Hatched area represents volume of contaminated soil that was previously excavated and treated on site. Suspected small areas of residual DNAPL are thought to remain within the loess infilling large channels and fractures at the top of bedrock.

The complexity of the fracture regime at the site makes detailed characterization of the nature and occurrence of NAPL difficult. Even dissolved-phase concentrations of contaminants may or may not be present in selected fractures a few meters apart. This is an example of applying EESA at its most basic level: examining the problem over a longer-term perspective, developing quick estimates of the benefits that occur from remediation, and comparing them to the costs of various overall remedial approaches.

BENEFITS OF REMEDIATION

A simple analysis identifies the following benefits that may accrue from remediation at the site; in each case, benefit estimates are overstated to skew the assessment toward active remediation as an outcome. Benefits examined were

1. Increase in property value at the site itself. The area is rural, and land values in the area are relatively low. Clean, the site is worth about US$0.50 million.
2. Uplift in the value of surrounding properties through removal of blight. Within a 2-km radius, the sum of property values is estimated at US$11 million by area realtors. Applying a 10% blight factor, remediation of the site would result in a one-time benefit of US$1.1 million.
3. Prevention of aquifer damage. Using simple modeling, the presence of the dissolved-phase plume at the site effectively eliminates about 200,000 m^3/year of potential abstraction (even though the aquifer is used only sparsely in the area and not at all for public supply). At 5% discount rate over 20 years and assuming a value for water of US$1.00/$m^3$, this equates to about US$2.5 million in lost aquifer potential.
4. Value of the wetland to which dissolved-phase contaminants may flow. Given the low concentrations expected under even worst-case conditions, a nominal value of US$0.1 million was assigned (see the refinery remediation example in this chapter).

Table 7.16 provides a summary of the possible benefits of remediation.

TABLE 7.16
Possible Remedial Benefits Summary (Millions of 2008 U.S. Dollars)

Benefit Category	Sum of Benefits Over 20 yr
Property value	0.5
Reduction in blight in neighborhood	1.1
Aquifer protection	2.5
Wetland protection	0.1
Total possible benefits	4.2

REMEDIAL APPROACH OPTIONS

Recognizing the difficulties involved in remediating low levels of dissolved-phase contamination and possible small concentrations of residual DNAPL (which could not be located) in the complex fractured rock environment, three main remedial approach options were identified as part of the evaluation process: (1) soil remediation (already completed) with MNA for groundwater over 20 years; (2) soil remediation (already completed) with selected in situ treatment of identifiable hot spots in groundwater and 20 years of MNA; and (3) soil remediation (already completed) with groundwater pump and treat to contain and reduce the mass of the dissolved-phase plume.

SIMPLE HIGH-LEVEL ENVIRONMENTAL AND ECONOMIC SUSTAINABILITY ANALYSIS

The 20-year life-cycle costs and benefits of each of the three remedial options being examined are provided in Table 7.17. Although the soil remediation has already been completed, the costs of soil remediation are included in each option to examine how they impact the decision.

   None of the options is economical; they do not result in a net increase in welfare for society. The combination of low property values (small benefits) and a complex and difficult to characterize and remediate aquifer (high costs) produces low BCRs below unity. Even if the value of the aquifer were increased by raising the value of water from US\$1.00/m^3 to US\$2.00/m^3, the costs of remediation are so high that all of the options remain uneconomic. Note also that here, the external costs of remediation were not included. External costs would only make all options even more uneconomic and unsustainable.

IMPLICATIONS

In this case, all options are uneconomic. The high cost of remedy for NAPL in fractured rock cannot be justified by the benefits that society will gain as a result of action. Even the soil remediation alone, which was accomplished with an energy-intensive thermal desorption system, was too expensive for the benefits created. This is a classic example of remediation driven by criteria in which risks have been

---

TABLE 7.17
Indicative Costs and Benefits, Base Case (20 yr, 5%) (Millions of 2008 U.S. Dollars)

Remediation Approach	PV Cost Soil	PV Cost Total	PV Benefit	Net Benefit	BCR
1: Soil remediation + MNA	3.0	4.5	2.85	−1.65	0.63
2: Soil remediation + in-situ DNAPL hot spot removal	3.0	9.7	3.6	−6.10	0.37
3: Soil remediation + pump-and-treat to contain dissolved-phase plume	3.0	13.9	4.2	−9.70	0.30

significantly overstated. A less-expensive soil treatment method may have brought the overall costs down to a point at which the proposition was economic and was worth exploring at the time. However, it is clear that further remediation, by deploying either option 2 or 3, is uneconomic and unsustainable.

In this case, a technical impracticability (TI) waiver would be both technically and economically justified, leading to selection of remedial approach 1. Similar issues in the United States have been granted TI waivers from the regulatory authorities. This analysis suggests that, in fact, TI waivers are really economic impracticability waivers. With enough effort and expenditure, using existing technology, it is highly probable that the low levels of VOCs at and off site could have been dealt with eventually. However, this would have required an unreasonable level of expenditure. It is not technology that limits our efforts in most cases, but our willingness to use scarce resources to achieve levels of remediation that may not be justified. EESA provides a way of rationally and objectively determining where and to what level remediation is warranted and where it is not.

EXAMPLE: NAPL IN A FRACTURED CARBONATE AQUIFER, UNITED KINGDOM

BACKGROUND

A disused MGP (manufactured gas plant) facility in the United Kingdom has caused significant contamination of the subsurface by coal tar compounds, including NAPLs. The site lies in the commercial center of a busy town, adjacent to a high-value residential neighborhood (Figure 7.5). The uppermost coarse-grained saturated gravel deposits are extensively contaminated by NAPL, which is very close to neutral density (Figure 7.6). Groundwater occurs at a depth of about 5 meters across the site. A small volume of the NAPL has penetrated a low-permeability layer and is present in small-aperture fractures within the upper horizon of the underlying carbonate aquifer to depths of up to 15 meters below ground level. The NAPL plume is stable and occurs mostly as residual saturation. The N-NAPL is composed predominantly of low-solubility PAHs.

FIGURE 7.5 Manufactured gas plants (MGPs) in Europe, many of which were built in the nineteenth century, are often sited within densely populated towns and cities.

FIGURE 7.6   Photo of coal tar in excavation at an MGP site during remediation. In this case, the NAPL is contained within an old gas holder base.

A risk assessment was undertaken that identified a series of risks associated with the presence of subsurface contamination on the site. First, the aquifer is extensively used throughout the region for public water supply (PWS), and an operating PWS well lies approximately 2 km down-gradient of the site. Trace levels of phenolic compounds have been detected at various times in this PWS well over the past several years. Groundwater monitoring and modeling have confirmed that these compounds are migrating from the site in very low concentrations. In addition, a small culverted river of poor quality runs past the site. The contaminated gravels on site are in hydraulic connection with the river, but again, the flux of contaminants to the river in the dissolved phase is low because of the low solubility of the N-NAPL. The site is derelict and empty, and its boundaries are completely fenced, allowing the public no access. A simple schematic cross section of the site and environs is provided in Figure 7.7.

REMEDIAL OBJECTIVE

The site owner wishes to remediate the site because of its high commercial value. The regulator and the operator of the PWS are concerned about contamination of the regional aquifer from contaminants on site. To date, the very occasional and low levels of contaminants detected at the PWS have not been a cause for inordinate concern by the well owner, primarily because the concentrations are below the drinking water quality standards, and the existing treatment system removes them. However, should conditions change, and the flux of contaminants to the wells increase, there is a risk that the PWS would have to shut down or substantially upgrade the existing treatment system. On this basis, the site owner sought to examine a range of possible remedial

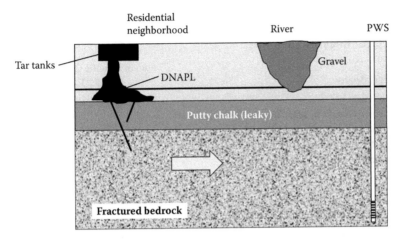

FIGURE 7.7 Conceptual cross-sectional schematic of the MGP site showing the extent of NAPL, the river, and PWS well. Groundwater flow in bedrock is from left to right, toward the well.

approaches that would achieve the objective of selecting a remedial approach that manages the risks present at the site in the most economic and sustainable way.

BENEFITS OF REMEDIATION

The direct financial benefit of remediation to the site owner is the value of the property itself (approximately US$14.0 million), which can be unlocked if remediation makes the site suitable for hard-covered residential or light commercial use. However, there are other substantial benefits that will accrue to other parties if remediation proceeds, depending of course on what remediation achieves. Sale and redevelopment of the site will immediately trigger an uplift in the value of properties surrounding the site through removal of blight. As discussed, this is a well-documented and empirically proven effect. Over 300 residential properties and apartments are situated within 0.5 km of the site.

If remedial action significantly reduced the mass flux of contaminants reaching the aquifer, this would result in avoidance of future possible damage to the aquifer itself and to water production from the PWS well. The PWS well is pumping 1.0 Mm^3 (millions of cubic meters)/year, and any significant increase in mass flux of contaminants to the well would likely bring it offline. Similarly, removal of NAPL on site would eliminate the source of future flow of contaminants to the nearby river, whose average annual flow is approximately 0.3 Mm^3. For the base case, it is assumed that replacing water would cost US$1.00/m^3. Later in the sensitivity analysis, TEVW estimates of up to US$2.00/m^3 are used.

A summary of the benefits that may be realized as a result of remediation over a 20-year planning horizon at a 5% discount rate is provided in Table 7.18. If all benefit categories could be realized, total benefits would be approximately US$34 million. However, not all benefit categories will be realized by each remedial approach, as shown next.

TABLE 7.18
Base Case Benefits Summary

Benefit Category	Sum of Benefits Over 20 yr (Millions of U.S. Dollars, 2008)
Property value	14.0
Reduction in blight in neighborhood (at 5% blight factor)	3.83
Aquifer protection (base case TEVW US$1/m³)	12.46
River protection (base case TEVW US$1/m³)	3.74
Total possible benefits	34.03

REMEDIAL APPROACH OPTIONS

Based on the risk assessment discussed, an understanding the overall social value of preventing damage to the various receptors, and the value of the site itself, a number of remedial approaches were identified to meet the objective. Table 7.19 lists each option and describes its basic components and its advantages and disadvantages.

INDICATIVE REMEDIAL COSTS

Table 7.20 presents indicative remedial costs for each of the remediation approaches identified. The remedial approaches are coded to indicate whether they address the risk source (S) or pathway (P) of the problem. Other options and other scenarios (for sensitivity analysis) are of course possible, but for the purposes of this example, the list has been limited to eight approaches. Both capital and operating and maintenance costs are provided for each approach. Note that capital costs are assumed to occur in year 1 (the first full year of the investment) and be complete within that year.

External costs of remediation, if applicable, have also been estimated and are included in the costs presented. External costs accounted for are generation of GHGs from thermal treatment, vehicle use, and power consumption. A base case cost of carbon was set at US$52/tCO$_2$e (tonnes CO$_2$ equivalent), increasing at 2% per year, based on current U.K. government guidance.[56] External costs of landfill were deemed to be included in the capital cost estimates, if applicable, under the assumption that the new landfill tax in the United Kingdom effectively accounts for the external social costs of landfill. The costs of road transport disamenity were also estimated assuming that the landfill was 100 km away from the site and that the HGVs used carry an average of 10 m³ of material each.

Each of the costs in Table 7.20 is assumed to represent the least-cost solution to achieve the particular approach. For illustration, two methods for achieving capture or control of contaminants in the shallow groundwater-bearing zone (P1 and P1 alternative [P1-ALT]) are presented in Table 7.20. The lowest-cost method is selected (the sparging barrier alternative [option P1], at PV of US$8.8 million, is considerably

TABLE 7.19
Remedial Approach Options

Approach Description	Relative Cost	Advantages and Disadvantages
Source Methods		
S1	Partial excavation and removal of main sources. On-site ex situ treatment. Removal and landfill of remaining material (10,000 m³). In situ remediation of NAPL in fractured bedrock using chemical oxidation (ISCO) technology.	Moderate to high	Pilot trials of ISCO show good success in oxidizing NAPL within fractured rock. On-site treatment focus reduces volume of material for off-site disposal to landfill.
S2	Full excavation of contaminants, including into uppermost part of aquifer, use of piling, maximize off-site disposal (40,000 m³).	High	Quickest and most complete remediation of sources; significant risk of disturbing aquifer during excavation and mobilizing some NAPL into the aquifer, possibly worsening situation at PWS.
S3	Partial excavation of NAPL-contaminated materials above water table only (10,000 m³); leave remaining contaminants in place and monitor.	Moderate	Removal of major concentrated sources allows site to be sold with some residual liability; leaves significant mass of contaminant in place.
Pathway Methods		
P1	Containment in gravels using air-sparge barrier along the down-gradient site boundary within gravels.	Moderate	Prevent further off-site migration of contaminants in gravel to river; does not deal with other hazards; perpetual management needed.
P2	Hydraulic containment of dissolved phase in bedrock aquifer through installation of a series of capture wells, and operation of a water treatment facility on site.	Moderate	Prevent further off-site migration of contaminants in aquifer to PWS; does not deal with other hazards; perpetual management needed.
Receptor Methods		
R1	Treat groundwater after abstraction at PWS.	Moderate	Directly prevents contaminants from site impacting delivery of water to PWS customers. Treats other contaminants from other sources impacting aquifer also; perpetual management required.
Institutional Management		
MNA	Monitored natural attenuation. Current data indicate that low levels of dissolved-phase contaminants are naturally degrading within the aquifer. No remedial action; monitor groundwater to prove attenuation is occurring.	Low	Does not actively remediate site to any degree; does not eliminate the risk that contaminant flux from the site could increase significantly in the future.

TABLE 7.20

Indicative Costs for Selected Remedial Approach Options (Millions of 2008 U.S. Dollars)

Approach	Capital Costs (CAPEX)	O&M Costs ($/yr)	20 yr, 5% PV Financial Costs	GHG Emissions (tCO$_2$e)	Road Transport (v-km)
R1	0.25	0.21	2.87	100/yr	—
MNA	0.5	0.10	1.75	10/yr	—
P2	3.5	0.35	7.86	200/yr	—
P1	3.8	0.40	8.78	100/yr	—
P1 – ALT	5.4	0.48	11.38	150/yr	—
P1 + P2	6.5	0.55	13.35	300/yr	—
S3	9.2	—	7.2	650	200,000
S1	13.9	—	14.4	1,200	480,000
S2	22.0	—	22.0	2,100	800,000

cheaper than the US$11.4 million for the pump-and-treat system [option P1-ALT]) and is carried forward in the analysis. The approach options in Table 7.20 are listed in order of lowest to highest capital expenditure (CAPEX).

Base Case Analysis

Based on this information, base case analysis can be undertaken for the remedial approach options considered. The base case assumes a 20-year life-cycle planning horizon and a 5% discount rate. Table 7.21 shows the results of these calculations. As discussed throughout this book, apportionment of benefits to the various options requires careful judgment of the anticipated effectiveness of the techniques which form the basis of the remedial approach. In many cases, this is straightforward. A zero benefit in the table indicates that the remediation approach does not address the damage to that receptor or does not allow that particular type of site redevelopment. However, in some cases, such as the prevention of damage to the river from the removal of the subsurface hot-spot sources of contamination in the shallow surficial materials (S3), this would require some level of professional judgment supported by groundwater modeling. Nevertheless, regardless of the level of technical analysis, the relative proportions of benefits realized must remain consistent and balanced between options.

So, for instance, in Table 7.21, S2 is deemed to eliminate 100% of the ongoing damage to the aquifer through aggressive removal of the NAPL in fractured bedrock through a concentrated in situ remediation program. In contrast, a system that captures contamination in the surficial groundwater and prevents it from reaching the river (option P1) does not protect the aquifer and so realizes no aquifer protection benefit. However, P1 is deemed to be 100% protective of the river, and that benefit would be accrued for as long as the system remains operational.

TABLE 7.21
Base Case 20-Year PV Costs and Benefits for Remedial Approaches
(Millions of 2008 U.S. Dollars)

Option	Total Financial Costs	Benefit Land Sale (Private)	Elimination of Blight to Neighboring Properties	River Protection	Value of Aquifer Protection	Total PV Benefits
R1	2.87	—	—	—	7.48	7.48
MNA	1.75	—	—	—	0.75	0.75
P2	7.86	—	—	3.74	—	3.74
P1	8.78	—	—	—	12.48	12.48
P1 + P2	13.35	—	—	16.21	12.48	28.69
S3	9.2	7.0[a]	3.875	1.25	1.25	13.38
S1	13.9	14.0	3.875	3.74	9.35	30.97
S2	22.0	14.0	3.875	3.74	12.48	34.95

[a] Assumes that site must be sold at a 50% discount due to the presence of residual deeper subsurface contamination. However, commercial redevelopment can take place, so blight is removed.

Figure 7.8 shows the calculated full social 20-year NPVs for each option (benefits minus costs). The S1 option is superior under base case conditions. Note the contrast with Figure 7.9, which shows base case NPVs over only 5 years. With a shorter-term view, S3 is most economic because the immediate returns to the problem holder from selling the site now dominate the analysis; other benefits that accrue to society

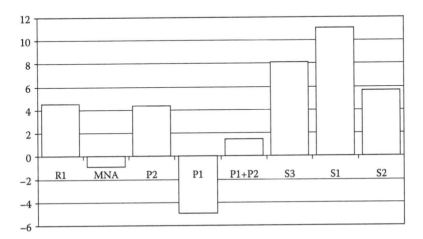

FIGURE 7.8 NPVs of each option under base case conditions, including all internal and external costs and benefits, over 20 years using a 5% discount rate. Option S1 (second from right) provides the most overall benefit. All options but two (MNA and hydraulic containment in bedrock [P1]) provide a net improvement in overall human welfare.

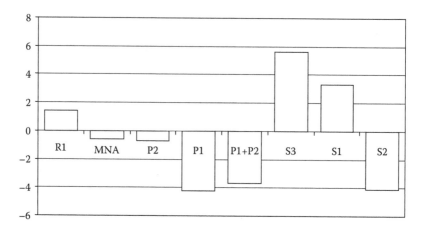

FIGURE 7.9   Base case NPVs with a 5-year planning horizon. The option that focuses on making the site fit for sale (S3) is most economic, despite being sold at a discount. Benefits that accrue to society over time are deemphasized, rendering a number of options uneconomic. Sustainability is not served with short-term planning horizons.

over time, such as protection of river and aquifer, are less important. This illustrates the importance of a longer-term view if sustainable outcomes are being sought.

With a 20-year perspective, remedial approaches MNA and P1 are NPV negative; they do not generate benefits sufficiently large to overcome their costs, even when wider benefits to the whole of society are considered. All of the other remedial approaches generate more benefits overall than they cost to implement. In this example, the analysis suggests that excavation and treatment of the bulk of contaminated soil, where the vast majority of the contaminant mass exists, is the most economic single remedial approach evaluated. This is driven by the relatively high value of the property for commercial development and the fact that all of the neighbors also benefit by elimination of blight on their property values. More aggressive source removal will also benefit both the river and the aquifer to varying degrees by removing a continuing source of contaminants.

Under base case assumptions, the overall value of the river and the degree of damage being inflicted upon it are too small to justify direct action aimed specifically at its protection. In this case, using conservatively high base case valuation parameters for the river, a PV of about $3.7 million is estimated. Thus, to be considered sustainable, remediation measures aimed specifically at protecting the river should not cost appreciably more than this figure in PV terms. If appreciably more is spent on this issue alone, then society as a whole is worse off. Other projects should be sought that provide higher benefits. However, the analysis also reveals that if combined with measures intended to manage risks to other, more valuable receptors, river protection can be achieved economically and sustainably. This conclusion is of course only valid to the extent that the value of the river assumed for the analysis encompasses all of the services it actually provides, and that value is stable over time. It is safe to say that both conditions are being violated. The full extent of the services that natural ecosystems

TABLE 7.22
Sensitivity Analysis Parameter Ranges
(Millions of 2008 U.S. Dollars)

Parameter	Low Value	High Value
Property value	$5 million	$25 million
Blight factor	0%	20%
Total economic value of water	$0.50/m^3	$2.00/m^3
Discount rate	0.03	0.10
GHG external cost	$52/tCO$_2$e	$100/tCO$_2$e
External cost of transport	$0.60/v-km	$2.40/v-km

provide to humankind are only slowly being revealed, and the inherent value of these ecosystems will increase over time as population growth, resource extraction, development, and pollution reduce their stock and health. This is directly reflected in society's WTP for those assets. This is explored in more detail in the sensitivity analysis section to determine if this kind of variability will affect the overall conclusions.

SENSITIVITY ANALYSIS AND DECISION MAKING

Because of the range of assumptions typically required for this type of analysis, sensitivity analysis is an important tool for testing the robustness of conclusions. As discussed in Chapter 3, the EESA pulls together and moves beyond the CBA results with sensitivity analysis to explore the consequences for decision making overall and to help guide an optimal decision. Benefit values, particularly, should be varied over the widest possible reasonable range to test how changes will affect decision making. Table 7.22 shows the ranges of parameters that are varied for the complete sensitivity analysis, following the method described in Chapter 3.

Figure 7.10 shows how the NPVs of the various options vary across the full range of values for TEVW, with all other factors fixed at base case values. This could reflect, for example, a wider anticipated trend toward an increasing value of water, driven by growing demand, resource degradation, and the impacts of climate change. Higher values for water have a significant impact on the analysis. More aggressive aquifer remediation efforts such as S1, S3, and P2 become increasingly more economic and sustainable. The more water is worth to society, the more remediation action that protects it can be justified.

Figure 7.11 shows the cumulative probability plot for each option under the full range of conditions explored in Table 7.22, in every direction. Option S1, identified as most economic in the base case analysis, fares well and is superior under the majority of conditions. This is further illustrated in Figure 7.12, which shows that S1 is the best choice under 82.5% of all conditions examined given equiprobable outcomes. S3 is the best option under 14% of conditions, namely, when the value of water is low and the revenue from property sale dominates the result. Figure 7.13 illustrates again the effect of short-term planning; looking only 5 years into the future, S3 is the best option under 86% of all conditions across the range of variables examined.

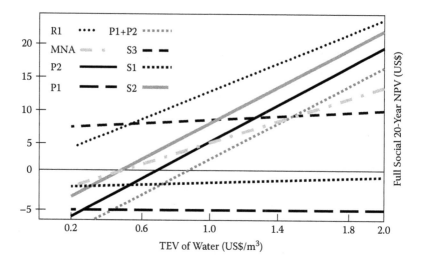

FIGURE 7.10 Sensitivity of NPV results for each option to change in the TEV of water, with all other parameters fixed at base case values. At the lowest TEV considered, S3 is marginally superior, but S1 rapidly becomes the most economic option as the TEV of water rises by virtue of its more comprehensive remediation of NAPL within the aquifer.

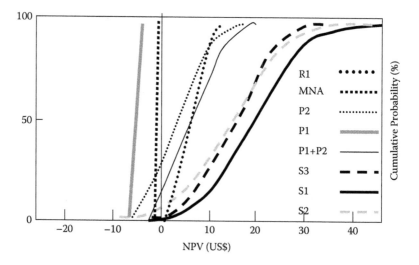

FIGURE 7.11 Cumulative probability plot for each option, showing the NPVs calculated for 1,000 random points across the full range of all parameters considered for each option. Option S1 is more NPV positive (further to the right) than the other options under the widest range of conditions.

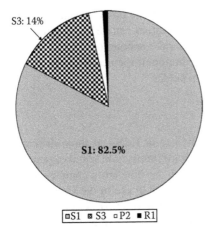

FIGURE 7.12 Chart showing the proportion of instances where options were the most environmentally and economically sustainable. S1 was most economic and sustainable over 82.5% of all conditions tested. S3 was next most economic, but only under conditions of low water value and high property value.

In this example, the site owners could clean up the site at a profit if no other stakeholder's interests were considered. This reflects reality at present in many countries, particularly in Europe, where high property values have driven considerable site remediation over the past decade. However, broader economic analysis reveals that some of this surplus (profit) could also be used to generate external benefits for

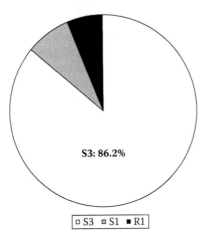

FIGURE 7.13 Proportion of instances where options were the most environmentally and economically sustainable over a 5-year planning horizon. With a short-term perspective, S3 is the most economic choice under 86% of all conditions. Only if TEV of water is very high is S1 better. However, such a short-term view eliminates much of the longer-term benefit to the aquifer and river that stems from options such as S1 and makes the option that maximizes immediate return (S3) appear to be most economic. A short-term economic perspective rarely delivers a truly sustainable outcome.

the community (and by extension the public image of the problem holder). In other words, sustainability in remediation can be identified and justified in an economic context. In this example, several combinations of approaches are in fact possible that would provide wider distribution of the economic benefits of remediation, albeit at higher cost to the problem holder.

DISCUSSION

This chapter examined the use of EESA at three different contaminated sites for three problems associated with immiscible liquid organic contaminants (NAPLs) in fractured rock systems, which are inherently complex and difficult to remediate. In the last of these examples (the MGP site in the United Kingdom), high urban property values combined with the risk of impacts to a major aquifer used by an important nearby PWS meant that remediation of NAPLs in fractured rock, although very expensive, was economic—it yielded overall benefits to society as a whole despite the requirement that the problem holder had to spend a considerable amount of money. Society as a whole would be better off as a result of remediation taking place.

In contrast, in the second example, involving a site situated in rural America atop a low-yield and little-used aquifer, remediation of very low concentrations of dissolved-phase contamination associated with small remnant volumes of immobile NAPLs in the fractured aquifer was not economic and not sustainable. The already-completed soil remediation had removed the continuing source of future dissolved-phase contamination and rendered the site suitable for limited commercial use. However, further remediation could not be justified based on the benefit to be gained. In both cases, active remediation of NAPLs in complex and highly heterogeneous fractured rock was expensive (over US$4 million and almost US$10 million, respectively). However, the environmental and economic sustainability of the two cases were starkly different.

In general, active remediation may be beneficial and warranted if property values are high and are expected to rise, aquifer use value is high, alternative water sources are scarce and an important nonuse value is threatened (such as groundwater recharging a sensitive and valued wetland ecosystem), and ecological assets at risk are unique and valued by society. In many cases, the anticipated benefits of remediation in complex technically challenging environments (such as fractured aquifers) will not justify the level of expenditure required to realize those benefits with currently available technology. To allow benefits to be more accurately quantified, studies are required on valuing aquifers in both the United States and Europe. It is also clear that continued research is needed to develop more effective and less-costly remedial techniques, particularly for more complex subsurface environments such as fractured rock systems.

For too long, decision making about remediation has been carried out with only half of the required information. Legislation and regulatory standards have driven problem holders to seek the cheapest possible way to meet the applied standards. The real overall benefits of remediation, however, have been largely ignored. EESA provides a robust way to examine the benefits to all parties from remediation and compare the overall economic sustainability of a variety of options across a range of possible future conditions to determine an optimal decision that will stand the test of time.

NOTES

1. Hardisty, P.E., R.A. Bracken, and M. Knight. 1998. The Economics of Contaminated Site Remediation: Decision Making and Technology Selection. *Geological Society Engineering Geology Special Publication*, 14: 63–71.
2. Hardisty, P.E., H.S. Wheater, P.M. Johnston, and R.A. Bracken. 1998. Behaviour of Light Immiscible Liquid Contaminants in Fractured Aquifers. *Geotechnique*, 48(6): 747–760.
3. Goist, T.O., and T.C. Richardson. 2003. Optimizing Cost Reductions While Achieving Long Term Remediation Effectiveness at Active and Closed Wood Preserving Facilities. *Proceedings of the NGWA Conference on Remediation: Site Closure and the Total Cost of Cleanup*, New Orleans, November, 2003. pp. 77–83.
4. Perakami, M.P., and P.B. Donovan. 2003. A Remedial-Goal Driven, Media-Specific Method for Conducting a Cost Benefit Analysis of Multiple Remedial Approaches. *Proceedings of the NGWA Conference on Remediation: Site Closure and the Total Cost of Cleanup*, New Orleans, November, 2003. pp. 23–31.
5. Hardisty, P.E., E. Kramer, E. Ozdemiroglu, and A. Brown. 1999. Economic Analysis of Remedial Objective Alternatives for an MtBE Contaminated Aquifer. *Proceedings of the Hydrocarbons and Organic Chemicals in Groundwater Conference*, NGWA, Houston, November, 1999. pp. 62–69.
6. Pearce, D.W., and J.J. Warford. 1993. *World Without End*. World Bank, Washington, D.C.
7. Hardisty, P.E., and E. Ozdemiroglu. 2005. *The Economics of Groundwater Remediation and Protection*. CRC Press, Boca Raton, FL.
8. Hardisty, P.E., R.A. Bracken, and M. Knight. 1998. The Economics of Contaminated Site Remediation: Decision Making and Technology Selection. *Geological Society Engineering Geology Special Publication*, 14: 63–71.
9. Hardisty, P.E., S. Arch, and E. Ozdemiroglu. 2008. Sustainable Remediation: Including the External Costs of Remediation. *Land Contamination and Reclamation*, 16(4).
10. Hardisty, P.E., and E. Ozdemiroglu. 2005. *The Economics of Groundwater Remediation and Protection*. CRC Press, Boca Raton, FL.
11. Ibid.
12. Maddison, D., D.W. Pearce, O. Johansson, E. Calthorp, T. Litman, and E. Verhoef. 1996. *The True Costs of Road Transport*. Earthscan, London.
13. Hardisty, P.E., S. Arch, and E. Ozdemiroglu. 2008. Sustainable Remediation: Including the External Costs of Remediation. *Land Contamination and Reclamation*, 16(4).
14. Ibid.
15. Hardisty, P.E., and E. Ozdemiroglu. 1999. *Costs and Benefits Associated with Remediation of Contaminated Groundwater: A Review of the Issues*. UK Environment Agency Technical Report P278. UK Environment Agency, London, UK.
16. Department for Food, Environment, and Rural Affairs (DEFRA). 2003. *A Study to Estimate the Disamenity Costs of Landfill in Great Britain*. Final report by Cambridge Econometrics with Eftec and WRc. DEFRA, London, UK.
17. Gawande, K., and H. Jenkins-Smith. 2001. Nuclear Waste Transport and Residential Property Values: Estimating Effects of Perceived Risks. *Journal of Environmental Economics and Management*, 42: 207–233.
18. Payne, B., Olshansky, S., and Segel, T. 1987. The Effects on Property Values of Proximity to a Site Contaminated with Radioactive Waste. *Natural Resources Journal*, 27: 579–590.
19. Hirshfeld, S., P.A. Veslind, and E.I. Pas. 1992. Assessing the True Cost of Landfills. *Water Management and Research*, 10: 471–484.

20. Department for Food, Environment, and Rural Affairs (DEFRA). 2003. *A Study to Estimate the Disamenity Costs of Landfill in Great Britain.* Final report by Cambridge Econometrics with Eftec and WRc. DEFRA, London, UK.
21. USEPA, Sediment Remediation Guidance for Hazardous Waste Sites, 2005. EPA-540-R-05-012. Office of Solid Waste and Emergency Response, United States Environmental Protection Agency, Washington, D.C.
22. Hardisty, P.E., and E. Ozdemiroglu. 2005. *The Economics of Groundwater Remediation and Protection.* CRC Press, Boca Raton, FL.
23. United Nations Environment Programme (UNEP). 2007. *Global Environmental Outlook 4: Environment for Development (GEO-4).* UNEP, Nairobi, Kenya.
24. Hardisty, P.E. 2009. The Social Cost of Landfill. *Proceedings of the 2009 Waste and Recycling Conference, Fremantle, September, 2009.* Waste Authority of Western Australia, Perth, Australia.
25. United States Environmental Protection Agency (USEPA). 2008. *Municipal Solid Waste in the United States: 2007 Facts and Figures.* USEPA, Office of Solid Waste, Washington, DC.
26. Australian Bureau of Statistics. 2007. *Australia's Environment: Issues and Trends, 2007.* ABS, Canberra, Australia.
27. Department of Environment, Water, Heritage and the Arts. 2008. *Waste and Recycling in Australia—Final Report.* Commonwealth of Australia, Canberra, by Hyder Consulting.
28. Eurostat. 2006. *Environmental Data Centre on Waste.* European Union, Brussels, Belgium.
29. United Nations Environment Programme (UNEP). 2007. *Global Environmental Outlook 4: Environment for Development (GEO-4).* UNEP, Nairobi, Kenya.
30. United States Environmental Protection Agency (USEPA). 2008. *Municipal Solid Waste in the United States: 2007 Facts and Figures.* USEPA, Office of Solid Waste, Washington, DC.
31. Australian Bureau of Statistics. 2007. *Australia's Environment: Issues and Trends, 2007.* ABS, Canberra, Australia.
32. Department for Food, Environment, and Rural Affairs (DEFRA). 2009. *Key Facts about Waste and Recycling: Municipal Waste Management in the European Union.* DEFRA, London.
33. European Commission. 2000. *A Study of the Economic Valuation of Environmental Externalities from Landfill Disposal and Incineration of Waste—Final Report.* EC Directorate General Environment, Brussels, Belgium.
34. Ibid.
35. Pearce, D., 1993. *External Costs from Landfill and Incineration.* U.K. Department of Environment, Her Majesty's Stationery Office, London, U.K.
36. Stern, N. 2006. *The Stern Review—The Economics of Climate Change.* Cambridge University Press, Cambridge, UK.
37. U.S. Energy Information Administration, 2009. Voluntary Reporting of Greenhouse Gases Program. Fuel and Energy Source Codes and Emission Coefficients. USEIA, Washington, D.C.
38. Department for Food, Environment, and Rural Affairs (DEFRA). 2003. *A Study to Estimate the Disamenity Costs of Landfill in Great Britain.* Final report by Cambridge Econometrics with Eftec and WRc. DEFRA, London, UK.
39. Roberts R.K., P.V. Douglas, and W.M. Park. 1991. Estimating the External Costs of Municipal Landfill Citing through Contingent Voluation Analysis—A Case Study. *Southern Journal of Agriculture,* 23(2), 155–165.
40. Hardisty, P.E., H.S. Wheater, D. Birks, and J. Dottridge. 2004. Characterisation of LNAPL in Fractured Rock. *Quarterly Journal of Engineering Geology and Hydrogeology,* 36: 343–354.

41. Hardisty, P.E., H.S. Wheater, P.M. Johnston, and R.A. Bracken. 1998. Behaviour of Light Immiscible Liquid Contaminants in Fractured Aquifers. *Geotechnique,* 48(6): 747–760.

42. Mackay, D.M., and J.A. Cherry. 1989. Groundwater Contamination: Pump-and-Treat Remediation. *Environmental Science and Technology,* 23(6), 630–636.

43. Cohen, R.M., and J.W. Mercer. 1993. *DNAPL Site Investigation.* C.K. Smoley–CRC Press, Boca Raton, FL.

44. Pankow, J.F., and J.A. Cherry (eds.). 1996. *Dense Chlorinated Solvents and Other DNAPLs in Groundwater.* Waterloo Press, Portland, OR.

45. Cohen, R.M., and J.W. Mercer. 1993. *DNAPL Site Investigation.* C.K. Smoley–CRC Press, Boca Raton, FL.

46. Kueper, B.H., and D.B. McWhorter. 1991. The Behaviour of Dense Non Aqueous Phase Liquids in Fractured Clay and Rock. *Groundwater,* 29(5), 716–728.

47. Hardisty, P.E., H.S. Wheater, P.M. Johnston, and R.A. Bracken. 1998. Behaviour of Light Immiscible Liquid Contaminants in Fractured Aquifers. *Geotechnique,* 48(6): 747–760.

48. Wealthall, G.P., B.H. Kueper, and D.N. Lerner. 2001. Fractured Rock-Mass Characterization for Predicting the Fate of DNAPLS. *Conference Proceedings of Fractured Rock 2001, Toronto, Ontario, Canada* (ed. B.H. Kueper, K.S. Novakowski, and D.A. Reynolds), USEPA, Washington, D.C. pp. 450.

49. Hardisty, P.E., H.S. Wheater, D. Birks, and J. Dottridge, 2004. Characterisation of LNAPL in Fractured Rock. *Quarterly Journal of Engineering Geology and Hydrogeology,* 36: 343–354.

50. Hardisty, P.E., H.S. Wheater, P.M. Johnston, and R.A. Bracken. 1998. Behaviour of Light Immiscible Liquid Contaminants in Fractured Aquifers. *Geotechnique,* 48(6): 747–760.

51. CL:AIRE, 2002. *Introduction to an Integrated Approach to the Investigation of Fractured Rock Aquifers Contaminated with Non-aqueous Phase Liquids.* Technical Bulletin TB1. CL: AIRE, London, UK.

52. Guswa, J.H., A.E. Benjamin, J.R. Bridge, L.E. Schewing, C.D. Tallon, J.H. Wills, C.C. Yates, and E.K. LaPoint. 2001. Use of FLUTe Systems for Characterisation of Groundwater Contamination in Fractured Bedrock. *Conference Proceedings of Fractured Rock 2001, Toronto, Ontario, Canada* (ed. B.H. Kueper, K.S. Novakowski, and D.A. Reynolds). USEPA, Washington, D.C. pp. 450.

53. Lane, J.W., Jr., M.L. Buursink, F.P. Haeni, and R.J. Versteeg. 2000. Evaluation of Ground Penetrating Radar to Detect Free-Phase Hydrocarbon in Fractured Rocks—Results of Numerical Modelling and Physical Experiments. *Ground Water,* 38(6), 929–938.

54. Schmelling, S.G., and R.R. Ross. 1989. *Contaminant Transport in Fractured Media: Models for Decision Makers.* EPA Superfund Issue Paper EPA/540/4-89/004. USEPA, Washington, D.C.

55. Taylor, T.P., K.D. Pennell, L.M. Abriola, and J.H. Dane. 2001. Surfactant Enhanced Recovery of Tetrachloroethylene from a Porous Medium Containing Low Permeability Lenses: 1. Experimental Studies. *Journal of Contaminant Hydrology,* 48: 325–350.

56. Department for Food, Environment, and Rural Affairs (DEFRA). 2007. *The Social Cost of Carbon and the Shadow Price. What They Are and How to Use Them in Economic Appraisal in the UK.* UK DEFRA, London.

8 Best Practice for the Twenty-First Century

SUMMARY

It is an important time for humanity. Across the planet, we face the simultaneous challenges of poverty, water scarcity, pollution, ecosystem degradation, and a changing climate. Each of these issues is being driven forward inexorably by a rapidly expanding population and steadily rising expectations of material prosperity around the globe. But, despite an almost universal espousal of the concept of sustainability—the idea that we want to safeguard the future for the next generations, allowing them to live at least as well as we have—the weight of evidence shows that we are accelerating in the wrong direction. Emissions of climate-damaging greenhouse gases (GHGs) continue to rise; per capita use of water, energy, and raw materials grows year after year; and the natural places of Earth continue to be eroded away at an alarming pace. A shocking gulf has opened between what we say we want for the future and what we do in the present.

The behavior that drives this growing disparity between aspiration and action lies firmly rooted within our economic system. Developed in an earlier time, our modern economic system is implicitly based on assumptions of inexhaustible natural resources, labor scarcity (not resource scarcity), and an unlimited capacity of the planet to assimilate waste and pollution. To this day, we continue to place no economic value on the natural world or any of the materials and services that it provides. These "externalities" remain almost exclusively unpriced, and because they appear to be free, we treat them as worthless—to be wasted, thrown away, consumed, burned up, and used without a thought. Even when we know, morally and ethically, that behavior should change, the day-to-day signals are not there to allow change to occur on a scale sufficient to make a difference.

Clearly, altruism is not enough. If we want to move toward a more sustainable world, we need to start pricing the real value of externalities within our decision making. A new notion of "best practice," fit and tailored for the twenty-first century, will require significant shifts in thinking across the whole spectrum of our activities, in business, industry, government, and our personal lives.

To achieve true sustainability this century, we will need to deploy technologies we have previously shunned as "too expensive"; we must change the way our economic system values life, the environment, and society and so drive fundamental changes in the ways that decisions are made. We will need to develop new policies that recognize the limitations of the free market, while preserving the undeniable benefits that markets bring in terms of efficiency and innovation and creating a new metric for success in our economies. Throughout the less-developed world, we must strengthen the institutions that protect society and the environment and eliminate the crippling affliction of corruption. We need to stress prevention in environmental and social protection while continuing to reclaim and heal the wounds we have made. We must act aggressively to reduce GHG emissions and reduce the risk of dangerous and unpredictable climate change, while beginning now the sad but inevitable task of adapting to a warming world to which we have already committed ourselves.

And in all of this, we must act now. We have wasted too much time on denial and obfuscation; there has already been too much procrastination. New laws and regulations that provide price-signals valuing the environment are needed. Industry can and must also play a key role in shaping a more sustainable future by bringing its resources, people, and innovation to bear to create a new standard for best practice in the twenty-first century.

TECHNOLOGY

When the term *best practice* is mentioned, particularly in industry, technology is often the first thing that comes to mind. As has been shown in the chapters of this book, there already exists a tremendous variety of technologies that have been developed to help improve the overall sustainability of humankind's activities—from all manners of energy efficiency systems, to renewable energy, emissions treatment equipment, and biological systems to filter and clear our air and water, among others.

At this point in history, humankind has at its disposal the technology it needs to design and execute virtually any kind of project in a way that can be far less destructive to the environment of the planet and more protective of our common social heritage than ever before. We also possess an increasing ability to repair the damage we have done in the past, through the application of reclamation and remediation technology and know-how. Across the entire project life cycle, a wide variety of tools, technologies, and products is now available (and in many cases being actively used) that allow industries of all kinds to operate in a far more benign and sustainable way than they do at present.

In short, technology is not the problem. Achieving sustainability in the twenty-first century does not depend on the discovery of some magic new piece of equipment that will solve our problems. We already have the technology we need to make substantial positive progress toward a more sustainable world. We can, today, massively reduce air emissions of all kinds, including GHGs, by deploying renewable energy, becoming more efficient in all we do, deploying clean public transport using electric and hybrid vehicles on a massive scale, and substituting cleaner forms of energy production. Many of the technologies needed for this transformation, such as thermal solar power discussed in Chapter 6, are not new; they have been around for a long time. Others, such as carbon capture and sequestration, discussed in Chapter 5, are newer and certainly require development and the kind of testing that can only come from doing—implementation at a large scale.

What is hampering our deployment of the sustainability-enhancing technologies already available is our perception about cost: We simply do not see the real cost of our current way of life and the hidden price of the environmentally and socially damaging ways to which we have become accustomed. Because we cannot see or feel the real cost of what we do now (in fact, much of our current unsustainable technological base is subsidized), and because our economic system does not explicitly value the benefits that environmental and social sustainability can bring, we see no reason to change, despite our growing appreciation that change is needed. There is no "price signal" that prompts us to respond. Until our economic system can explicitly account for these external costs and benefits, best-practice technology cannot be deployed on a scale sufficient to affect real improvement in environmental and social sustainability on the planet.

MANAGEMENT AND DECISION MAKING

The technology exists today to start us on the journey toward a more sustainable world. In the same way, huge strides have been made in understanding how to manage, control, and regulate industries and industrial production for improved sustainability. Regulatory systems have been developed that allow industries to operate within a free market bounded by laws and regulations that protect society. In many countries these have worked well over the last few decades to deliver improvements. Risk-based approaches allow responses to be gauged to the level of expected damage. Despite this, progress overall remains slow, and on a global scale we continue to lose ground and even accelerate in the wrong direction. Clearly, there is significant room for improvement in how we manage for sustainability.

THE NECESSARY EVOLUTION OF THE ENVIRONMENTAL IMPACT ASSESSMENT

The widespread adoption of the environmental impact assessment (EIA), and its more recent evolutionary descendent the environmental, social, and health impact assessment (ESHIA), into the mainstream of environmental legislation and regulations worldwide has been a major step in providing a more balanced view to development of all kinds. Almost every nation on the planet now has some form of regulatory requirement for the identification of environmental baseline conditions and the assessment of predicted impacts of a new development on the local environment. Most regulations and guidance also require development of some form of environmental management plan designed to mitigate predicted impacts. Much of the regulation, especially in less-developed countries, is modeled to some degree on World Bank guidelines.

The spread of the EIA concept around the globe has undoubtedly been a good thing, particularly in raising awareness within the financial and investment community of environmental issues and liabilities associated with major projects. However, in practice, EIAs have tended to be executed largely outside the core of the project design team. Impact assessments are typically conducted by external consultants who are presented with a ready-made project design and asked to determine the impacts of that design and suggest mitigation measures if possible. The ultimate focus of the exercise has become, in many instances, to secure regulatory approval for the project rather than actually to influence the scale, shape, and concept of the project itself. In fact, in many developed economies the EIA process is now officially called the "approvals" process. This process can also lead to a "green-colored-glasses" effect, by which impacts of development are underpredicted and the effectiveness of proposed mitigation measures overestimated. By extension, once the EIA document has been approved, little effort typically tends to be exerted toward verifying that predicted levels of damage are being realized or exceeded, or that proposed mitigation measures are being followed or are proving effective. This has led to a growing number of examples worldwide of projects with approved EIAs that have impacted the environment far more than ever predicted.

The world needs and deserves an environmental safeguarding process that ensures that new projects are conceived, designed, and executed to be as environmentally, socially, and economically sustainable as possible from the outset. We need a process by which environmental baseline data are fed into project concept selection, and predicted possible impacts are used to influence all aspects of project design and technology selection as part of an integrated process.

The role of EIA needs to be expanded, along the current World Bank best-practice guidelines, to explicitly include full consideration of project options from a complete environmental, social, and economic perspective. This would represent a major step forward in the necessary evolution of conventional EIA toward a more cooperative, conciliatory, and integrated planning and decision-making tool. The environmental and economic sustainability assessment (EESA) is an ideal companion to a new generation of EIA, by which project options can be examined from a full life-cycle environmental, social, and economic perspective to add real value to the process for all parties.

REGULATORY CAPACITY DEVELOPMENT

In societies with well-developed regulatory systems and capacity, environmental protection measures can be put in place, delivering significant positive results. However, much of the industrial growth and resource industry expansion of this century is expected to occur in parts of the world where regulatory governance and capacity are not well-developed. In many of these countries, regulatory structures are still not mature enough to manage and direct environmental protection to the degree required.

International development institutions have started to put significant effort into developing the regulatory capacity and capability of less-developed countries, helping them to protect their natural and social resources. However, much of this effort is modeled on the systems and thinking used in the richer Organization for Economic Cooperation and Development (OECD) nations—thinking based on a lopsided traditional financial view of the world, in which prices are distorted through perverse subsidies and do not reflect the value of environmental and social damage.

In much of the developing world, not only is regulation, in many cases, not sophisticated enough to deal with the wide range of issues being faced, but enforcement ability is hampered by lack of capacity, expertise, training, funding, and quite often corruption.

REVEALING THE REAL COST OF CORRUPTION

The role of corruption as a causal factor in environmental and social degradation in many less-developed countries of the world, and indeed even in developed countries, cannot be ignored.[1] Unfortunately, when corruption starts to dictate decision making, the environment and society are invariably the losers.

A road that should have been diverted around a designated nature reserve is given permission to cut through the most ecologically valuable parts of its extent. Drilling wastes that should have been properly collected and treated are dumped into water courses or unlined pits and simply covered over, hiding evidence of the crime. A new megaproject is quickly rushed through the approvals process, bypassing the usual environmental and social impact process, with economic development and job creation as the rationale, despite clear environmental and social impacts (this type of thing still occurs regularly in developed nations such as Canada, Australia, and the United States). In each case, short-term personal and corporate gain have been put ahead of the protection of the heritage of the nation, the long-term well-being of society, and the natural capital account of the country.

Public tolerance of corruption would certainly diminish if society was aware of the real costs and benefits of this unscrupulous behavior. Bribe payers often see the practice in purely financial terms—paying $1 million dollars now to a corrupt official allows the savings of $20 million of additional expenditure on environmental protection that otherwise would have been required. How would they change their view if they knew that the value of the environmental asset that was damaged or destroyed was worth ten times the amount they saved? And, how would society now consider the bribe taker's choice if they knew that for $1 million dollars they had sacrificed over $100 million dollars in the country's natural assets—their natural capital?

The economics of corruption are always far worse for the country than for the company. In the end, regulatory sophistication and superior technology are of little use if unscrupulous operators and governments (for it takes two to engage in the dark waltz of corruption) can sacrifice the environment and society in favor of short-term gain. Unfortunately, there are all too many examples from around the world of significant environmental degradation caused by corruption. Understanding the real social value of the damage caused by corruption can play an important part in preventing this type of behavior and rallying public opposition against myopic short-term decisions.

INTO THE FUTURE

Now Is the Right Time

The time is right for a fundamental shift in the way industry examines and manages its environmental and social impacts, allowing it to contribute toward a more sustainable future for everyone on the planet. If industrial projects actually create more damage than the value of the positive services they provide to humanity, then surely these activities are in the end futile at best and damaging at worst. What is needed is a rational balanced way to objectively quantify the value of both sides of the equation in a way that all stakeholders can see and agree on, a transparent accounting of the value of the products that these industries produce for the benefit of humankind, and the damage to the life-sustaining environment we all share that is caused in the course of the exploitation, production, and use of those products. At present, this is done only rarely. Currently, most decisions that directly affect the environment, and the communities and societies that depend on it, are made without rational, explicit, and objective consideration of the value of social and natural assets affected. Costs of labor, machinery, fuel, and other inputs are explicitly taken into account. The products generated in the market are valued—but nothing more. Acknowledgment of external issues—the costs and benefits to the environment and society associated with clean air, clean water, biodiversity, natural ecosystem protection—is provided in qualitative terms, often as a footnote to the decision, but the "economics" of the project do not consider these other factors.

Once companies, regulators, and the public are able to quantify the value of the environment, then the true economics of the project or proposition can be examined. This will illustrate the value of the environmental damage that would have been caused under a "business-as-usual" scenario and, by extension, the opportunities that exist to cost-effectively reduce that damage. Finally, the costs of environmental protection, which companies rationally seek to minimize under current practice, can more productively be optimized. The optimization of environmental spending by industry would represent a paradigm shift in the way sustainability has heretofore been considered.

From Remediation to Prevention

Currently, our legal and economic system places a strong emphasis on remediation. If damage is done, blame is apportioned, restitution is sought, and repairs are made. In environmental and social terms, an emphasis on remediation rather than prevention is in the best interests of business: Pollution and environmental degradation go on until the damage is deemed unacceptable, at which point action is contemplated. This does two things. First, because dealing with the issue is pushed into the future, the firms and individuals responsible for the damage are often no longer around. It is the public that must bear the costs of remediation, either because the companies responsible for the damage are no longer solvent or because the damage is a result of so many combined individual discharges that no single industry can be held responsible. Second, from the perspective of the firm, remediation has a big financial advantage over prevention: The costs of dealing with the problem are pushed back in time. So, while up-front capital investment on mitigation and prevention will have an impact on project finances now, remediation costs that occur in the future are heavily discounted and appear quite small in present value terms (if indeed any provision is made at all). By delaying expenditure, remediation appears to be the cheaper alternative, especially with the high discount rates used in the private sector and if the external benefits of prevention are not recognized. With the current economic system skewed so heavily in favor of short-term outcomes, it is not surprising that we continue to stress cure over prevention.

But, remediation of environmental and social damage is usually much more expensive than the equivalent preventive measure (site contamination is a good example) in real terms. In addition, prevention of damage will usually result in higher benefits to society due to the inherent irreversibility of many types of environmental and social impact. Over typical one-generation planning horizons, prevention is usually a more economic strategy than remediation. The reality, of course, is that prevention is better than cure; we know this intuitively from personal experience. Moving toward a more sustainable world will require significantly more emphasis on prevention and recognition of the hidden but very real cost of deferral.

Climate change is a powerful example. Preventing the worst of climate change means substantially decarbonizing the global economy over the next three decades. But, the task is monumental and expensive, so much so that many are now saying that the effort will be too great, that we will not be able to afford it, that it will cripple our economies. But, climate change is irreversible. There is no remediation option here. If we persist with business as usual, we should be very sure about our choice because there is no going back to the way things were. All we will be able to do is adapt. And chances are that adapting to climate change will be a lot more expensive, in every way, than preventing it. This is a choice we all have to make, but in the end, we have to make it together because only collaborative action on a planetary scale will prevent the irreversible ravages of climate change.

Unfortunately, a substantial and ongoing effort is also required to remediate, reclaim, and otherwise repair the significant environmental damage that has been wrought over the past several decades. A stronger emphasis on prevention will require continuing development of the trend toward implementation of robust environmental management and loss prevention systems (which is now well under way in many industries) and an understanding of the fundamental economics of environmental and social protection. The challenge industry now faces is to move away from legacy issues (dealing with these in the most economic way possible) and to focus squarely on prevention—ensuring that the next generation of legacy issues is never created in the first place. In doing so, industry also needs to come to the understanding that the most economic solution is rarely the cheapest, but that equally, spending more does not always bring commensurately better results.

Future Value Trends

In the future, the value of all environmental assets and natural resources is expected to rise inexorably as population growth drives increased demand and inevitable scarcity. The price of water will continue to climb toward a true reflection of its total economic value (TEV; which is also expected to rise) as available freshwater supplies continue, unfortunately, to be overexploited and contaminated and as the effects of climate change alter hydrological regimes across the planet. Biodiversity, of little relevance to decision makers in the nineteenth century, will continue to emerge as a dominant issue in the twenty-first century. The "fraying fabric of life" on Earth is more vulnerable now than at any time in the past few millennia, straining under the combined effects of overexploitation, land clearing, burning, and climate change.[2]

The huge value of ecosystem services to humanity needs to be recognized within our economic systems and increasingly will be. People will be less and less willing or able to put up with the impacts of industrial activity: health effects, loss of amenity, environmental damage, nuisance, odor, and pollution. Whereas in the past there may have been a tendency to see this type of impact as a necessary part of "progress"— undesirable but inevitable, to be endured, in return for a job and income—in the future people will increasingly challenge the necessity of these impacts and demand that industry does more.

This has certainly been the trend over the past few decades in the developed world, and it is likely to accelerate. Evidence of this attitude is now being seen across the developing world also, in places such as China, where public dissent over environmental issues is increasingly frequent despite the consequences of speaking out against the system. People now see that environmental and social damage affects them and their families in a very real and palpable way. Increasingly, they understand that this damage is simply a transfer of cost from industry to them—they are suffering so that industry may profit. Armed with an appreciation of the value of those transfers in monetary terms, people will increasingly demand more of industry and mandate their governments to place even greater pressure on those who generate unsustainable damage.

Given the conditions on the planet today, the twenty-first century will undoubtedly see continuous and significant increases in the values placed on natural and environmental resources, which will translate into higher costs to industry. Industry will certainly pass these costs on to their customers. In the near future, people will start to pay more for everything. The key is that they should benefit overall.

Toward a New Metric of Success

At present, it is the production and consumption of goods and services that are the measure of success in society. For industry and business, profit, measured in the conventional sense (private benefit minus private costs), remains the driving force behind decision making. But, the twenty-first century will gradually bring about a change in what we define as success. Including the costs of environmental and social damage in the macro- and microeconomic equations will have a profound impact on how companies, individuals, and governments behave. Over time, a genuine progress indicator (GPI) will come to replace the antiquated notion of gross domestic product (GDP), and corporate accounting will increasingly include environmental and social costs as real costs of doing business. These things will occur simply because they must. The current measures of success are inaccurate and misleading and must be replaced if we are to meet the massive challenges that face us.

Industry Can Lead the Way and Benefit in the Process

Rising costs for what, hitherto, have been external to the considerations of industry are not necessarily bad news. Companies, with the participation of other key project stakeholders, can compare the costs of various environmental and social protection measures to the monetary benefits that they will produce. This will allow them to seek an optimal balance between the two rather than simply arguing, as they have in the past, for the lowest-cost solution. As shown in the previous chapters, cheaper does not always mean better, and perhaps just as importantly, spending more is not always a guarantee of improved environmental protection or sustainability. Once a rational basis for analysis is provided, sustainability can rise from being what many consider to be a soft, intangible, "touchy-feely" issue to something quantitative that can be injected into mainstream decision making. This in turn will unlock the true potential of many other environmental and sustainability-oriented tools and systems, including a new generation of EIA, and emerging environment-in-design methods, which allow environmental issues to directly influence project design.

By moving away from conventional financially dominated decision making and looking at the real overall economics of choices over the longer term, industry can better optimize operations, increase efficiency, reduce waste, and improve its relationship with society, customers, neighbors, and regulators. Shareholders and employees will also benefit by a longer-term, more balanced view of operations that is designed to improve the sustainability of the business as a thriving, profitable entity.

SUMMARY

Making these changes will not be easy, however. Doing what we know, what has worked before, over decades of accumulated experience, is engrained in our individual and organizational memories. It is comfortable; we know the routine. But, it is not working. Albert Einstein, Benjamin Franklin, and others have all been attributed, defining insanity as "doing the same thing over and over again and expecting different results." Avoiding collective insanity in the twenty-first century requires that we realize that prosperity must be measured by more than just material wealth, that the wealth we do have needs to be more equitably shared, and that unless we value other things explicitly within our economic system, we will continue our hurtling journey into the dark reaches of an unsustainable and ultimately perilous future.

NOTES

1. Sachs, J. 2008. *Common Wealth: Economics for a Crowded Planet.* Allen Lane, London.
2. World Watch Institute. 2001. *The State of the World, 2001.* Earthscan, London.

Index

"f" indicates material in figures. "n" indicates material in footnotes. "t" indicates material in tables.

Milton Keynes UK
Ingram Content Group UK Ltd.
UKHW021626071024
449327UK00020BA/1209